# OIL HISTORY

## A SELECTED & ANNOTATED BIBLIOGRAPHY

COMPILED BY

E. RICHARD NEFF

INTERNATIONAL ASSOCIATION
OF DRILLING CONTRACTORS
HOUSTON, TEXAS

1995

Printed in the United States of America

International Association of Drilling Contractors
15810 Park 10 Place, Suite 242
Houston, Texas 77084-5134

Library of Congress Cataloging-in-Publication Data

Neff, E. Richard

Oil History: A Selected & Annotated Bibliography.

Includes Index.

1. Petroleum 2. History-Petroleum 3. Bibliography-Oil
4. Reference-Petroleum.

95-79319 1995

ISBN 0-9648314-0-6

FIRST EDITION

Printed by D. Armstrong, Inc.
Houston, Texas

# CONTENTS

Foreword

Introduction

Preface

Part I - Histories . . . . . . . . . . . . . . . . . . . . . . . . . . . . . . . . 1

Part II - Biographies . . . . . . . . . . . . . . . . . . . . . . . . . . . . . 123

Part III - Companies . . . . . . . . . . . . . . . . . . . . . . . . . . . . . 273

Author Index . . . . . . . . . . . . . . . . . . . . . . . . . . . . . . . . . . 357

Subject Index . . . . . . . . . . . . . . . . . . . . . . . . . . . . . . . . . 367

# Foreword

The International Association of Drilling contractors has long recognized the value of maintaining a record of the past while moving forward to meet industry's present and future meeds. In meeting our goals we have published references of interest to the industry which are not available elsewhere.

A classic example of this is IADC's Drilling Manual (formerly the Toolpusher's Manual). Now in its eleventh edition, it is distributed world-wide and found on rigs, in field offices, drilling engineering offices and in company and public libraries.

When George Williford brought the present manuscript to our attention, he pointed out that the last such reference was published in 1960. Although still a valuable reference, it is long out of print and out of date, as much has been written in the intervening years. As a consequence, it is in dire need of replacement and Neff's work does that admirably.

Upon reviewing the compiler's background in the industry, his reputation as a collector/dealer and observing the meticulous work put in to the manuscript's presentation, we agreed to publish and distribute this book. It is our belief it will be a valuable reference to the industry, academicians and to the general reader interested in the history of this great business.

A. Lee Hunt, Jr.
President
International Association of Drilling Contractors

# Introduction

It is with great pleasure that I have been involved with the IADC in the publication of the eleventh edition of the Drilling Manual and now this bibliography of petroleum histories. Both works are unique and fill specific needs of different facets of the industry.

Richard Neff is a petroleum geologist and out-of-print book dealer specializing in oil industry publications, more specifically oil history. After attaining his Master's Degree in Geology from the University of Oklahoma he worked for major oil companies in the southwest for thirteen years prior to becoming an independent consulting geologist. He is the author of Hydrocarbon Quest; One Man's Search For Oil And Gas, an autobiographical journal of oil exploration over thirty years, privately published in 1990. He is also the author of One Hundred Good Oil Books: A Collectors Check List, which was privately published in 1994.

I have known Richard Neff as a fellow collector of books on the petroleum industry and a competitor in the out-of-print book business. During this time, I have developed a sincere respect for his knowledge of the subject and his professionalism in his business endeavors. In those few titles in this work he has not personally reviewed, he fully credits his sources(s) and he has provided an unusually detailed and lively description of those he has reviewed. Furthermore, he notes the general availability of each title for the reader's benefit. It is perhaps one of the most complete petroleum bibliographic references written. I am confident the reader will not only find this work a useful reference but enjoyable reading as well.

Every effort has been made to be all inclusive in the compilation of these titles. Naturally, there will be some titles which have been omitted and the IADC would welcome being made aware of them for future revisions.

George A. Williford
Editor

# Preface

Oil history books make fascinating reading. The written history of the oil industry and its people is like all history, with its different writers putting different spin on the same events based on their own perspectives. I have been reading oil history since my college days of the late 1950's. Utilizing my twenty percent student discount, I first acquired a couple of University of Oklahoma Press books. For a total of around $10.00, I bought Ruth Knowles' THE GREATEST GAMBLERS; William Connelly's THE OIL BUSINESS AS I SAW IT; and Carl Rister's OIL TITAN OF THE SOUTHWEST. These three rank high on the list of quality oil history books, each coming from a totally different perspective. Knowles weaves her history around the big independent wildcatters with their remarkable exploratory success. Connelly begins with autobiographical accounts of early Pennsylvania oil and later describes working with Harry Sinclair as they become involved in the Teapot Dome scandal. Rister's book is a respected scholarly work on oil history and developments in the southwest with much attention to chronological events and those who made them happen. It is unfortunate that all of the books included in this bibliography are not of such high quality, however several are even better.

I have been buying and selling oil-related books for some time now and have reviewed around 4,000 different titles. Oil fiction is entertaining and technical books are informative, however history books provide even more... a window to the past which in turn provides a key to the future. We can learn more by observing other peoples triumphs and failures than by floundering through life, making mistakes and learning everything the hard way.

The simple criteria used for inclusion in this bibliography is that the book be an oil history or oil-related book of historic interest still in circulation (including reprints). Excluded books include many obscure pamphlets, government investigations, political works, faux fiction, reprints from periodicals and technical books. These are reserved for a future undertaking.

This book is intended to be of use as an up-to-date bibliographic reference for those in research, historians and collectors, but it is perhaps of even greater value as a history itself. It was designed to supplement the oil history section in Swanson's 1960 bibliography and also to pique interest in the subject. Along with a short review of the book, I provide some information on various editions and printings of the more important ones.

A few oil book bibliographies are in circulation , however none of them deal exclusively with history and the latest is thirty-five years old. Hardwicke, in 1937, published his PETROLEUM AND NATURAL GAS BIBLIOGRAPHY with 1,400 items dealing with all phases of the industry, about 150 of which are oil history. Giddens, in 1941, published his THE BEGINNINGS OF THE PETROLEUM INDUSTRY. Part II contains numerous references to early Pennsylvania oil history. DeGolyer, in 1944, published his BIBLIOGRAPHY ON THE PETROLEUM INDUSTRY, with 12,000 items, mainly articles dealing with all phases of the oil industry. In 1960, Swanson published his A CENTURY OF OIL AND GAS IN BOOKS. It was prepared in conjunction with the API, and contains over 2,000 well annotated listings of books and important articles. It is the bible of oil book dealers and collectors. Surprisingly, it contains less than 300 oil history and biography items. The above authorities are regularly cited throughout this book. In addition to the above, Yergin has provided an eighteen page bibliography of oil titles in his excellent book, THE PRIZE.

E. Richard Neff
July, 1995

# PART I - HISTORIES

**A1. (MONTANA) ABC Publicity Service (compilers), BILLINGS BECKONS, The Company, 1921, 56pp.**
Excellent promotional brochure for the town. Much information on the four new oil fields which began production in 1919. Devils Basin, Cat Creek, Soap Creek and Elk Basin contained forty-one wells and they are all listed through May 22, 1921. Petroleum geologists working the area relate the history of the development in those four areas which surround Billings. Included are some great old ads and photos of the small town which are provided to entice oil and other people to move there. The softbound first edition is scarce.

**A2. (GENERAL) Abrams, Kathleen Crotty (editor), THE STORY OF THE AMERICAN OIL INDUSTRY: BRINE TO BONANZA, The Woodlands, Pioneer Publications, (1981), 430pp.**
Many oil companies contributed to this industry-promoting history. One-third of the book is a compact and excellent oil history with many good old photos not published elsewhere. The history takes you from Pennsylvania and the early Appalachian play to California and then through developments in Oklahoma and Texas. Much on the landmark Spindletop discovery which essentially freed the country from Standard Oil domination. Early offshore exploration is covered and optimistically projected into the future. The remainder of the book is oil and service company histories from those unique outfits such as Boots & Coots to major oil companies like Shell Oil. Seventy-five full-color company biographies cover the famous and the obscure. Some of the more prominent are: Gulf, Superior, Hunt, Ethyl, HNG, Geosource, and Standard of California. The hardbound first edition is becoming scarce.

**A3. (ALASKA) Alaska Geographic, ALASKA'S OIL/GAS AND MINERALS INDUSTRY, Anchorage, AK, Alaska Geographic, 1982, 216pp, Volume 9, Number 4.**
Prepared by the publisher's staff for members of the Alaska Geographic Society, the book traces early history of exploration and development of minerals of the state. Emphasis is on oil and gas. Included are historic and modern photos along with maps and charts. The softbound first edition is common.

**A4. (WYOMING) Allen, Frederick Lewis, ONLY YESTERDAY, New York, NY, Harper & Brothers, 1931, 370pp.**
The author was editor of Harper's Magazine and this was his first book. He covers the sensational, largely political stories of the 1920's and creates a newsreel of the times, allowing it to pass before your eyes in a well written format. Much of his story concerns the Teapot Dome oil scandal which involved President Harding, Albert Fall, Andrew Mellon, Harry Sinclair, Edward Doheny and the Continental Trading Company. Allen's matter of fact style fits the tone of the 1920's. A widely circulated and still popular history. Cited in Swanson on page 149. The hardbound first edition is common.

**A5. (ALASKA) Allen, Lawrence J., THE TRANSALASKA PIPELINE: SOUTH TO VALDEZ, Seattle, WA, Scribe Publishing, 1976, 151pp.**
Coffee table book. It opens with historic pre-pipeline photos of Alaska, its natives and industry followed by impressive color photos of the scenery and people working at the pipeline construction sites. Brief history and dialogue accompanies the photos. Numerous oil companies and businesses contributed to the book. The hardbound first edition is common.

**A6. (GREAT BRITAIN) Alvarez, A., OFFSHORE: A NORTH SEA JOURNEY, Boston, Houghton Mifflin Co., 1986, xii, 190pp.**
An excellent history and graphic description of the North Sea oil industry as seen through the eyes of a poet-novelist who, because of a fortunate contact made at Shell Oil, became intimately familiar with the subject. In order to gain material and insight, he interviewed divers, toolpushers, helicopter pilots, engineers, crane drivers, scaffolders and roustabouts, as well as geologists, administrators and businessmen. This is a very well prepared and readable book. The hardbound first edition is common.

**A7. (PENNSYLVANIA) American Association of Petroleum Geologists, HISTORY OF THE PETROLEUM INDUSTRY SYMPOSIUM, Tulsa, OK, AAPG, (1989), 84pp.**
Not only a brief, well prepared history of the early Pennsylvania oil industry, this is also a guidebook-roadlog for three interesting field trips. The roadlog takes you to many of the historic sites in the Titusville and

Oil City area. Some geology is included in the log, along with visits to the former homes of Colonel Drake and Coal Oil Johnny among others. This is a first class presentation for layman and oil industry personnel alike, with many photos. The softbound first edition is common.

**A8. (GENERAL) American Bar Association, LEGAL HISTORY OF OIL AND GAS: A SYMPOSIUM, Chicago, IL, The American Bar Association, (1939), vii, 302pp.**
From the title page: "Consisting of 11 papers which, as a whole, give the legal history of the regulation of the oil and gas industry in the principal producing states in the interest of conservation; and which, at the same time, necessarily contain much instructive data and interesting discussion relating to the political and economic phases of governmental regulation of an industry". It is that. Cited in Swanson on page 64. The hardbound first edition is common.

**A9. (GENERAL) American Petroleum Institute, AMERICAN PETROLEUM INSTITUTE QUARTERLY: CENTENNIAL ISSUE, NY, API, 1959, 72pp.**
The special issue contains several excellent articles. "Birth of an Industry" by Giddens, gives his expert historical perspective on the early Pennsylvania oil industry. "The Well Shooters" by Dolson, tells of the early well stimulation technique of exploding nitroglycerine torpedoes. "Three Fabulous Decades" by Nevins, covers the rapid expansion years of the petroleum industry years from 1900 to 1930. "The Blue Chip Sea" by James Clark, relates the story of offshore oil which was just then coming into its own. "The War on Waste" by Stanley Clark, covers conservation efforts. Twenty-three other articles by well-known authors concern every phase of the industry from exploration to marketing. Very well done with good photos and drawings. The softbound first edition is scarce.

**A10. (GENERAL) American Petroleum Institute, HISTORY OF PETROLEUM ENGINEERING, Dallas, TX, The Institute, 1961, ix, 1241pp.**
This is a major compilation, conducted over a nine year period and presented here in five parts, covering every phase of petroleum engineering. Numerous contributions by the top oil minds of the day which include James Clark, Everette DeGolyer, C.A. Warner, Shuman,

Brantley, Vance, Swigart and Hardwicke. Primary emphasis is on part two, "Drilling and completing wells" and part three, "Producing and handling oil and gas". Part five is titled "Impact upon society" by chairman of the editorial board, D.V. Carter. Several individual profiles are given of the primary contributors. Significant bibliographies accompany the various papers and the work is well illustrated with drawings and photos. Definitive. The hardbound first edition is scarce.

**A11. (GENERAL) American Petroleum Institute, PETROLEUM: IN LEGEND, IN HISTORY AND TODAY, NY, API, 1930, 15pp.**
The small pamphlet briefly covers oil history from ancient to 1930. This is the second of the Oil Industry Series, aimed at layman readership. Lots of firsts are included, such as the first oil well, first boomtown, etc. Several drawings. Not cited in Swanson. The softbound first edition is common.

**A12. (GENERAL) American Petroleum Institute, PETROLEUM: THE STORY OF AN AMERICAN INDUSTRY, NY, API, (1930), 118pp., Revised 1949 edition.**
From the foreword: "..answers to the many old and new questions regarding the origin, source, and uses of petroleum and its products, as well as to furnish factual information about the complex but highly efficient operations of the petroleum industry". This is a brief and simplified story which is well illustrated with maps and drawings. More propaganda for the layman. The softbound first edition is common.

**A13. (CANADA) Anderson, Allan, ROUGHNECKS & WILDCATTERS, Canada, Macmillan of Canada, 1981, 303pp.**
The book provides glimpses of the Canadian oil industry through time, including its major fields, the Beaufort Sea, Athabasca tar sands, and more. From the title page: "Hundreds of first-hand exciting stories by Canadian rig builders, drillers, mudmen, jughounds, scouts, landmen, geologists, and many others...an informal popular history of the oil patch". Generally in chronological order, most of the stories are from the Calgary Edmonton area and the tellers are identified in the appendix except for a few who wished to remain anonymous. Many historic and modern photos are included. Fun reading. The hardbound first edition is common.

**A14. (GENERAL) Anderson, G.S., ROMANCE OF ROYALTY, Fort Worth, TX, The Author, 1928, 62pp.**

Exceptional promotional work with interesting photos of some spectacular wells, blowouts and fires. Also photos of wealthy Black families who had become rich through royalty payments due from their mineral rights. The author provides recommendations on buying oil royalty. He describes the terms and the possibilities for profit. From the conclusion: "...to the uninitiated, to the one that might scoff at oil royalty investments, or those that are tired and sick of the ordinary garden variety of oil stocks, the writer wants to impress that an investment in carefully selected oil royalties offers an exceptional opportunity for steady and substantial returns". The hardbound first edition is very scarce.

**A15. (MIDDLE EAST) Anderson, Irvine H., ARAMCO, THE UNITED STATES AND SAUDI ARABIA: A STUDY OF THE DYNAMICS OF FOREIGN OIL POLICY 1933-1950, NJ, Princeton University Press, (1981), xiii, 259pp.**

This is an historic look at developments which set the stage for massive changes in the balance of industrial power in the world which would take place twenty-five years down the road. Much on the roll of the Seven Sisters in controlling world oil prices. Even more on the U.S. and European government's complacency to this trade restraining policy. A large and useful bibliography is included. The hardbound first edition is common.

**A16. (ALASKA) Andrasko, Kenneth, ALASKA CRUDE: VISIONS OF THE LAST FRONTIER, Boston, Little, Brown and Co., (1977), vii, 152pp.**

Excellent black and white photos by Marcus Halevi, a photographer with an eye for story-telling through pictures. Pictures and narrative of the pipeline construction characters as they coped with the environment (physical and social) which went along with this major project. From the preface: "Alaska is still in the early stages of bright boom, and is a kinetic remembrance of the way it must have been in the first colonies, the migration West, and the Gold Rush". An unusual look at the boom phenomenon. The softbound first edition is common.

**A17. (GENERAL) Arnold, Ralph and William J. Kemnitzer, PETROLEUM IN THE UNITED STATES AND POSSESSIONS, NY, Harper & Brothers Publishers, 1931, xxi, 1052pp.**
A presentation and interpretation of the salient data of geology, technology, and economics of petroleum in each state and possession treated according to the conventional major field divisions. Excellent compilation of oil information from state and federal sources. Non-producing states are also covered in the same format: Historical, Geological, Stratigraphy, Structure, Physiography and Geography, Land Classification, Technical Review and Future Possibilities. Quoting from Swanson: "Excellent survey of development of U.S. oil fields and areas with detail not found elsewhere, with maps". Cited in Swanson on page 31; Hardwicke page 98, Giddens page 97. The hardbound first edition is scarce.

**A18. (VENEZUELA) Arnold, Ralph, George A. MacReady and Thomas W. Barrington, THE FIRST BIG OIL HUNT: VENEZUELA 1911-1916, NY, Vantage Press, (1960), 353pp.**
This comprehensive pictorial and written work covers the five year period by the men who were there. Arnold was an oil expert and much published author. A large loose map of the country and oil areas is included along with several others, which are bound in. This important work on early Venezuela oil is a classic. The hardbound first edition is scarce.

**A19. (PENNSYLVANIA) Asbury, Herbert, THE GOLDEN FLOOD: AN INFORMAL HISTORY OF AMERICA'S FIRST OIL FIELD, NY, Alfred Knopf, 1942, xv, 324pp.**
Asbury was a much published writer who dealt largely in non- fiction sensational expose's of sin, murder and mayhem in places like New York, New Orleans and Chicago. This book is a real change of pace for him, however chapter VII is titled "Sin Among the Derricks". He gives a good account of the events leading up to the Drake discovery and then chronologically documents the developments which followed. Most of the important events and characters are described along with a particularly good account of Coal Oil Johnny Steele. Historic photos and reproductions are included along with a useful bibliography. Highly readable. Cited in Swanson on page 149. The hardbound first edition is common.

**A20. (TEXAS) Atlantic Richfield Company, THE FABULOUS EAST TEXAS OIL FIELD, Los Angeles, CA, Atlantic Richfield, Circa 1975, 17pp.**
This is a brief history of the largest (lower 48) oil field in the United States. The pamphlet is done in black pictorial wrappers with historic sepia tone photos accompanying the short history. A first class effort. The softbound first edition is becoming scarce.

**A21. (GENERAL) Bacon, Raymond Foss and William Allen Hamor, THE AMERICAN PETROLEUM INDUSTRY, NY, McGraw-Hill Book Co., Inc., 1916, xvi, 963pp.**
This is the first of two major two-volume works by the much published authors. Generally styled as a textbook, this work was widely circulated, as was AMERICAN FUELS, a more technical work published by the same authors in 1922. Swanson calls this a: "Classic reference book, representing comprehensive survey of American petroleum industry as of publication date. History, finding, drilling, production and refining of oil, with section on shale oil". It also contains an extensive Glossary of Bitumenology. Cited in Swanson on page 178; Hardwicke page 7; Giddens page 87. The hardbound first edition is scarce.

**A22. (GENERAL) Ball, Max W., THIS FASCINATING OIL BUSINESS, Indianapolis, Bobbs-Merrill Co., 1940, 420pp.**
This is one of the top general oil industry books, giving everything from history to technology. Swanson calls it a "comprehensive, authoritative and highly readable description of oil business in all its branches". Ball is a much published and respected oil geologist who brings it all together in an easy reading format. Maps and drawings are included. Cited in Swanson on page 178. The hardbound first edition is common.

**A23. (GENERAL) Baroid, THE HISTORY OF DRILLING MUD: A BRIEF HISTORY OF WELL DRILLING PRIOR TO COMPLETION OF THE SPINDLETOP WELL IN 1901, Houston, TX, Baroid Sales Division, (1954), 23pp.**
Excellent pamphlet covering equipment, techniques and procedures for drilling. Progress is followed from the ancient Chinese methods to spring pole, to cable tool, and finally rotary drilling. The unidentified author was probably Jack D. Rittenhouse. Photos and drawings of the equipment are

included as well as copies of patents for the various drilling and coring devices. Not cited in Swanson. The softbound first edition is scarce.

**A24. (MEXICO) Barron, Clarence W., THE MEXICAN PROBLEM, Boston, Houghton Mifflin Co., 1917, xxv, 136pp.**
Written during the height of WW I, Barron makes a case for what he believes will be the salvation of Mexico. After reviewing Mexico's erratic and corrupt governments of the past, he tries to find a solution to the Mexican problem. He cites the work of Doheny and Canfield as they developed the Tampico oil fields, causing the building of the modern cities of Tuxpan and Tampico. Both of these are shown as examples to the remainder of the chaotic country. Barron gives a hard look at Mexico's social, political and economic problems, concluding that international cooperation in developing Mexico's resources is the only salvation. He further concludes that it will only happen under the guidance of a stable government which can guarantee property rights. Cited in Swanson on page 51; Hardwicke page 24. The hardbound first edition is scarce.

**A25. (TEXAS) Bartley, Ernest R., THE TIDELANDS OIL CONTROVERSY: A LEGAL AND HISTORICAL ANALYSIS, Austin, University of Texas Press, 1953, 312pp.**
Although academic in nature the book still provides interesting reading. Woodrow Wilson once wrote that every great internal crisis in the course of the nations history has turned on an issue of opposed state and national rights. Some things never change. The dispute lies between the state and feds concerning submerged and offshore lands of Texas, Louisiana and California. This book was written as the U.S. House of Representatives agreed to execute a quitclaim deed to the states to certain tidewater lands. Large appendix with tables of cases is included. Cited in Swanson on page 21. The hardbound first edition is becoming scarce.

**A26. (WYOMING) Bates, J. Leonard, THE ORIGINS OF TEAPOT DOME: PROGRESSIVES, PARTIES, AND PETROLEUM, 1909-1921., Urbana, University of Illinois, (1963), viii, 278pp.**
From the dust jacket: "This first book revealing the origins of the Teapot Dome scandal finally puts that complex, infamous event in its proper historical perspective". Bates pursues the political sectionalism between East and West also the clash between those who would develop public oil

lands and those who would conserve them. The highly political corruption scandal is described in great detail. Ultimately Albert Fall paid with jail time although his co-conspirators, Edward Doheny (Elk Hills lease) and Harry Sinclair (Teapot Dome lease) were acquitted. The developed leases were, of course, confiscated by the government. A useful bibliography and several photos are included. The hardbound first edition is becoming scarce.

**A27. (TEXAS) Benton, Minnie King, BOOMTOWN: A PORTRAIT OF BURKBURNETT, Wichita Falls, Privately Published, (1972), 65pp.**

Excellent historic photos are included in this town history. The story begins with early ranching and farming of the area. Although Corsicana Petroleum completed an oil well just west of Burkburnett in 1912 for 45 barrels per day, the area did not come alive until 1918. A group of farmers joined together to drill the Fowler wildcat number one near the Red River, specifically located as per Mrs. Fowler, who had a vivid oil premonition. The shallow Burkburnett field now covers about sixty square miles. Probably the wildest and most frantic of all boomtowns, the peaceful, modern day Burkburnett does not resemble the past. The hardbound first edition is common.

**A28. (MEXICO) Bermudez, Antonio J., THE MEXICAN NATIONAL PETROLEUM INDUSTRY, Stanford, CA, Stanford University Press, 1963, xvii, 268pp.**

The author was Director General of the Mexican national petroleum company from 1947 through 1958. He largely discusses problems he encountered during his administration and how he handled them, not the least of which was maintaining public relations and trying to improve them with the U.S. The Mexican expropriation of U.S., British and Dutch oil interests following the 1938 government decree had been pretty well settled by 1947, the year Bermudez took office. However, he still had to deal with poor public image from that, plus the separate acquisition of Gulf Oil Co's interests, which for some reason (not made clear) had not been expropriated with the others. Probably his most difficult problem involved cancellation of the private oil contracts signed with Mexican citizens. This policy was to have been a big part of the reform. The contracts and the policy were canceled. The Cities Service, Sinclair, and Standard Oil groups settled in 1942, and finally in 1947, the Royal

Dutch-Shell group which owned Mexican Eagle, settled for $130,000,000. The total settlement is reported here as $200,000,000. William Kemnitzer helped in translating the original Mexican version to English. The hardbound first edition is becoming scarce.

**A29. (WYOMING) Bille, Ed, EARLY DAYS AT SALT CREEK AND TEAPOT DOME, Casper, The Author, 1978, vii,170pp.**
Exceptional photo coverage of the Salt Creek boom and subsequent events. Many newspaper articles are reproduced here, containing stories of the characters who played a part in this great oil field. The nearby Teapot Dome oil field and its federal leasing scandal of the Harding administration is also well covered through narrative, photos and print-journal reproductions. Front endpapers are a map of the history and romance of Wyoming from the 1920's and the back endpapers are a modern satellite photo of the area with the Salt Creek & Teapot Dome anticlines clearly visible. Includes many photos and maps. The hardbound first edition is becoming scarce.

**A30. (TEXAS) Black, F. Michael, MIRANDO CITY: A NEW TOWN IN A NEW OIL FIELD, Laredo, TX, Laredo Publishing Co., (1972), 93pp.**
Story of how a seasoned lawyer-politician named Killam came to Laredo from Indian Territory in 1920 and after several tries, drilled the discovery well for the Mirando Valley Field, and later the Schott field south of Mirando City. He was soon known /$Ç1 as the greatest wildcatter in South Texas. Having had enough of oil, he decided to go into real estate and promoted the town of Mirando City from scratch. Includes promotional material, history, early photos by the author plus maps and drawings. The softbound first edition is scarce.

**A31. (GENERAL) Blakey, Ellen Sue (editor), OIL ON THEIR SHOES: PETROLEUM GEOLOGY TO 1918, Tulsa, OK, A.A.P.G., 1985, 192pp.**
Excellent documentation of early petroleum geology. In the early 1900's, it was not a very exact science and oil operators and companies were skeptical about its value. The families of early geologists supplied oil hunting stories and many photos which are generally not published elsewhere. Oil company field parties and exploratory work from around the world is documented up to the formation of the American Association

of Petroleum Geologists in 1917. The A.A.P.G. was the first organization to officially recognize the critical part played by petroleum geologists in finding oil and gas. The hardbound first edition is common.

**A32. (GENERAL) Blakey, Ellen Sue (editor), TO THE WATERS & THE WILD: PETROLEUM GEOLOGY 1918 TO 1941, Tulsa, OK, A.A.P.G., 1991, 207pp.**
Second in a planned series of three photographic chronicles of petroleum geology around the world. During this important exploration period, geology had become the number one technique for finding oil. Famous and obscure geologists are portrayed here in history and photos as they practiced their trade. Worldwide oil exploration was an "Indiana Jones" type adventure and the geologists' quest is represented here by men such as Weeks, Dott, Kirk, Owen, Oppenheim and Engleman. The priceless photos are candid and become the essence of this fine history. The hardbound first edition is common.

**A33. (MIDDLE EAST) Blandford, Linda, OIL SHEIKHS, London, Weidfeld and Nicolson, (1976), x, 286pp.**
Blandford presents the history of the Arab world from the birth of Muhammed in 571 to the assassination of King Faisal in 1975. She delves into all facets of Arab life but generally concentrates on their tradition of subjugating women. Not pretty. It seems that no one really cared when there were only poor Arabs, but the oil-rich high-profile modern day Arab incited her to this research and revelation. The hardbound first edition is common.

**A34. (GENERAL) Bone, J.H.A., PETROLEUM AND PETROLEUM WELLS, Philadelphia, J.B. Lippencott & Co., 1865, vi, 153pp.**
An early history of the oil industry in the U.S. Subtitle: "What petroleum is, where it is found, and what it is used for; where to sink petroleum wells, and how to sink them. With a complete guidebook and description of the oil regions of Pennsylvania, West Virginia, Kentucky and Ohio". The author personally observed many of the events and had published his work in other journals. He claims this to be an unbiased and reliable work. Much information not published in book form is included here. Cited in Swanson on page 32. Swanson also cites the 92 page 1st edition, also issued in 1865. All hardbound editions are rare.

**A35. (TEXAS) Bosworth, Allan, OZONA COUNTRY, NP, Privately Published, ND, xiv, 238pp.**
Ozona is sheep and goat raising country in a sleepy little ranching-farming community in Crockett County of West Texas (Named for Davy of course). With a population of 2,000, the town has a disproportionate number of wealthy people...farmers, ranchers and oil people. The town was founded in 1891, and when the author moved there as a clerk in 1907, it was still a frontier town. Chapter fifteen documents the oil and gas history. Leasing activity in 1922 led to the first well in 1925, the #1 Powell which produced 25 barrels per day. Several good sized oil fields were found while Ozona survived the depression. Natural gas was found later and the town is now nearly surrounded by one of the largest gas fields in the U.S. The hardbound first edition is common. **/$**

**A36. (PENNSYLVANIA) Botsford, Harry, THE VALLEY OF OIL, NY, Hastings House, (1946), viii, 278pp.**
The author covers events leading up to the 1859 Drake discovery and for a few years after that, as Pennsylvania went wild with oil fever. His stories are unusual and some are not covered in the more academic early oil history works. Many people assume this book is fiction since the author provides no source notes, footnotes or bibliographical references. However, the characters and events were certainly real, only a little embellished. Much on the 1865 Pithole boom and bust along with the incredible logistical problems encountered in trying to transport crude oil from the various areas to primary refining sites. A popular book and fun reading. Cited in Swanson on page 150. The hardbound first edition is common.

**A37. (TEXAS) Bowden, J.J., UNCERTAIN RICHES: THE DISCOVERY AND EXPLOITATION OF THE YATES OIL FIELD, Austin, TX, Eakin Press, 1991, vii, 186pp.**
Bowden is an attorney and has done much research on the litigation which has persisted for years over rights in the tremendous Yates Field of West Texas. He provides much detail on the development drilling and complex lease ownership. His narrative is comprehensive concerning the history of this unique field and the characters involved. Coverage is up to the 1991 Texas Supreme Court decision that the Midland Group had conspired to defraud the state of its rightful share of production. The state

sought 162 million dollars in damages. It is probably not over yet. The hardbound first edition is common.

**A38. (GENERAL) Bowles, Charles E., THE PETROLEUM INDUSTRY, Kansas City, MO, Schooley Stationery & Printing, 1921, xv, 189pp.**
The author was a statistician and publicity director for the Independent Petroleum Association and also editor of their monthly magazine. This early work, as he saw it, was written to help people, from within and outside the petroleum industry, understand its history and how it works. It is a general look at the industry, including a description of oil fields, methods used, and oil finance. Included are two exceptional, large fold-out tables, one of the U.S. oil production by state and year, the other of World oil production by country and year, also some historic photos of oil wells and oil fields. Cited in Swanson on page 186. The hardbound first edition is common.

**A39. (TEXAS) Bowman, Bob, THIS WAS EAST TEXAS: AN ANTHOLOGY OF GHOST TOWNS, Diboll, TX, The Angelina Free Press, 1966, 21pp.**
The book is printed on sepia paper and has excellent photos. Bowman covers eighteen ghost towns in some detail and another thirty are given cursory mention. The historically significant "Oil Springs" site is well described since it was where Barret did his 1866 drilling. His well was non commercial, but in 1887, the area exploded as Texas' first boom town. Unfortunately, it went bust just as fast. The pamphlet is very well done. The softbound first edition is common.

**A40. (GENERAL) Brantley, J.E., HISTORY OF OIL WELL DRILLING, Houston, TX, Gulf Publishing, (1971), xxiii, 1525pp.**
The author was a well-respected drilling contractor with world wide experience. He was also an historian of note. The book is not overly technical but contains great detail of text and is well illustrated. Everything you always wanted to know about oil well drilling is in this book. From pre-historic to modern technology along with the history of men and events which were critical to development of the oil industry. The monograph is three inches thick and definitive. It contains a useful bibliography along with many photos and drawings. The hardbound first edition is becoming scarce.

**A41. (PENNSYLVANIA) Bristow, Arch, OLD TIME TALES OF WARREN COUNTY, Meadville, PA, The Author, 1932, 389pp.**
The author disclaims this book as a history, however he did considerable research and provides many valuable stories which are not found in scholarly works. Consequently, they are only as reliable as the teller's memory. From the title page: "A collection of the picturesque and romantic lore of early days in Warren County, Pennsylvania". About one third of the book is oil related, which is expected since Warren County lies near the center of the earliest oil developments. Very interesting and enjoyable reading. Cited in Swanson on page 150; Hardwicke page 29; Giddens page 88.

**A42. (PENNSYLVANIA) Brown, George W, OLD TIMES IN OILDOM, Oil City, PA, The Derrick Publishing Co., 1911, 251pp.**
This review was prepared from the above cited fourth, revised, and last edition. The first 1909 edition contained only 79 pages, the second 1910 edition had 172pp, the third 1910 edition had 196pp. Brown provides a rambling series of very readable reminiscences over the fifty year life of the oil industry, 1895-1909. Not much biography on Brown, although he was born in 1829 within ten miles of the Drake discovery well at Titusville. He claimed to have personally observed the first two days production from the well. He also mentions smelling the crude odor from Oil Creek, years before the wildcat well was drilled. Brown was basically a lumber yard manager who only dabbled in oil a few times, with little success. He relates interesting first-hand stories of the severe logistical problems which plagued the industry not only in drilling and completing the wells, but in transporting oil to market by teams and barges in wooden barrels. He was a newspaper reporter for a time and interviewed the famous "Coal Oil Johnny" Steele shortly after he had apparently spent a million and a half dollars in oil money. Johnny told Brown that sharpers took advantage of his inexperience and cleaned him out, thus squelching the rumors that Johnny had blown it all on high living. The thirty-nine chapters contain reference to lots of dollars and barrels, but provide virtually no dates or notes. There are fascinating stories about places like Parker City, the most interesting of the oil boom towns in Pennsylvania. John Galey of Spindletop fame started his oil career just outside of Parker City. Brown tells of railroads, robbers, oilfield accidents and oil speculators. Forty-four biographical sketches are provided at the end of the book, including one on John D. Rockefeller. These were the most

influential people operating within the Oil Regions at the time. Excellent first-hand accounts of everyday life during the birth and early development of the oil industry. One of the best. Cited in Swanson on page 150; Hardwicke page 30; Giddens page 88. The softbound first edition is rare, all hardbound editions are scarce.

**A43. (GENERAL) Byles, Axtell J., AMERICA'S PETROLEUM INDUSTRY: A REVIEW AND COMMENTARY, NJ, Guild of Brackett Lecturers, 1938, 28pp.**
The author was president of the API and delivered this address at Princeton in 1937. It is compact and styled like a Newcomen pamphlet. Byles covers every facet of the oil industry from a brief history through the negative effect taxes were having on the producers at the time. Not cited in Swanson. The softbound first edition is common.

**A44. (LOUISIANA) Carter, Hodding, JOHN LAW WASN'T SO WRONG: THE STORY OF LOUISIANA'S HORN OF PLENTY, Baton Rouge, Esso Standard Oil, 1952, 105pp.**
The title comes from John Law, the Scotsman who controlled the French treasury for Louis XV, who had Louisiana as part of his colonial enterprise. This is a comprehensive promotional book on Louisiana with much on the oil and gas industry. Chapter 10 deals with the oil history which began nine months after the Spindletop discovery in Texas. A wildcatter named Heywood brought in a gusher near the town of Jennings. Louisiana became one of the top oil producing states shortly thereafter. Much on Esso's part in exploration, production, refining and transportation. A twenty- two page chronology of the state history and industry is found at the back of the book 1519-1952. Contains maps and photos. Not cited in Swanson. The softbound first edition is common.

**A45. (GENERAL) Carter, Worrall Reed, BEANS BULLETS AND BLACK OIL: THE STORY OF FLEET LOGISTICS AFLOAT IN THE PACIFIC DURING WORLD WAR II, Washington, Government Printing Office, (1953), xix, 482pp.**
The foreword is by Secretary of the Navy, Kimball and the introduction is by Admiral Spruance, "Who helped turn back the Japanese at Midway and later took from them the Gilberts, Marshalls, Marianas, Iwo Jima, and Okinawa". Many maps and photos along with the detailed war history from Alaska to Australia. Each campaign and its distribution of fleet tankers and fueling ships in the Pacific Theater is chronicled. Logistical

superiority won the war in the Pacific and the tanker fleet was integral. Cited in Swanson on page 150. The hardbound first edition is scarce.

**A46. (TEXAS) Carson, Ira (compiler), OZONA, TEXAS: CROCKETT COUNTY 1891-1966 DIAMOND JUBILEE, Ozona, TX, Crockett County Historical Society, (1966), viii, 90pp.**
Ozona motto "The Biggest Little Town in the World", is the only town in the gigantic Crockett County of West Texas. The town was founded about 1891. Much detail on the town with chapter eleven covering the oil industry. Small oil production was found in 1920 and followed in 1925 by discovery of World, Watt, and Todd Deep fields. Flush gas production was discovered in 1962 and spread into a wildfire development which continued through the 1980's. Ozona is the major gas field which later became one of the largest in the U.S. by areal extent. The softbound first edition is common.

**A47. (ALASKA) Chasan, Daniel Jack, KLONDIKE '70: THE ALASKAN OIL BOOM, NY, Praeger Publishers, (1971), 184pp.**
The author was a staff writer for the New Yorker Magazine. He gives a fair amount of history from the giant 1968 Atlantic Refining Company oil discovery on Naval Petroleum Reserve #4 at Prudhoe Bay. He is even more interested in the ins and outs of the growing political and environmental controversy. One of the most popular books on the Alaska boom, although it was written long before the full impact of the oil field would be known. The hardbound first edition is common.

**A48. (GENERAL) Clark, J. Stanley, THE OIL CENTURY: FROM THE DRAKE WELL TO THE CONSERVATION ERA, Norman, University of Oklahoma Press, (1958), xxii, 280pp.**
From the title page: "It is a story of the quick grab for mineral riches; of unpredictable results in times when geology had yet few or no applications; of wild-flowing wells and insufficient storage and pipe line facilities; ....In short, it is the record of the greatest bonanza of them all". All major characters and events are reviewed up to and including the problem of imports which, by the late 1950's were beginning to seriously affect the U.S. oil industry. A well organized, well written, widely circulated and popular history. Cited in Swanson on page 150. The hardbound first edition is becoming scarce.

**A49. (TEXAS) Clark, James A., GLADYS CITY. TX, Reprint from Beaumont Enterprise-Journal newspaper (1976). 13pp.**
A series of James Clark columns as they appeared in the Beaumont Enterprise-Journal regarding the Spindletop history. Much of the history is centered on Patillo Higgins' Gladys City Oil Co. and his persistent efforts to get oil from the low-rising hill near Beaumont, called Spindletop. The softbound first edition is becoming scarce.

**A50. (GENERAL) Clark, James A., THE CHRONOLOGICAL HISTORY OF THE PETROLEUM AND NATURAL GAS INDUSTRIES, Houston, TX, Clark Book Company, (1963), vi, 317pp.**
Edited by Charles Warner, famous Texas oil historian. Six thousand entries trace the history of petroleum from 6000 BC to 1962. Brief paragraphs describe, by year and date, important oil events pertaining to scientific and technological development within the exploration, drilling, production, transportation, refining, marketing and research fields. Ten years of research were required to prepare this major reference book. Impressive. The hardbound first edition is becoming scarce.

**A51. (TEXAS) Clark, James A. and Michel T. Halbouty, SPINDLE-TOP, NY, Random House, (1952), xvi, 306pp.**
One of three oil books granted full bibliographical status in Jenkins' BASIC TEXAS BOOKS, Number 30. From Jenkins: "This is the best account of the great discovery of oil near Beaumont dome, where oil became an industry". From the dust jacket: "The true story of the oil discovery that changed the world...of the events leading up to it and the boom days that followed." The Drake discovery at Titusville, Dad Joiner's discovery near Kilgore, and the Atlantic discovery at Prudhoe Bay are three other events which created major changes in the U.S. oil industry and in the country. The Lucas discovery at Spindletop, near Beaumont was equal to the others in all respects. The Spindletop story has so many individuals and companies involved in its discovery and later development that the detailed telling could easily create volumes. Of considerable importance is the fact that born from the wildcat were major oil companies, Texaco, Gulf, Sun and Humble. Other independent oil companies and individual fortunes were also created from the bonanza.

Historically important individuals involved are Patillo Higgins (the idea man), Doc Lloyd (scientific staff), Captain Anthony Lucas (driller-expediter), Guffey & Galey (more science and expediting), the Mellon family (money), The Hammil family (drilling-technical), and many others. Clark and Halbouty wrote a popular, easy-reading and exciting history of the field from their intimate knowledge of the people and events. Reprinted in 1984 by Gulf Publishing. Cited in Swanson on page 150. The hardbound Beaumont and H&P editions are scarce. The hardbound first edition is becoming scarce. Others are common.

**A52. (TEXAS) Clark, James A. and Michel T. Halbouty, THE LAST BOOM, NY, Random House, (1972), 305pp.**

This one is also mentioned in Jenkins' BASIC TEXAS BOOKS on page 79 as a superb book about the discovery. From the dust jacket "It was all of the oil booms in history rolled into one". This is the story of the East Texas field, told by Texas' premier oil historian. It is the definitive work on the history of the Black Giant, which was discovered in early October 1930. A fascinating story from Dad Joiner, with his trials, tribulations and persistence to drill on the Bradford farm, to proration, to the big inch pipeline. As Spindletop before it, East Texas became the vehicle which made small oil companies large, and which made individual speculators and investors wealthy enough to start their own oil companies. H.L. Hunt was the biggest name to come out of the field as he built an empire on Dad Joiner's success. There are many more, some of whom are still enjoying income from the field. All of the characters and all of the events are covered and provide exciting reading. Reprinted in 1984 by Shearer. The hardbound first edition is becoming scarce.

**A53. (PENNSYLVANIA) Clark, Mabel K., TITUSVILLE: AN ILLUSTRATED HISTORY, Cambridge, MD, Western Publishing, 1976, 192pp.**

A bicentennial project, the book is a narrative of the coming of settlers and of their struggles in the development of the area. First in the area were Jonathan Titus and Samuel Kerr, during 1796. Much on the lumbering and oil history of this unique city. The author of this well-composed and illustrated book is a native Titusville school teacher and historian. The hardbound first edition is becoming scarce.

**A54. (OKLAHOMA) Clinton, Fred S., FIRST OIL AND GAS WELL IN TULSA COUNTY, Oklahoma City, Oklahoma Historical Society, 1950, pp312-332.**

A special reprint from THE CHRONICLES OF OKLAHOMA, Volume XXX, Number 3. Frontispiece is a photo of the monument erected at Red Fork on U.S. Hwy 66 dedicated March 23, 1950. The well was completed on June 25, 1901 and was drilled by two doctors, Bland and Clinton. It brought first nationwide publicity and an oil boom to Indian Territory. Subsequent rapid industrial development made Tulsa "Oil Capital of the World." A well prepared commemorative pamphlet, although historians later discredited the Clinton-Bland claims to fame and properly gave it to they Heydrick family. Not cited in Swanson. The softbound first edition is common.

**A55. (ALASKA) Cohen, Stan:, THE GREAT ALASKA PIPELINE., Missoula, MT, Pictorial Histories Publications., (1988), 134pp.**

This is an excellent, although brief history of oil in Alaska. From first reports of natural oil seeps to early developments and production in 1907 to the establishment of Naval Petroleum Reserve #4, to the Prudhoe Bay discovery and finally the pipeline. Color as well as black and white photos along with reproduced newspaper headlines emphasize the fabulous history of Alaska Oil. The importance of the pipeline cannot be overstated in that it currently (1995) accounts for about twenty-five percent of the domestic U.S. oil production. Maps are included. The softbound first edition is common.

**A56. (OKLAHOMA) Collings, Ellsworth and Alma Miller England, THE 101 RANCH, Norman, University of Oklahoma, (1937), xxx, 255pp.**

The ranch was founded in the 1890's by Colonel George W. Miller, in two short decades the 101 Ranch achieved legendary status. Spreading over 110,000 acres of land in the Ponca Indian country of north central Oklahoma, it became a showcase for experimental agricultural methods under the management of the Colonel's sons. E. W. Marland had his early oil exploration success on the ranch and eventually developed several good sized oil fields there. The Millers lost the ranch after the crash of 1929. Numerous reprints. Not cited in Swanson. The hardbound first edition is scarce.

**A57. (GREAT BRITAIN) Collins, George B., WILDCATS AND SHAMROCKS, N. Newton, KS, Mennonite Press, (1976), 90pp.**
This is a chronicle of the search for oil and gas off the Irish coast. Collins (of Irish decent) was a successful oil company attorney in the U.S. for years and became interested in doing something positive for his ancestor's homeland. He was convinced that offshore Irish oil was a viable project based primarily on onshore showings previously encountered in a long string of dry holes. After eighteen years and countless more dry holes, he was finally rewarded by Marathon's 1975 offshore discovery at Kinsdale Gas Field. Persistence pays again. The hardbound first edition is common.

**A58. (PENNSYLVANIA) Cone, Andrew, and Walter R. Johns, PETROLIA: A BRIEF HISTORY OF THE PENNSYLVANIA PETROLEUM REGION, ITS DEVELOPMENT, GROWTH, RESOURCES, ETC., FROM 1859 TO 1869, NY, D. Appleton and Company, 1870, iv, 652pp.**
The word brief in the title is misleading, as this is a massive work, containing everything that was known about oil in the first ten years of the industry's existence and particularly around its starting place. The author did considerable research in detailing the earliest recorded mention of oil and or asphalt and brings it up to the current scene in Pennsylvania. Oil had been known around the world for centuries, as it was found in seeps, springs, pits and in salt wells. Drake and his business associates went a step further and actually drilled for oil, finding it in good quantities, starting the industry which today touches everyone's lives. Oil was being sold from the Oil Region as early as 1803, largely skimmed from Oil Creek. Before that, Indians had mined for it and used it for something or other. Good detail is given on early refining methods which produced kerosene for lighting. Also, much information is provided on drilling technique and equipment. The accompanying problems, and procedures for overcoming them, such as control of flowing wells and handling the associated gas are covered. Over 300 pages of detailed descriptions of individual farms and wells includes: ownership, location (and how to get there), depth, cost, pay thickness, number of wells flowing, pumping and dry. Cone also gives his impression of future development possibilities on certain farms. Lots of statistics. A chapter titled Flood, Fire and Financial Disaster gives details of those common problems. Floods, both natural and from engineered pond freshets usually

created havoc on Oil Creek as the barges loaded with barrels attempted to float down river to the Allegheny. Fires were often ignited from careless people as oil and gas constantly leaked from producing facilities. Wild speculation and price volatility routinely created financial chaos. Many pages of statistics on oil and product production, storage and consumption are presented with other useful tables. This is a much sought oil book. Venango County Historical reprinted the book in 1984. Cited in Swanson on page 150; Hardwicke page 30; Giddens page 88. The hardbound first edition is very scarce.

**A59. (TEXAS) Conger, Mrs. Joe, McCAMEY, CHILD OF BLACK GOLD: 1925-1975 50TH ANNIVERSARY, McCamey, TX, Privately Published, (1975), 96pp.**

Fifty years of history for this little West Texas oil town. McCamey is located in the center of the Permian Basin and surrounded by shallow oil production. Not much oil is mentioned in the book, but McCamey probably would not have lasted long without it. Typical city history prepared by a proud civic committee. The hardbound first edition is becoming scarce.

**A60. (PENNSYLVANIA) Conn, Frances G. and Shirley Sirota Rosenberg, THE FIRST OIL RUSH, NY, Meridith Press, (1967), xi, 141pp.**

The authors did a considerable amount of research prior to presenting this early Pennsylvania oil history. The book is in compact form, but manages to touch on many major characters and incidents. Several photos are also included. This is not one of the top early oil books, but worth reading to learn the author's idea of what oil history events are the most important and interesting. The hardbound first edition is common.

**A61. (TEXAS) Cotten, Kathryn, SAGA OF SCURRY, San Antonio, TX, The Naylor Co., (1957), 165pp.**

Snyder is the Scurry County seat and its history is revealed along with several of the other smaller county towns. Much detail is provided on the important citizens. The author spends little time on oil but has unusual photos of Walter Lechner and E. E. Fogleson with their early well. Scurry is one of three, billion- barrel West Texas counties and when this was written, oil production had reached 200,000 barrels per day. Many photos are included. The hardbound first edition is scarce.

**A62. (GENERAL) Crew, Benjamin J., A PRACTICAL TREATISE ON PETROLEUM, Philadelphia, Henry Carey Baird & Co., 1887, xxiv, 508pp.**
The treatise comprises its origin, geology, geographical distribution, history, chemistry, mining, technology, uses and transportation, together with a description of gas wells, the application of gas as a fuel, etc. Quoting from Swanson: "Author connected with petroleum refining from early days; died before manuscript complete. Work completed by brother, J. Lewis Crew, and others". This is a very comprehensive early work. Usually bound in the back is a thirty-one page Baird & Co. book catalog. Cited in Swanson on page 178; Hardwicke page 8, Giddens page 89. The hardbound first edition is very scarce.

**A63. (PENNSYLVANIA) Crum, A.R. and Dungan, A.S. (editors), ROMANCE OF AMERICAN PETROLEUM AND GAS: VOLUME I, NY, Romance of American Petroleum Company, (1911), 363pp.**
The author died shortly after completing this major work which largely deals with the first fifty years of oil development in Pennsylvania. Later the Titusville play extended to the massive Bradford field and into New York. Lima, Ohio then boomed and later, California. Crum provides an excellent photo-biographical section of the major players up to 1911 along with information on their companies. Historic photos of places and events are also included in this comprehensive work, which essentially reviews everything that was known of American oil history at the time. There are 246 pages of the History and Romance of Oil followed by 110 pages of biographical sketches. A second volume was completed by Crum's associates ten years later. Cited in Swanson on page 151; Hardwicke page 30. The hardbound first edition is very scarce.

**A64. (GENERAL) Curtin, D. Thomas, MEN OIL AND WAR, Chicago, Petroleum Industry Committee, (1946), 360pp.**
Author's statement: "The writer found absorbing interest and drama in the oil transportation revolution by which the Middle West and the Gulf States rushed oil overland to save the East...when the East rushed tankers to save the United Nations". Largely a history of the logistical organization and operations of the petroleum industry in fifteen Middle Western states comprising District No. 2 of Petroleum Administration for War. Curtin was a war correspondent and prepared this volume for the

National Archives. Amazing story of how the war needs were met. Includes photos, maps and graphs. Cited in Swanson on page 151. The hardbound first edition is common.

**A65. (PENNSYLVANIA) Darrah, William Culp, PITHOLE, THE VANISHED CITY: A STORY OF THE EARLY DAYS OF THE PETROLEUM INDUSTRY, Gettysburg, PA, Privately Published, (1972), viii, 252pp.**
From the dust jacket: "Pennsylvania's fabulous ghost town. Born in the wilderness, 1865; died 1877". The famous boomtown grew to 20,000 people and within a few short years was back to wilderness again. The United States Petroleum Company started it all in 1865 with their well on the Frazier farm which flowed 250 barrels per day. The price of crude was $8.00 per barrel and their company stock increased eight fold almost overnight. There are great photos of the original well and other confirmation wells which were drilled the next week. Lots of stories. Good photo-documentation of the town growing, then shrinking. Included is a useful bibliography. The hardbound first edition is common.

**A66. (GENERAL) Davis, William, IT'S NO SIN TO BE RICH: A DEFENSE OF CAPITALISM, NY, Thomas Nelson, (1977), 264pp.**
The author discusses the rich of the past, of today, and of the future. Much on Rockefeller, Getty and the oil rich Arabs. Somewhat right-wing conservative effort, with humor. The hardbound first edition is common.

**A67. (BIBLIOGRAPHY) DeGolyer, Everette and Harold Vance, BIBLIOGRAPHY ON THE PETROLEUM INDUSTRY, College Station, TX, Agricultural and Mechanical College of Texas, 1944, Bull 83, Vol 15, No. 11, xxxii, 730pp.**
This is a major bibliographical effort, consisting of 12,000 entries which were amassed by the famous geologist and book collector, DeGolyer, with the aid of the Texas A&M library staff. The format is difficult for anyone other than a serious oil industry researcher to use. It is essentially a comprehensive list of periodical articles. The bibliography is descriptive and not annotated. It is cited in Swanson on page 184. The hardbound first edition is scarce.

**A68. (GENERAL) DeGolyer, E., THE ANTIQUITY OF THE OIL INDUSTRY: WITH COPIOUS NOTES AND REFERENCES BY THE AUTHOR, Godley, TX, The Peripatetic Press, 1947, 20pp.**
Facetious look at the beginnings of the oil industry by one of its best known bibliophiles. Biblical quotes as to Noah being the first oilman and Job being the first geologist. Many important oilmen are quoted in the book with their reactions to the twelve page first 1946 edition of this DeGolyer spoof. Not cited in Swanson. This printing was limited to 300 copies and all softbound editions are very scarce.

**A69. (GENERAL) DeGolyer, E., THE DEVELOPMENT OF THE ART OF PROSPECTING: WITH SPECIAL REFERENCE TO THE AMERICAN PETROLEUM INDUSTRY, Princeton, NJ, Princeton University Press, 1940, 38pp.**
This booklet is an address delivered before Princeton University on December 22, 1939, in the Cyrus Fogg Brackett Lectureship in Applied Engineering and Technology. DeGolyer, the brilliant explorationist and historian provides good detail of oil prospecting from before the Drake discovery, up to 1939. He reviews geological thinking from the early days to the most modern efforts used in determining the best place to drill an exploratory well. He was a pioneer in utilizing the exploratory geophysical tools of seismic and gravity meter in America and provides that first-hand history. His thrust and conclusion generally is that subsurface mapping through the use of all available well information and other geological-geophysical tools is the best technique available for modern prospecting. He did miss the boat when he indicated that the use of reflection seismography had about run its course as other geophysical tools had, by 1939. It is still (1995) the premier exploration tool due to the incredible sophistication and power available through computer processing. Cited in Swanson on page 3. The softbound first edition is scarce.

**A70. (PENNSYLVANIA) Dolson, Hildegarde, THE GREAT OILDORADO, THE GAUDY AND TURBULENT YEARS OF THE FIRST OIL RUSH: PENNSYLVANIA 1859-1880, NY, Random House, (1959), xi, 300pp.**
Dolson was a very enthusiastic novice writer and became caught up in the excitement of oil history. She relates the most interesting and unusual,

often undocumented stories along with the actual history as it has been accepted over the years. She puts her own spin on the facts and makes the area come alive again for the reader. This work is not too serious, but it is fun reading. Cited in Swanson on page 151. The hardbound first edition is common.

**A71. (TEXAS) Eason, AL, BOOMTOWN: KILGORE, TEXAS, Kilgore, TX, Chamber of Commerce, Nodate, ix, 144pp.**
A good history of the city which was one of the wildest boomtowns. It is located right in the middle of one of the largest oil fields in the world. The author writes of the rangers and lawmen along with the girls and bars in the first section. In the second, he gives the history of Kilgore and in the third, presents biographical sketches on Kilgore pioneer families. Photos of downtown Kilgore with the wells almost on top of each other are excellent. Also there is an unusual photo of Mrs. Lou Della Crim in a rocker on her front porch with three pumping wells and derricks in her yard. The Bateman & Cain well was drilled on her land and extended the East Texas field nine miles from the original discovery. The hardbound first edition is common.

**A72. (CALIFORNIA) Easton, Robert, BLACK TIDE: THE SANTA BARBARA OIL SPILL AND ITS CONSEQUENCES, NY, Delacorte Press, (1972), xvi, 336pp.**
From the dust jacket: "When offshore well A-21 blew out, it started a chain of events that reached the highest levels of government, and repercussions persist to this day". A major work on a touchy subject. Was it an accident, or negligence, or a combination of unlucky events? This unfortunate incident remains the environmentalist's rallying point, anytime the subject of offshore oil comes up. One of the best of many books on the oil spill, this one includes an accurate chronology. The hardbound first edition is common.

**A73. (PENNSYLVANIA) Eaton, Reverend S.J.M., PETROLEUM: A HISTORY OF THE OIL REGION OF VENANGO COUNTY, PENNSYLVANIA, Philadelphia, J.P. Skelly & Co., 1866, x, 299pp.**
Continued subtitle: "Its resources, mode of development, and value: embracing a discussion of ancient oil operations; with a map, and illustrations of oil scenes and boring implements". Eaton, in writing this early documentation of the oil industry, says he: "Aims at nothing more

than a mere popular description of the oil region, of the early history of petroleum operations, of the manner of boring oil wells, bringing the oil to the surface, and preparing it for use". Oil Creek flows southward through the county to join the Allegheny River. Oil city and Franklin are in the center of the county and Titusville is just outside the north border. Eaton methodically recounts all that was known of oil at the time. The first accounts of oil in Venango County were from Indians who utilized the oil long before the 1800's for either medicine, ceremonies, water proofing, or something we cannot even imagine. Blanket skimming was used by white settlers up to the early 1850's to supply the small medicine market. In 1855, Bissell and his partners thought the small market had big potential. They proceeded to have the famous Yale chemist, Dr. Benjamin Silliman, Jr., study the liquid and prepare a report of its potential commercial uses. He touted the distillation of kerosene which he believed could conceivably replace whale oil and coal oil as an illuminant. Coal oil was expensive to process and they were running out of whales. American ingenuity saved the day. Eaton describes the origin of petroleum, and everything else mentioned in the book title. He does not spend much time with the individual wells and farms, however he has put together a wealth of information, which, if assimilated by a newcomer to the area, would have provided enough oil knowledge to go into the business in any phase that appealed, be it leasing, drilling, refining, transporting, supply or as many chose, promoting. A fold out map of Venango County, showing rivers, towns and important oil producing farms is included. Important book. Venango County Historical reprinted the book in 1984. Cited in Swanson on page 151; Hardwicke page 32; Giddens page 90. The hardbound first edition is very scarce.

**A74. (MIDDLE EAST) Edmonds, I.G., ALLAH'S OIL: MIDEAST PETROLEUM, Nashville, TN, Thomas Nelson, Inc., (1977), 160pp.** The author grew up in several oil boomtowns in the American Southwest and has since published numerous books on culture in Asian and African countries. This is a fairly unbiased look at the Middle East from first oil in 1908 to after OPEC and the embargos. From the DJ: "..compelling story of Mideast oil and the part it has played in modern history. Today, with Arab states brandishing the oil weapon in defense of their foreign policies, I.G. Edmonds' revealing study is must reading for everyone concerned with the world's future". Contains a detailed chronology of events. The softbound first edition is common.

**A75. (GENERAL) Egloff, Gustav, EARTH OIL, Baltimore, Williams & Wilkins Co., 1933, xi, 158pp.**
Egloff gives the history of exploration and drilling for oil around the world. He discusses most aspects of the industry, with particular attention to refining and the uses of petroleum products. Much on his ideas of where future oil resources will come from. Quoting from the conclusion: "In view of the known crude oil reserves, the vast amount left underground by present production methods, the undiscovered fields that no doubt exist, and modern refining, an oil shortage, despite the gloomy prophets, is far in the future". Cited in Swanson on page 187; Hardwicke page 9. The hardbound first edition is becoming scarce.

**A76. (MIDDLE EAST) El Mallakh, Ragaei, QATAR: DEVELOPMENT OF AN OIL ECONOMY, NY, St. Martin's Press, (1979), 183pp.**
Very good coverage of the oil history of this tiny country. The strategic Arab state is located between Kuwait and Iraq and the Strait of Hormuz. The country first exported crude oil in 1947 by Petroleum Developments Limited, which was owned by Shell, Standard, Mobil and the French Company. Gulbenkian family was in for five percent. Somewhat political, the chapters cover finances and economic policies involved in the industrialization of the country. The hardbound first edition is common.

**A77. (TEXAS) Elder, Jack & Caleb Pirtle III, THE GLORY DAYS, Austin, TX, Nortex Press, 1986, iv, 60pp.**
Impressive photo-essay on Kilgore and the East Texas Field. Printed entirely in sepia. The history is short but the photos of the derricks, almost touching, are impressive. Some are aerial shots and most are of downtown city blocks taken while the steel derricks were still up. Elder took most of the photos himself during the 1930's. The hardbound first edition is becoming scarce.

**A78. (GENERAL) Erni, Henri, COAL OIL AND PETROLEUM: THEIR ORIGIN, HISTORY, GEOLOGY AND CHEMISTRY, ETC., Philadelphia, Henry Carey Baird, 1865, viii, 196pp.**
This book was not available for review by Neff, but is regarded as important. Quoting from Swanson: "Author was Chief Chemist to Dept. of Agriculture; articles originally written for Washington D.C.

newspaper". Cited in Swanson on page 151; Hardwicke page 32; Giddens page 90. The hardbound first edition is very scarce.

**A79. (GENERAL) Ethyl Corporation (editors), ETHYL NEWS: OIL CENTENNIAL ISSUE, MAY-JUNE 1959, NY, The Corporation, 1959, 40pp.**
History of oil, starting with an early 1755 map by Evans which indicated petroleum at several sites in Pennsylvania and near the 1859 Drake drillsite. From the foreword: "..articles relating to the birth, early development and growth of oil, it is designed to pay tribute to pioneer oilmen and events...appeared in previous issues of this magazine over a period of almost fifteen years". Only four pages on Ethyl. Ethyl also funded the reprinting of several rare oil publications such as the Evans map and Gale's WONDER OF THE NINETEENTH CENTURY. Interesting article on Ethyl's extensive collection of early 1900's "Dime Novels" with oil plots, and their relationship to oil history. The softbound first edition is becoming scarce.

**A80. (Great Britain) Fairhall, David, and Philip Jordan, THE WRECK OF THE AMOCO CADIZ, NY, Stein and Day, (1980), 248pp.**
The book is touted as the first to look at a major oil pollution incident in detail and to set out its findings for the general and concerned reader. From the cover: "The story of the greatest ecological disaster at sea and what it bodes for the future". From the DJ" "..wrecked on the coast of Brittany in March 1978, deep-laden with 220,000 tons of crude oil, it was the largest disaster of its kind the world had yet seen". The hardbound first edition is common.

**A81. (GENERAL) Fanning, Leonard M., MEN, MONEY AND OIL: THE STORY OF AN INDUSTRY, Cleveland, The World Publishing, (1966), 188pp.**
One of the most respected and prolific of oil authors, Fanning here tells stories of powerful oil figures from Rockefeller to Mossadegh along with many others in between. He discusses the economic, social and political implications of the struggle for power through oil. The hardbound first edition is becoming scarce.

**A82. (GENERAL) Fanning, Leonard M., OVER MOUNTAINS, PRAIRIES AND SEAS: THE OIL TRADE, NY, McGraw-Hill Book Co., (1968), 128pp.**

The last of Fanning's many excellent oil industry books and one of the best. This is an excellent photo-enhanced history of oil transportation and trade from the first barrels moved from Drake's well to the modern supertankers. Much on the WW II tanker sinkings off the Atlantic Coast as well as the importance of the Suez crisis to oil shipping. The hardbound first edition is becoming scarce.

**A83. (GENERAL) Fanning, Leonard M., THE RISE OF AMERICAN OIL, NY, Harper & Brothers, 1936, ix, 221pp.**

Quoting Swanson: "Oil history and achievement of oil industry in meeting needs of advancing civilization". This was a very popular non technical account of the part played by petroleum in the life of this country. It was a widely circulated book, heralded by oil companies and organizations for the way it portrayed the oil industry in history and today. Written in an easy reading style using adventure and entertainment as part of the story. Highly respected oil author. Second 1948 edition has 178pp. Cited in Swanson on page 31; Hardwicke page 9; Giddens page 90. The hardbound first edition is becoming scarce.

**A84. (ALASKA) Fears, Jerry, BOOM, CASH AND BALDERDASH: A DIFFERENT LOOK AT FAIRBANKS DURING PIPELINE CONSTRUCTION, Fairbanks, AK, That New Publishing Co., (1978), 175pp.**

Fears is a television news director. He prepared this book as a compilation of his many commentaries (selected) which chronicled the difficult period in Fairbanks while the 800-mile 48-inch oil pipeline was being constructed. The hardbound first edition is common.

**A85. (CANADA) Finnie, Richard, CANOL: THE SUB-ARCTIC PIPELINE AND REFINERY PROJECT CONSTRUCTED BY BECHTEL-PRICE-CALLAHAN FOR THE CORPS OF ENGINEERS UNITED STATES ARMY 1942-1944, San Francisco, Bechtel-Price-Callahan, 1945, ix, 210pp, Large.**

This is a picture history of one amazing wartime construction project detailed in fifteen chapters from the idea stage through completion. About a hundred and fifty good sized, black and white and color photos of the project are impressive as to quality and subject. At the start of WW II, Japan was harassing the west coast and hanging around the Aleutian Islands. The U.S. and Canada decided that an inland fuel supply route to Alaska would be advisable to keep the air force in business in case Japan cut off shipping entirely. Norman Wells field on the Mackenzie river in far northern Canada was capable of 3,000 barrels per day and was virtually the only source. Bechtel and others were contracted to build the pipeline from the field and also a refinery which would deliver aviation gas and other fuels. Two years later it was done, however the threat had passed and the project functioned only sporadically, becoming a well-intentioned debacle. The hardbound first edition is very scarce.

**A86. (GENERAL) Floherty, John J., FLOWING GOLD: THE ROMANCE OF OIL, NY, J. B. Lippencott Co., (1945), 255pp.**
Prior to this effort, the author had written some twenty books on various industries from a first-hand perspective. His books are generally targeted at high school seniors who would probably choose careers in industry, business or government. Here he describes all phases of the oil industry based on his personal contacts and interviews while depicting the romance of the oil industry. His stories are unusual, as they include little-known facts and men behind a great industry. The author's own photos and drawings are included to show the modern oil industry in action. A very popular book. Second 1957 edition has 223pp. Cited in Swanson on page 187. The hardbound first edition is becoming scarce.

**A87. (GENERAL) Forbes, Gerald, FLUSH PRODUCTION: THE EPIC OF OIL IN THE GULF-SOUTHWEST, Norman, University of Oklahoma Press, 1942, xi, 253pp.**
Forbes was widely published in the oil field, being a journalist, historian and editor of many Texas and Oklahoma newspapers. This is generally a social history of oil where he integrates all phases of the industry into his story. According to Swanson it is: "Largely the development of oil and gas in Gulf-Southwest; histories of selected major fields; oil field lore". Styled as a dialogue between himself and an exploration geologist, the book is reasonably informative. Cited in Swanson on page 151. The hardbound first edition is becoming scarce.

**A88. (GENERAL) Forbes, R.J., STUDIES IN EARLY PETROLEUM HISTORY, Netherlands, E.J. Brill, 1958, ix, 199pp.**
Many authorities believe this and two other pre-Titusville oil histories by Forbes are incomparable. Quoting Swanson: "Petroleum and its usage in the ancient world, Middle Ages, and up to Drake well; early Galician, Rumanian, and Russian production and refining, with statistics. A series of separate papers by a distinguished petroleum scholar". This is definitely a different look at oil history. Well illustrated with photo-plates and drawings. The other Forbes titles are: MORE STUDIES IN EARLY PETROLEUM HISTORY, 1959 and BITUMEN AND PETROLEUM IN ANTIQUITY, 1936. Cited in Swanson on page 151; Hardwicke page 32. All hardbound editions are Scarce.

**A89. (CALIFORNIA) Franks, Kenny A. and Paul F. Lambert, EARLY CALIFORNIA OIL: A PHOTOGRAPHIC HISTORY, 1865-1940, College Station, Texas A&M University Press, (1985), xiii, 243pp.**
Franks and Lambert do their usual exceptional job of presenting oil history with an array of historic photos. This is number four of the Montague History of Oil Series and complements Rintoul's three detailed California photo histories. More than three hundred photographs portray the development of California's rich and colorful petroleum industry from 1865 to 1940. Frontispiece is a photo of the Lakeview gusher at North Midway Field. The well is flowing 90,000 barrels per day and shown in mirror-effect across a large temporary lake of oil. Much on Standard of California and Union Oil of California in their early days. The areas are reviewed by geologic sub-province or field as Santa Maria Basin, Los Angeles Basin, Coalinga and Northern California Basins, McKittrick-Kern River, Midway-Sunset and Elk Hills. The hardbound first edition is common.

**A90. (OKLAHOMA) Franks, Kenny A., OKLAHOMA: THE LAND AND ITS PEOPLE, Helena, MT, American and World Geographic Publishing, 1994, 104pp, Oklahoma Geographic Series No. 1.**
The much published authors do another fine job on their state. From the promotion material: "This popular geography gives an overview of the geology of Oklahoma, then moves to chapters on each area of the state. History includes Indian Territory, Oklahoma Territory, the Unassigned

Lands, Indian Removal Acts, land runs, the oil industry and more". The book is well illustrated with historic and current photos. Oil is pervasive in the six general area chapters, although the chapter titled "Black Gold in the Red Man's Land" details the earliest oil history in northeastern Oklahoma with particular emphasis on the Osage. The softbound first edition is common.

**A91. (LOUISIANA) Franks, Kenny A., EARLY LOUISIANA & ARKANSAS OIL: A PHOTOGRAPHIC HISTORY, 1901-1946, College Station, Texas A&M University Press, (1982), xiv, 243pp.**
Countless historic photos are shown from the 1901 Jennings field discovery in Louisiana to the giant El Dorado and Smackover fields of Arkansas in the 1920's. The authors insert a fair amount of history into this pictorial chronology of the significant oil and gas area. Much on Bennedum and Trees, Kerr-McGee and Standard Oil in the role of developing the numerous Louisiana and Arkansas oil fields. Bibliography. The hardbound first edition is common.

**A92. (OKLAHOMA) Franks, Kenny A:, RAGTOWN: A HISTORY OF THE GREATER HEALDTON-HEWITT OIL FIELD, Oklahoma City, Oklahoma Heritage Association, (1986), viii, 139pp.**
Just west of the city of Ardmore lies one of the major oil areas in the U.S. The area was known to have active oil seeps and surface asphalt from pre-history. Drilling encountered uneconomical oil in this area in the 1880's, however the first big development started in 1906 at Sho Vel Tum, north of Healdton (locally called Ragtown). Wirt Franklin and Samuel Apple, two Attorneys from Ardmore formed Crystal Oil Company and acquired some 6,000 acres around a dry hole plugged in 1888 which had oil showings. They discovered the Healdton field in 1913 with their wildcat flowing 100 barrels per day. Ragtown had already become a regional supply and boomtown, but this discovery set off more frantic development. Good history and photos of a prolific oil area. The hardbound first edition is becoming scarce.

**A93. (OKLAHOMA) Franks, Kenny A., THE OKLAHOMA PETROLEUM INDUSTRY, Oklahoma City, Oklahoma Heritage Association, (1980), xvi, 283pp.**
Franks presents a complete history of Oklahoma oil with much detail and many photos. This book spawned many more of his excellent picture

histories. It can be used as a companion book to his "Early Oklahoma Oil". Coverage is of most important boomtowns, oil companies and famous characters, from the unbelievably wealthy Osage Indians to the residual philanthropies left by the great oil finders. Comprehensive. The hardbound first edition is common.

**A94. (OKLAHOMA) Franks, Kenny A., THE OSAGE OIL BOOM, Oklahoma City, Oklahoma Heritage Association, (1989), xiv, 180pp.** Great early photos of the Indians and the oil men. Franks gives much history of the Osage from their forced move to the 1.5 million acre reservation through the early oil development. Exploratory and development activity stemmed from the "Blanket Lease" which was exploited by H.V. Foster. The modern day map of the reservation gives evidence to the fact that this is one of the most prolific oil areas in the country. All of the key companies and men on both the Indian and oilman side are profiled in this fascinating saga. Major early players were Marland (Conoco), Skelly, Phillips, ITIO (Cities Service), Gypsy (Gulf), Roxana (Shell) and Minnehoma (Getty). A yearly breakdown of Osage oil royalty and lease payments from 1901 to 1950, reveals an impressive $300,000,000. The legal, moral, and ethical problems created by this giant wealth have been enormous and much has been written about the greed-induced robberies, murder and general mayhem it spawned. The hardbound first edition is becoming scarce.

**A95. (OKLAHOMA) Franks, Kenny A., THE RUSH BEGINS: A HISTORY OF THE RED FORK, CLEVELAND AND GLENN POOL OIL FIELDS, Oklahoma City, Oklahoma Heritage Association, (1984), xiii, 163pp.**

The cities of Tulsa and Red Fork were in competition to determine which would become the industrial capital of Northeastern Oklahoma. A big oil strike at Red Fork in 1901 got the area going and was shortly followed by discovery of Glenn Pool near Tulsa in 1905. Red Fork and Glenn Pool are largely within Creek Indian lands, while the Cleveland field lies in Osage country. Galbreath, Getty and Sinclair are among many important oil industry pioneers and characters profiled in this picture history of another fabulous oil area. Includes maps. The hardbound first edition is becoming scarce.

**A96. (OKLAHOMA) Franks, Kenny A., A HISTORY OF THE DIAMOND JUBILEE 1907-1982: YOU'RE DOIN' FINE, OKLAHOMA!, Oklahoma City, Oklahoma Historical Society, 1983, x, 198pp.**
Of course the jubilee recognized oil as the key to economic growth of the state. Drilling, pumping, pipelining and seismograph operations were demonstrated for the public during the big show. Governor George Nigh appointed Jack Conn as chairman of the commission for honoring the seventy-fifth year of statehood. It was a major success, as the commission undertook 1,466 historical projects designed to instill pride in the state, the results of which are evident. The hardbound first edition is common.

**A97. (OKLAHOMA) Franks, Kenny A., Paul F. Lambert and Carl N. Tyson, EARLY OKLAHOMA OIL: A PHOTOGRAPHIC HISTORY, 1859-1936, College Station, Texas A&M University Press, (1981), xiv, 245pp.**
This is number two of the Montague Series. Early Oklahoma oil discoveries and the boomtowns that followed them are traced chronologically. First oil on the Osage & Creek reservations near Bartlesville then Red Fork & Glenn Pool near Tulsa. Good coverage of E.W. Marland, the 101 Ranch, oil-rich Indians, oil company housing and primitive drilling equipment. The 325 photos also include important discovery wells and many blowouts. The picture history is well constructed and impressive. It ends with the spectacular ITIO discovery and development of the Oklahoma City field with much on the famous Wild Mary Sudik blowout. Includes maps. The hardbound first edition is common.

**A98. (GREAT BRITAIN) Fraser, Duncan, EAST COAST OIL TOWN, BEFORE 1700, Montrose, Scotland, Standard Press, 1974, 122pp.**
The author gives a well illustrated 800 year history of the small Scottish fishing village of Montrose. Oil was aggressively being pursued in the North Sea and the little town was caught up in the excitement and culture shock. The author saw what was going to be the passing of a way of life and published this fine history so the world could see the village as it had been before oil. As a base of operations for the offshore drilling rigs, deep ports were dredged for the supply ships and warehouses were built, changing the landscape and seascape. Much history with excellent photos

of relics and maps from as far back as the Bronze Age. The hardbound first edition is becoming scarce.

**A99. (GENERAL) Frey, John W. and H. Chandler Ide (editors), A HISTORY OF THE PETROLEUM ADMINISTRATION FOR WAR 1941-1945, Washington, GPO, 1946, xviii, 463pp.**
From the forword: "It is the history of a unique experience, dealing with an unprecedented program of government-industry cooperation". Swanson describes the effort as: "Operations of Petroleum Administration for War and its predecessor agencies. Authoritative account written at time by major participants. Important statistics not found elsewhere". Harold Ickes guided the administration for about five years and selected the best minds in the country from industry and government to do the job. The PAW had an excellent record of supplying fuel to Navy, Air Force and Army troops wherever they were located. Patton would probably not agree with that statement. The PAW was motivated and had power to issue orders, which they did effectively. The Big Inch pipeline, tanker sinkings off of the Atlantic coast, gasoline rationing, and the massive production of 100 octane aviation fuel were their main problems, along with constant logistical complications of moving fuel and oil to support fighting around the world. Much history with a large appendix, photos and charts. Cited in Swanson on page 152. The hardbound first edition is becoming scarce.

**A100. (TEXAS) Fuermann, George, HOUSTON: LAND OF THE BIG RICH, NY, Doubleday & Co., 1951, 255pp.**
Author Fuermann, a transplant from New York, was a daily columnist for the Houston Post. He lived in Amarillo, Laredo, and San Antonio before settling in Houston. From the dust jacket: "From oil, real estate, cotton, cattle, shipping, and chemical industries, the inhabitants of Houston have accumulated such an abundance of wealth that philanthropy is a profession, and currency is as expendable as Kleenex." Much on oilmen, particularly McCarthy and Cullen. Pictorial endpapers and inserts of paintings of Houston scenes. Fun reading. The hardbound first edition is common.

**A101. (TEXAS) Fuermann, George, RELUCTANT EMPIRE, NY, Doubleday & Co., 1957, 284pp.**
Fuermann's second book about Texas. He is entranced by the phenomenon of Texas. Here he concentrates on the big names and the big

cities as he discusses the pluses, minuses and eccentricities in an interesting and fun format. From the dust jacket: "...I have tried to show Texas in terms of mentality, politics and culture, of strengths and weaknesses, to separate facts from popular conceptions..." The book is somewhat controversial, but good reading. The hardbound first edition is common.

**A102. (OKLAHOMA) Galbreath, Frank, GLENN POOL...AND A LITTLE OIL TOWN OF YESTERYEAR, Tulsa, Frank Galbreath, 1978, x, 177pp.**
The author wrote this book at age seventy-six, at the urging of friends. He had witnessed much of Oklahoma's growth and along with his father was a significant contributor to it. Galbreath's Dad, Robert and brother Herman were farm boys seeking the frontier when they participated in the 1889 Oklahoma land run. Later excited by the Red Fork pool discovery, Robert and some business associates drilled #1 Glenn, which at 1467 feet discovered a giant oil field containing 300,000,000 barrels. It was almost entirely Robert's persistence that made it happen. In his personal observations of the limestone outcroppings on the Glenn farm, he had taken note of the small oil globules which appeared in the open rock pores. From that, he was convinced that an underground oil pool was waiting to be discovered there. An excellent photo section is included. The hardbound first edition is common.

**A103. (PENNSYLVANIA) Gale, Thomas A., THE WONDER OF THE NINETEENTH CENTURY: ROCK OIL IN PENNSYLVANIA AND ELSEWHERE, Erie, PA, Sloan & Griffith, 1860, 80pp.**
According to Swanson: "This first book about petroleum following the Drake well was written to satisfy public desire for more information about rock oil, its origin, geology, production, costs, uses, history, prospects; invaluable eyewitness descriptions of early oil wells". The compact and comprehensive work told most of what was known about petroleum and the incipient industry at the time. Cited in Swanson on page 152; Hardwicke page 32; Giddens page 166. The first edition is rare, and a 1952 Ethyl Corporation reprint is scarce.

**A104. (TEXAS) Gard, Wayne, THE FIRST 100 YEARS OF TEXAS OIL & GAS, Dallas, TX, Texas Mid-Continent O&GA, (1966), 40pp.**
Gard is a well known western history writer and dedicates this pamphlet to Warner, the great Texas oil historian. Much of the pamphlet is

abstracted from Warner's acclaimed "Texas Oil & Gas Since 1543". This is a brief, accurate account of Texas oil history with all of the main characters and events replayed along with historic photos, celebrating 100 years since Barret's discovery. Often included is a four page postscript up-dating 1966 to 1973. The softbound first edition is becoming scarce.

**A105. (GENERAL) Gibson, George R., A LAMPFUL OF OIL, NY, Harpers New Monthly, Jan 1886, pp235-257.**
Much of what was known about the worldwide oil industry in 1886 is encapsulated in this short article which had been reprinted from Harpers. It is mostly about the Oil Regions of Pennsylvania, but also contains a surprising amount of information on the Russian oil industry which was thriving at Baku. Engravings of the Nobel refinery and an oil steamer operating on the Caspian Sea are included. Reproduction of the engravings is exceptional, as are the drawings of oil equipment and a geological representation of the drillers log from a Bradford Field oil well. Cited in Giddens on page 170. The special reprint edition is becoming scarce.

**A106. (PENNSYLVANIA) Giddens, Paul H., EARLY DAYS OF OIL: A PICTORIAL HISTORY OF THE BEGINNINGS OF THE INDUSTRY IN PENNSYLVANIA, Princeton, NJ, Princeton University Press, 1948, vii, 149pp.**
One of the best picture-history books available. This book was widely circulated and has been reprinted several times. It is complementary to Giddens' "Birth of the Oil Industry", described later, and together they give the most comprehensive picture of early Pennsylvania oil available. Much of the content is the unique photographs taken by John A. Mather, the first and best oil photographer of the oil regions, to whom Dr. Giddens dedicates the book. Cited in Swanson on page 152. The hardbound first edition is becoming scarce.

**A107. (PENNSYLVANIA) Giddens, Paul H., THE AMERICAN PETROLEUM INDUSTRY: ITS BEGINNINGS IN PENNSYLVANIA!, NY, The Newcomen Society, 1959, 32pp.**
A Newcomen address delivered before the society at Erie, PA, in 1953 by famous oil historian, Dr. Giddens. He gives his usual thorough history of

oil with emphasis on the social and economic impact it has had on us. Giddens concentrates on important oil characters and events through the Civil War period, at which time Pennsylvania had around 3,000 producing wells. Cited in Swanson on page 152. All softbound editions are becoming scarce.

**A108. (PENNSYLVANIA) Giddens, Paul H., THE BEGINNINGS OF THE PETROLEUM INDUSTRY: SOURCES AND BIBLIOGRAPHY, Harrisburg, PA, Pennsylvania Historical Commission, 1941, v, 195pp.**
Part I contains copies of fifty-three letters written to Dr. Brewer from 1854 to 1872, plus nineteen letters written to Bissell from 1855 to 1859. Much on Pennsylvania Rock Oil Company of New York. Part II is a summary of Dr. Giddens several years of Pennsylvania oil research in the form of a complete bibliography of early history of the oil industry including articles from newspapers and magazines. This is a comprehensive collection of early oil history material and an often cited bibliography of early Pennsylvania oil which is annotated with interesting comments when Giddens thought it appropriate. Cited in Swanson on page 152. The hardbound first edition is scarce.

**A109. (PENNSYLVANIA) Giddens, Paul H., THE BIRTH OF THE OIL INDUSTRY, NY, The MacMillan Co., 1938, xxxix, 216pp.**
The twenty-six page introduction by Ida Tarbell is very insightful and educational. She was writing her autobiography at the same time and was very reflective on the early oil industry. She consulted heavily with Dr. Giddens as he prepared this book which covers Pennsylvania up to 1870. Oil had been dug for and skimmed from creeks for centuries before Colonel Drake drilled his discovery well. It was used largely for medicinal purposes. Samuel Kier refined oil and sold kerosene as an illuminant. Two young enterprisers, Bissell and Brewer had the idea to exploit this product and got the famous chemist from Yale, Dr. Silliman to write a report describing potential uses of oil. Good report. Drake drilled the first oil well for these two at Titusville, on Oil Creek. The well flowed its first oil on August 27, 1859. Subsequent leasing, drilling, transporting, refining and marketing of oil was rapid and chaotic. Numerous historic photo-plates by Mather are scattered throughout the book. Excellent bibliography. Cited in Swanson on page 152; Giddens page 91. The hardbound first edition is becoming scarce.

**A110. (PENNSYLVANIA) Giddens, Paul H. (editor), PENNSYLVANIA PETROLEUM 1750-1872: A DOCUMENTARY HISTORY, Titusville, Pennsylvania Historical Commission, 1947, xv, 420pp.**

Copies of many important early documents relating to petroleum and to the start of the Pennsylvania oil industry. The author explains the significance of the material and its sources, which includes many letters and reports. Most of the original material is now held at the Drake Museum at Titusville of which the author was curator while compiling this book. This is a primary research tool for the early oil historian. Well organized into five sections, the following: Part I covers the period before the Drake discovery well; Part II is previously published first-hand accounts of the Drake well by Brewer, Bissell, Townsend, Drake, Smith (the driller) and many others; Part III includes other letters and documents relating to the Drake well; Part IV is later Drake correspondence of interest; Part V is various other contemporary accounts of the oil regions from 1859 to 1872 such as the big wells, transportation, the oil exchange, Pithole boomtown and general living conditions there. Cited in Swanson on page 152. The hardbound first edition is scarce.

**A111. (TEXAS) Gillispie, Ora (editor), THE OVERTON STORY: A CENTURY OF PROGRESS, Overton, TX, Overton Centennial Corp., 1973, 152pp.**

Overton was a nice little town in Rusk County, Texas which suddenly became a boom town in 1930 as East Texas came alive with oil prospectors, their equipment and followers. Drillers and their derricks marched right through the town, completing prolific oil producers, one after the other. Much information on Overton pioneer families. The softbound first edition is common.

**A112. (OKLAHOMA) Glasscock, C.B., THEN CAME OIL: THE STORY OF THE LAST FRONTIER, NY, Bobbs-Merrill, (1938), 349pp.**

Glasscock writes in clear and simple terms of how the Eastern Indians were forced from their lands into Oklahoma, then the Western Indians came under similar circumstances. He tells of the government give-away of much of the promised Indian lands during the five land-runs starting in 1889, which was aimed at settling the territory for whites. Chapter nine is entitled Then Came Oil. The oil springs and seeps scattered around Oklahoma led oilmen down from Pennsylvania and by 1889 the first commercial oil was produced from Cherokee land. Cudahy Oil Company and ITIO Company began developing shallow oil at Bartlesville in 1897 and the rush was on. Cattle, industry, statehood, cities and their interrelations are covered. This book is generally recognized as one of the best and most readable early Oklahoma oil histories. Cited in Swanson on page 153. The hardbound first edition is common.

**A113. (GENERAL) Goralski, Robert and Russell W. Freeburg, OIL & WAR: HOW THE DEADLY STRUGGLE FOR FUEL IN WW II MEANT VICTORY OR DEFEAT, NY, William Morrow and Co., (1987), 384pp.**

From the dust jacket: "...the important role oil played in World War II provides a new view of the forces that controlled the greatest war in history...and a stunning analysis of the importance of oil in terms of world peace for the years to come". This is an anecdotal narrative about the important role oil played in the war. It is well illustrated with maps and charts. The book explodes several myths such as the general belief that Hitler's mechanized army was well supplied as it made the lightning strikes of the late 1930's and early 1940's. The facts dispute this, in that his tank, troop, and supply vehicles were shut down in the middle of campaigns more often than not, having to scavenge fuel from the locals. Hitler spent years developing his plan of fuel independence which was largely based on his synthetic oil (from coal) plants which in 1938 were producing only 2,000 barrels per day, but by the early 1940's had a projected capacity of 86,000 barrels, half of their requirements then. Germany's largest fiasco was their invasion of Russia which was stymied by a lack of fuel supplies. Japan also had no oil resources and began their part of the war essentially to conquer oil rich nations in the Pacific to gain reserves. A case can be made that they lost the war of the Pacific because of lack of fuel and supply. The hardbound first edition is common.

**A114. (TEXAS) Graham, Hal, et al (editors), SHOWCASE OF THE PERMIAN BASIN VOLUME I, Lubbock, TX, Plains Publishers, 1981, 240pp.**

At the very height of the late 1970's, early 1980's oil boom, Midland oil people decided to publish this glorious tribute to their town. Great color photos of first class oil operations and equipment fill the pages. Excellent individual and company profiles along with pieces of the Permian Basin history. Optimism was rampant. This is the Charter Edition. See the second edition for more information. The hardbound first edition is becoming scarce.

**A115. (TEXAS) Graham, Hal, et al (editors), PERMIAN BASIN SHOWCASE VOLUME II, Lubbock, TX, Plains Publishers, 1982, 224pp.**
Unlike Volume I, issued the year before, optimism is gone and will be reserved for a time in the future, not for today. The first chapter is entitled Return to Realism in 1982. In six months, drilling activity had slumped 40% and the price of crude had turned ominously downward. Amazing changes took place over a one year period. Even more amazing is the fact that by 1994, over 65% of those people interviewed and advertising in volume I are no longer in business. Most left in 1986. The hardbound first edition is becoming scarce.

**A116. (CANADA) Gray, Earle, THE GREAT CANADIAN OIL PATCH, Toronto, Canada, Maclean-Hunter Limited, (1970), 355pp.**
The author was editor of Oilweek Magazine and deeply involved in recording and reporting information on the expanding Canadian oil industry. This was the second of his books and it is a first class effort. He recounts the history and predicts the future. Much on the argument that Canada was producing oil prior to Drake's discovery. Production in Canada developed very slowly compared to the U.S., but in 1914 Alberta's first oil boom began spreading. Then in 1920, Norman Wells field was found in the Mackenzie River country of the far north. This discovery led to a historic fiasco called Canol which was a massive, outrageously expensive pipeline project designed to supply fuel to Alaska navy bases during WW II. Japanese submarines were the main threat as they had been wreaking havoc up and down the West Coast. The Canol pipeline was determined to be unnecessary and functioned only for a short while. The giant Leduc field was discovered in 1947 and it was followed by Pembina and Rainbow, which along with others became an integral part of the Canadian exploration success story. Gray looks optimistically at the future of the Athabasca Tar Sands and the Canadian Arctic region, both of which were showing tremendous potential. This is a very well organized and easy reading history for both laymen and oil people. The hardbound first edition is becoming scarce.

**A117. (OKLAHOMA) Gregory, Robert, OIL IN OKLAHOMA, Muskogee, OK, Leake Industries, 1976, 90pp.**
Gregory reviews Oklahoma oil history and provides great photos to enhance it. He focuses several chapters on Tulsa as it rose from a small farming town to an oil center. Major topics are as follows: J. Paul Getty who with his Dad got rich in Oklahoma; Phillips Petroleum with its phenomenal success and the boomtowns thus created; E. W. Marland, Ponca City and the Conoco saga; Bill Skelly, the tough and shrewd oilman who began Skelly oil at age 41 and turned it into a major oil company; The Osage Indians and their oil generated wealth; Thomas Gilcrease and his headright in the Glenn Pool. The remaining chapters cover the big booms of Greater Seminole area and Oklahoma City. A classy well-done book. The hardbound first edition is common.

**A118. (TEXAS) Griffin, John Howard, LAND OF THE HIGH SKY, Midland, TX, First National Bank, (1959), xv, 212pp.**
The First National Bank was proud of its city and its brand new bank building and showed it by commissioning this fine history. Griffin did much research of the Midland area and its people with interviews of the ranchers, farmers, oilmen and bankers. From cotton, cattle and oil, the economy maintained an uneven growth, but it was going to grow much more. Little did the author and sponsor suspect that FNB would become the largest independent bank in the U.S. by 1981 with over a billion dollars in deposits. According to the dust jacket, the book: "...is an epic of the land and of its effects, good and bad, on the people; and of their effects, good and bad, on it". A well written chronology with source notes. The hardbound first edition is becoming scarce.

**A119. (TEXAS) Haley, James L., TEXAS: FROM SPINDLETOP THROUGH WORLD WAR II, NY, St. Martin's Press, (1993), xii, 292pp.**
Largely a pictorial treatment of Texas during the first part of the 20th century depicting the effect Spindletop had on the state. Previously the economy was agriculture based and subsequently it became economic and industrial based. Oil is pervasive throughout the work with good concentration on the Spindletop phenomenon and other boomtowns. Much on various political characters and the role of Texas during both wars. A chapter gives detail on the East Texas oil field and the conservation methods enacted to stem the hot oil flow. Good rendition of

the Red River bridge controversy between governors Sterling of Texas and Bill Murray of Oklahoma. Federal and state forces clashed over a boundary dispute. A significant Texas bibliography is included. The hardbound first edition is common.

**A120. (TEXAS) Hamill, Curtis G., WE DRILLED SPINDLETOP, Houston, TX, The author, 1957, 39pp.**
Hamill wrote this pamphlet at age eighty-five. He had outlived his two brothers and all of the others involved in drilling the famous Lucas gusher, discovery well for Spindletop oil field. Much has been written about the field but this first-hand rendition gives details of actual drilling operations. Several previous unsuccessful drilling attempts were stopped short of the 400 foot level because of unconsolidated sands which wreaked all kinds of havoc on the hole. Young Hamill developed a revolutionary technique (drilling mud) which sealed the sands and finally allowed drilling through the bad zone and another one at 640 feet. He describes how they almost abandoned the well at 870 feet where they had a slight show of oil, but were unable to get a successful test. Evidently, Mrs. Lucas told the Hamills that they had better honor the contract and drill to 1200 feet, or else. They set pipe at 1060 feet and took off for Christmas, 1900. Drilling resumed January 1, 1901, and on the tenth, the well blew in. Curtis Hamill was on the derrick scaffold board when the oil erupted and was temporarily blinded by the mud and oil. Fortunately, he groped his way to safety. Wiping his eyes, he observed the spectacle of oil spurting 100 feet above the derrick. He gives details of their capping technique which had the well under control by January 21, 1901. Details of building the rig and other early rotary drilling operations are also related along with the parts played by Hamill's brothers Jim and Al. Important work. Cited in Swanson on page 153. The softbound first edition is scarce.

**A121. (MEXICO) Hamilton, Charles W., EARLY DAY OIL TALES OF MEXICO, Houston, Gulf Publishing Co., (1966), 246pp.**
From the dust jacket: "...reflections of the early days of the oil industry in Mexico bring to life again one of the most colorful and important periods in Mexican history. He was active in the oil business in Mexico from 1912 to 1922, the days of the great oil boom..." This presentation by Hamilton is similar to a geologist's journal as it relates oil history in first-hand biographical form. It is well written and readable. The hardbound first edition is common

**A122. (TEXAS) Hanna, Betty Elliott., DOODLE BUGS AND CACTUS BERRIES: A HISTORICAL SKETCH OF STEPHENS COUNTY, TX, Nortex Press., (1975), ix, 143pp.**

This history of Stephens County is told by the daughter of a pioneer county family. She incorporated verbal history with diaries and other documents collected over the years. Forty-one chapters, each a separate story, give the chronological history of people and events in a colorful, informative style. Chapters 27 through 31 are all oil. A 300 barrel discovery near Caddo in 1916 stirred further interest and some large wells were completed within the year as the play extended over Stephens County. The Stoker #1 near Breckenridge city limits came in during 1919 and the boom was on. By the end of 1920, 1730 wells had been drilled in the county and the population of Breckenridge shot up from 1,500 to 30,000. A chaotic situation prevailed there as with all boom towns. There was the constant threat of fire because oil was spilled everywhere. Occasional explosions caused by carelessness of the nitro shooters was also a real hazard. Crooked oil promoters accompanied by every form of vice were there in force. The chamber of commerce still gets inquiries about oil company stock certificates from long-defunct companies, issued during the boom. Daily production surged to over 120,000 barrels in 1921 while crude was bringing $3.50 per barrel. The hardbound first edition is becoming scarce.

**A123. (GENERAL) Hansl, Proctor W., YEARS OF PLUNDER: A FINANCIAL CHRONICLE OF OUR TIMES, NY, Harrison Smith and Robert Haas, 1935, 312pp.**

This book was not available for review by Neff, but is regarded as important. Quoting from Swanson: "Chapters on John D. Rockefeller, Standard Oil Company, leasing of Elk Hills, and Teapot Dome naval petroleum reserves". Cited in Swanson on page 153. The hardbound first edition is very scarce.

**A124. (BIBLIOGRAPHY) Hardwicke, Robert E., PETROLEUM AND NATURAL GAS BIBLIOGRAPHY, Austin, University of Texas, 1937, xii, 167pp.**

This early oil book bibliography was considered definitive, although it only contained 1,400 books at the time of printing. Hardwicke was an attorney and his field is well covered. The format is very easy to use and he tells whether each book was in print, and if so, what the retail price was. Descriptive and occasionally annotated when the author had a sidelight to report. Cited in Swanson on page 184; Giddens page 96. The hardbound first edition is becoming scarce.

**A125. (GENERAL) Hardwicke, Robert E., THE OILMAN'S BARREL, Norman, University of Oklahoma Press, (1958), xi, 122pp.**
Hardwicke is a well-known Fort Worth attorney who has been associated with the oil industry for many years and published several important books and papers. From the dust jacket: "A unique and fascinating story of the oil industry's unit of measure...full of legends and lore of the oil business". Keir, Drake, Nellie Bly, and Ben Hogan are all involved in a very well presented history of the subject. Forty-two gallon wooden barrels were commonly in use during the 1700's and 1800's as a convenient size for whiskey, beer, whale oil and other items, allowing five percent tolerance over the round number of forty. After several years of research, Hardwicke put together a lucid story and provides a useful bibliography of this barrel which is not what it appears to be. Although the forty-two gallon barrel has been the common measurement for crude oil and many crude oil products, it does not really exist except as a unit of measurement, the barrel itself having been abandoned long ago. Unusual subject, but someone had to do it. Cited in Swanson on page 179. The hardbound first edition is common.

**A126. (PENNSYLVANIA) Harris, Charles H., HISTORY OF THE VENANGO OIL REGIONS: SHOWING WHERE PETROLEUM IS FOUND, Buffalo, NY, 1866, 108pp.**
This book was not available for review by Neff, but is regarded as important. Quoting from Swanson: "Accounts of original oil properties and companies, statistics of production, sketches of towns and cities in oil regions, leaders of the industry". Cited in Swanson on page 34; Hardwicke page 33; Giddens page 91. The first edition is very scarce.

**A127. (TEXAS) Harter, Harry, EAST TEXAS OIL PARADE, San Antonio, Naylor Co., 1934, 220pp.**
Excellent early rendition of the East Texas oil field written about four years after its discovery. Harter's oil parade refers to the parade of people

who came to East Texas because of Dad Joiner's incredible Daisy Bradford #3 discovery well. The first part is concerned with history of the land on which the oil field now lies as it had been owned by the original Mexican grantee, Juan Ximines. The second part recalls the saga of Dad Joiner, his associates and investors as they persisted through almost total industry ridicule to bring in the wildcat well. In a later section, much detail is provided on the political problems, largely concerned with handling the market glut which dissolved the crude price structure of the world. The Texas Railroad Commission, the feds, the independent producers, and the majors were all at odds on how to prorate or apportion production. It took almost four years of constant court battle as well as military intervention to finally get an equitable settlement. A large section called Panorama consists of several stories relating to oil royalty in the field, prediction of the life and future of the field, and finally a small piece on Dad Joiner's doings after he sold his interests to H.L. Hunt....he kept on wildcatting of course. Cited in Swanson on page 39; Hardwicke page 34. The hardbound first edition is scarce.

**A128. (USSR) Hassmann, Heinrich, OIL IN THE SOVIET UNION: HISTORY-GEOGRAPHY-PROBLEMS, NJ, Princeton University Press, 1953, xvi, 173pp.**
Translated from German by Alfred Leeston, prominent oil author, to which he added many notes, the latest statistics, a bibliography and an index. Eighteen maps are included in this work which reviews the birth of the Russian oil industry, its development, oil-producing districts, oil problems, demand, and supply. Foreword by DeGolyer praises the work. Cited in Swanson on page 57. The hardbound first edition is becoming scarce.

**A129. (PENNSYLVANIA) Hatch, Vernelle A. (editor), ILLUSTRATED HISTORY OF BRADFORD, McKEAN COUNTY, PA., Bradford, PA, Burk Brothers, 1901, 261pp.**
This book is not listed in Swanson, however it is an excellent town history containing numerous old oil field photos of Bradford. Photos are of everyone in town who was of any importance as well as their homes. Oil Well Supply Co. and Boviard plants are shown with their employees. From the title page "Containing sketches and illustrations of the village in the early days of its existence, together with photographs and biographies of the present time". The hardbound first edition is scarce.

**A130. (CALIFORNIA) Hatfield, Reginald W., A HISTORY OF THE NAVAL PETROLEUM RESERVES AND OF THE DEVELOPMENT OF THE PRESENT NATIONAL POLICY RESPECTING THEM, Los Angeles, CA. Privately Published, 1944, 180pp.**
The author was one of Harry Sinclair's attorneys during the Teapot Dome investigations and trials. Solid detailed information showing how the government developed its policy regarding and resolving the fiasco. Four exhibits are included. An impressive color map of Elk Hills field showing ownership and status of leasehold as of 1938, including the various illegal canceled leases is enclosed. Cited in Swanson on page 157. The hardbound first edition is very scarce.

**A131. (GENERAL) Hatley, Allen G. (editor), THE OIL FINDERS: A COLLECTION OF STORIES ABOUT EXPLORATION, Tulsa, OK, A.A.P.G., (1992), xv, 134pp.**
A collection of world-wide exploration geology stories. From the Standard Oil "Rover Boys" in Timbuctu to Hunt oil in Yemen. It is hard to put this book down with the thrill of the oil chase going on. Very popular book which includes historic photos as well as maps. The softbound first edition is scarce.

**A132. (TEXAS) Helgesen, Sally, WILDCATTERS: A STORY OF TEXANS, OIL AND MONEY, NY, Doubleday & Co., 1981, x, 198pp.**
Author Helgesen is a contributing editor to Harpers. From the dust jacket: "...profiles three generations of oil tycoons: such pioneer oilmen as Clint Murchison, Sid Richardson, and Monty Moncrief...men who played by their own rules and clinched multi- million dollar deals with a handshake". This is an up-beat history of these oilmen, written at the height of the modern oil boom. The hardbound first edition is common.

**A133. (RUSSIA) Henry, James Dodds, BAKU, AN EVENTFUL HISTORY, London, Archibald Constable & Co., 1906, xviii, 256pp.**
This book was not available for review by Neff, but is regarded as important. Quoting from Swanson: "History of Baku and oil and gas developments, refining, leading companies and personalities; about half the book is devoted to a detailed account of the 1905 massacre of Armenians by Tartars; chapter on Bedim, Black Sea refining town for Baku oil". Cited in Swanson on page 57; Hardwicke page 34. The hardbound first edition is very scarce.

**A134. (PENNSYLVANIA) Henry, J.T., THE EARLY AND LATER HISTORY OF PETROLEUM: WITH AUTHENTIC FACTS IN REGARD TO ITS DEVELOPMENT IN WESTERN PENNSYLVANIA, Philadelphia, James B. Rodgers, (1873), 607pp.**
Further subtitle tells all "The oil fields of Europe and America. Gas Wells. Spiritual Wells. Oil Well Shafts. Petroleum Products. Oil Companies. Pipe Line Statistics. Early Modes of Transportation. Flowing Wells of 1861 to 1864. Pit Hole in 1865. The Lubricating Oil District, also, Statistics of Product Export, and Consumption, with prices of Oil from 1859 to 1872, etc.". The author was a newspaperman at Titusville and wrote with an intimate familiarity of his subject. He knew the people involved and their stories. The book is bound as two volumes in one. Volume I is mainly first-hand history dealing with the Drake well, but also describing several other early important oil lease developments. Statistics, photos and illustrations are included. Volume II includes histories of certain oil towns along with biographies of the important oil men of the day. The only two easily recognizable characters are Drake and Bissell. The book has been reprinted by Burt Franklin of New York, and others. Cited in Swanson on page 154; Hardwicke page 34; Giddens page 91. The hardbound first edition is rare.

**A135. (TEXAS) Heritage (The Magazine), THE OIL INDUSTRY IN TEXAS: BRINGING FORTH A NEW ERA, Austin, Texas Historical Foundation, 1994, Volume 12, Number 3, 30pp.**
This is an exceptional issue of the quarterly magazine, dedicated to events surrounding the 1901 Spindletop oil discovery. The discovery is regarded as establishing the oil industry in Texas and diminishing the pervasive influence of Standard Oil. Four feature articles include: Beaumont, Where Oil Became An Industry, by Christine Sanders; Midland's Emergence as the West Texas Oil Capital, by Betty Orbeck; Museum Celebrates Early Days of the Oil Industry in Texas, by Christy Marino; Captain George Washington O'Brien and the Gladys City Company at Spindletop, by Christine Sanders. The softbound edition is common.

**A136. (NEW YORK) Herrick, John P., BOLIVER, NEW YORK: PIONEER OIL TOWN, Los Angeles, Ward Ritchie Press, 1952, x, 150pp.**
Recounted here are short stories and reminiscences of the author, publisher and editor, Herrick. He reviews the early oil discoveries around Bolivar and the resulting booms and busts. Sketches of individual oil

pioneers are also provided. The history begins with extension of the Pennsylvanian Allegheny Valley oil play to the nearby town of Richburg in 1881. People swarmed the area and an oil drilling frenzy took place. Bolivar went from almost zero to 4500 citizens in less than a year. At this writing in 1952, Bolivar was back to around 1000 people and the shallow oil production was still helping support the area. The hardbound first edition is becoming scarce.

**A137. (NEW YORK) Herrick, John P., EMPIRE OIL: THE STORY OF OIL IN NEW YORK STATE, NY, Dodd, Mead & Co., (1949), xx, 474pp.**
Herrick was an oilman and publisher who did considerable research over the years to document and chronicle the history of oil in New York. This book is his report on the men and the firms involved in exploring for and developing these oil fields. The Hall Farm Corporation #1 well was completed in 1865 flowing seven barrels per day. The stockholders rebelled at such a poor producer, having been told they would get a gusher. Job Moses ended up with all of the stock and methodically began development of the unspectacular oil field. A slow start for the state which would eventually produce around 200 million barrels and much natural gas. A very comprehensive and readable work with amusing vignettes along the way. Cited in Swanson on page 154. The hardbound first edition is becoming scarce.

**A138. (MIDDLE EAST) Hewins, Ralph, A GOLDEN DREAM: THE MIRACLE OF KUWAIT, London, W.H. Allen, 1963, 317pp.**
The third of Hewins' important oil history books, all of which deal with the Middle East. His biographies of Gulbenkian and Getty are tributes to individual accomplishments, amidst centuries old Arab and non-Arab conflicts. A Golden Dream is a similar triumph...that of a small politically created nation of Arabs which was to become second only to Saudi Arabia in world oil reserves. Kuwait is situated between Iran, Iraq and Saudi Arabia, all more unstable and petulant than they are. Hewins delves into Kuwait's ancient history and brings us up to the oil age of 1962 where Persian Gulf exports were beginning to dominate world trade. Britain was virtually dependent on Kuwait oil at this time. Hewins is critical of almost everyone involved in exploration and development of Kuwait and the Neutral Zone oil concessions, particularly the inept politicians and greedy oil companies. Much on Frank Holmes who caused

Saudi oil to be found and developed. Also, Archie Chisholm who expedited the Kuwait developments. The author closes with the quote: "This is an inflammable corner of the world. One day somebody may put a match to it". As usual with Hewins, no notes, sources or index, just pure Hewins....take him or leave him. The hardbound first edition is becoming scarce.

**A139. (OKLAHOMA) Heydrick, L.C., RED FORK DISCOVERY: JUNE, 1901, Wichita Falls, TX, The Author, (1953), 116pp unnumbered.**

This is a most unusual report organized in ten sections, wherein Heydrick documents and details the fact that his family promoted and drilled the historic Red Fork discovery well near Tulsa in 1901. Newspaper accounts and generally accepted history to 1953 allowed that promoters Clinton and Dr. Bland promoted and drilled the well. Also much erroneous published information was circulated prior to this. Heydrick's case was well taken and confirmed in Rister's monumental "Oil, Titan of the Southwest". This historic report is largely typewritten and contains photocopies of the more important letters and instruments. Publication limited to seven copies. Not cited in Swanson. The hardbound edition is rare.

**A140. (GENERAL) Hidy, Ralph W. (editor), OIL'S FIRST CENTURY: PAPERS GIVEN AT THE CENTENNIAL SEMINAR ON THE HISTORY OF THE PETROLEUM INDUSTRY, MA, Harvard Graduate School, (1960), vi, 147pp.**

Papers presented at Harvard on November 13-14, 1959 are reproduced here and include the premier oil history authors. Forbes writes on Eastern Europe; Beaton on pre-Drake refining; Giddens on significance of the Drake well; Larson on the rise of big business in oil; and Pratt on business history and the search for oil. Many more are included in this excellent compilation. The hardbound first edition is common.

**A141. (TEXAS) Higgins, Patillo, PROSPECTUS: THE TRUE HISTORY OF THE BEAUMONT OIL FIELDS, Beaumont, TX, Higgins Standard Oil Company, Ltd., 1902, 71pp.**

This book was not available for review by Neff, but is regarded as important. Quoting from Swanson: "History of Spindletop by its original promoter, who subsequently leased to Capt. Anthony F. Lucas, operator of the successful discovery well. Maps, geological sections, well logs; views of Beaumont during oil boom; account of Higgins' real estate company, the Gladys City Oil, Gas & Mfg. Co. Cited in Swanson on page 154; Hardwicke page 34. The softbound first edition is very scarce.

**A142. (TEXAS) Hoffman, William, TEXAS - A YEAR WITH THE BOYS, Dallas, Taylor Publishing, (1983), 205pp.**
Author Hoffman previously took on oil industry people in three of his books, specifically Paul Mellon, Jack Grimm, and David Rockefeller. Hoffman leaves his comfortable New York to spend the year with some of Texas' most eccentric characters. After his previous episode looking for the Titanic with Jack Grimm, he concluded that Texas and Texans are a subject which can sell a book. He travels the state with a nut named Gosdin and manages to meet several other obscure and bizarre characters who are continually wheeling and dealing. Much of the time is spent with Grimm in Abilene. Hoffman touches on all of the unusual Texas personalities such as Billie Sol, Bunker Hunt, Ross Perot, Eddie Chiles and Cullen Davis to name a few. Hoffman gives an exaggerated impression of Texas and Texans. These characters and their method of operating were basically the result of boom times and they would have done the same things in Nebraska if that was where the action was. Most had moved on by 1986. The hardbound first edition is common.

**A143. (GENERAL) Holbrook, Stewart H., THE AGE OF MOGULS, Garden City, NY, Doubleday & Company, 1953, x, 373pp.**
Rockefeller is the common thread throughout Holbrook's story as the other Robber Barons are constantly being compared to him, except for Mellon. The Mellon story is different in that he controlled banks and companies and businesses, including Gulf Oil, and was publicly the least known, although probably the wealthiest of the bunch. John Gates the compulsive gambler who fell into half ownership of Texaco is one of the most fascinating stories...a smart gambler? The book is not cataloged in Swanson, but has much on Standard Oil. Good coverage of the other famous moguls as well. The hardbound first edition is common.

**A144. (GENERAL) Hollyman, Thomas, THE OILMEN: THEIR ENERGY MOVES A NATION, NY, Rinehart & Co., (1952), 91pp unnumbered.**
Two hundred photographs of Shell Oil employees at work in various jobs of the oil industry including the fields, pipelines, refineries and service

stations. Hollyman was the photographer who spent four months traveling across the U.S. to take photos for the book. It is an early 1950's snapshot of the oil industry and very well done. Cited in Swanson on page 187. The hardbound first edition is common.

**A145. (TEXAS) House, Boyce, I GIVE YOU TEXAS: 500 JOKES OF THE LONE STAR STATE, San Antonio, TX, The Naylor Co., (1943), 100pp.**
House is in his element here and presents jokes on every conceivable funny and otherwise Texas situation. The book is named after his weekly syndicated news column and radio show. Most of the oil related jokes are about Gib Morgan, either inspired or told by him. The hardbound first edition is common.

**A146. (TEXAS) House, Boyce, OIL BOOM: THE STORY OF SPINDLETOP, BURKBURNETT, MEXIA, SMACKOVER, DESDEMONA AND RANGER, Caldwell, Idaho, The Caxton Printers, Ltd., (1941), 194pp.**
One of the top oil history books by the famous author-historian-journalist. Here, he covers several of the most famous Texas oil booms. From the dust jacket: "...of poor croppers and Negroes catapulted to vast wealth, of thousands of dollars changing hands in gambling joints, of leases that became magic carpets for their owners, of engineering feats that tested the strength of brawny men. These and thousands of other incredible happenings make up the amazing story of the great American oil boom. ...the whole story is presented in all its vivid, roaring reality by one who saw the great oil boom come and go". House had more first-hand information on Texas oil boom towns than anyone. He is also a good storyteller which makes this book his most popular. Cited in Swanson on page 39. The hardbound first edition is scarce.

**A147. (TEXAS) House, Boyce, OIL FIELD FURY, San Antonio, The Naylor Company, (1954), vii, 142pp.**
House was editor of the Eastland Daily Oil Belt News and writes first hand of the aftermath of the Texas oil booms of the 1920's. Mostly they are stories of the Ranger boom. Factual accounts which are sometimes humorous but mostly poignant. He had written several Ranger books and stories prior to this effort, but in this book he shows a great sensitivity to the times and to the people who for a brief moment were on top of the

world. Cited in Swanson on page 40. The hardbound first edition is becoming scarce.

**A148. (TEXAS) House, Boyce, ROARING RANGER: THE WORLD'S BIGGEST BOOM, San Antonio, TX, The Naylor Company, (1951), 122pp.**
This book could have also been titled "The World's Biggest Bust", although Pithole, Pennsylvania might contest that. There are many good photos of the town before the boom and of the important characters during the rapid growth period. A gusher near the sleepy town of Ranger in October of 1917 stimulated a three and a half year boom which forced the city population from a few hundred to over 30,000. House uses his rapid fire, hit and run, vignette method of telling the history of this boom which he observed first-hand and reported on a daily basis. His epilog tells of the modern Ranger which by 1951 had settled back to 5,000 residents and had become as stable as any Texas town. Authentic. Cited in Swanson on page 39. The hardbound first edition is becoming scarce.

**A149. (TEXAS) House, Boyce, THE RANGER STORY: A BRIEF HISTORY OF RANGER, TEXAS, Ranger, TX, The Ranger Times, ca 1965, 16pp.**
This pamphlet covers the entire boom town history from the Ranger Field discovery well, #1 McKlesky in 1917 through eventual stabilization. It was probably reprinted from House's newspaper articles over a period of time. Presented in eight parts with numerous photos. The softbound first edition is common.

**A150. (TEXAS) House, Boyce, WERE YOU IN RANGER?, Dallas, TX, Tardy Publishing Co., (1935), xxii, 210pp.**
House recounts history from the discovery of oil on John McClesky's farm, south of Ranger on October 22, 1917, through the boom that followed. This was Boyce House's first book on the Eastland County, Ranger boom and was to be followed by three more historic works on the subject. This one is the local history with one anecdote following another. Countless good stories in a you are there format with lots of dialogue. Subtitle following Were You In Ranger? .. "or Cisco, DeLeon, Dublin, Leeray, Merriaman, Eastland, Desdemona, Breckenridge?" Cited in Swanson on page 39; Hardwicke page 34. The hardbound first edition is scarce.

**A151. (CANADA) House, J.D., THE LAST OF THE FREE ENTERPRISERS: THE OILMEN OF CALGARY, Toronto, Macmillan of Canada, (1980), ix, 230pp.**
House is not happy with the way oil companies have manipulated the Canadian government over the years, and through a general examination of the history and inner workings of the industry, concludes that the situation is unhealthy. Tax breaks and other legislative concessions are condemned. The Canadian history of government intervention and regulation of private industry is similar to that of the U.S., although it has worked in different ways. He compares the Canadian system with that of USSR and Mexico and concludes that something in-between is necessary. He includes a map and tables. The softbound first edition is common.

**A152. (MISSISSIPPI) Hughes, Dudley J., OIL IN THE DEEP SOUTH: A HISTORY OF THE OIL BUSINESS IN MISSISSIPPI, ALABAMA, AND FLORIDA, 1859-1945, Jackson, University Press of Mississippi, (1993), xi, 267pp.**
From the dust jacket: " The first wildcatters in the Southeastern states...began drilling around the turn of the century, but success eluded them. Except for a few insignificant shallow gas fields in northern Alabama, all early wells...more than two hundred of them by 1925 proved to by dry". Hughes documents the business history which really started with the 1939 discovery of the Tinsley Oil Field in Yazoo County, Mississippi and then expanded through the WW II years. Included are historic photos and maps. The hardbound first edition is common.

**A153. (TEXAS) Humble Oil & Refining Co., 250,000,000 YEARS WITH TEXAS, Houston, TX, Humble Oil, (1936), 31pp.**
This pamphlet was prepared as part of the Texas Centennial Celebration. From the cover: "A story of oil and geology told by the geologic exhibits in Humble's Hall of Texas History, Texas Centennial Exposition, Dallas, Texas, 1936. This excellent Texas oil history includes impressive photos of Humble's relief maps and true scale cross-sections, showing surface and subsurface at East Texas, Powell, Spindletop and Big Lake. Often bound in with TWICE TOLD TALES OF TEXAS, published at the same time by Humble. Not mentioned in Swanson. The softbound first edition is common.

**A154. (OHIO) Humphrey, William D., A BRIEF HISTORY OF OIL & GAS IN FINDLAY TOGETHER WITH AN ACCOUNT OF THE CELEBRATION OF JUNE, 1937, Findlay, The Author, (1940), vii, 160pp.**

Fort Findlay was established during the war of 1812 and a pioneer town grew around it. In 1836 Richard Wade discovered gas in his water well at ten feet. He piped it to his fireplace and used it for heating and cooking. Dr. Oesterlen determined that Findlay was built on a large gas field. In 1874 Findlay Gas Light Company lighted the streets of the town, first by artificial gas then in 1885 by natural gas. The good doctor drilled and found three separate pay zones to 1092 feet which supplied the 250 mcf required by the city then. The Findlay gas boom was on and was followed by associated oil production as well. Cited in Swanson on page 41. The hardbound first edition is becoming scarce.

**A155. (TEXAS) Hurley, Marvin, DECISIVE YEARS FOR HOUSTON, Houston, Houston Magazine, (1966), 414pp.**

Hurley examines the post WW II era of America's sixth largest city. From its founding in 1836 by the Allen brothers, New York real estate promoters, and a very slow start, to the modern oil capitol of the world. Civic progress during the first 100 years was erratic at best. The author analyzes Houston's growth in what he describes as the critical years (Pre WW II to the mid-1960's). The city's future was decided during that time according to Hurley. The hardbound first edition is common.

**A156. (GENERAL) Ickes, Harold L., FIGHTIN' OIL, NY, Alfred A. Knopf, 1943, x, 174pp.**

The book was written by Ickes while he was Secretary of the Interior and Chairman of Petroleum Administration for War under Roosevelt and right in the middle of the WW II. It is an excellent propaganda piece and an informative tool for the public. It must have distressed our enemies (who surely read it). It is an appeal for support in conservation of oil. Much on the PAW and the extraordinary construction of Big Inch pipeline. Big Inch carried oil from the Southwestern oil fields to the East Coast which then supplied European markets and the allied armies which had become cut off by Germany, both from land and sea. From the cover: "The inside story of the relationship between our petroleum supply & the state of global war told by the man charged with the responsibility of delivering oil on time to the battle lines & the home front". The book usually includes a hand addressed presentation slip from Mr. Boyd of the War Council. Cited in Swanson on page 154. The hardbound first edition is common.

**A157. (CALIFORNIA) Independent Petroleum Association of America, VALLEY OF OIL, Oildale, CA, The Oildale News, (1964), 143pp.**
Much research was involved in determining the true beginnings of the California oil industry. John Hamilton was the voted winner by the IPAA. In 1864 he dug the first producing oil well. Not only that, he built the first refinery, distilled kerosene, bottled it and sent it to market. This special issue commemorates the first 100 years of California oil. A number of early oil articles are presented covering such topics as the tar pits at Los Angeles, Buena Vista, the first oil company, the unions, discovery and location of the oil fields, towns oil built, etc. Background on thirteen important oil companies is included with details on how they got to be big in California. Union, Standard, Shell, PG&E, Richfield and Tidewater are among the prominent. An exceptional photo section is included. The hardbound first edition is scarce.

**A158. (GENERAL) Irvin, Samuel P., THE OIL BUBBLE, Franklin, PA, The Author, 1868, 70pp.**
The oil industry had collapsed and the author had just lost a hard-gained fortune in the business. He takes a little time to reflect on the general state of the 1868 oil industry and the people involved in it. He states what he believes are the reasons for the ups and downs of crude oil prices. Had modern day oil industry people read this enlightening treatise prior to the January 1986 price collapse, they would not have been quite so surprised. History does repeat itself and most of us are doomed to suffer the consequences. One of several rhyming quotable quotes from the booklet: "Riches, like insects, while concealed they lie, Wait but for wings, and in their seasons fly". Cited in Swanson on page 135. The softbound first edition is rare.

**A159. (GENERAL) Johnson, Arthur Menzies, THE DEVELOPMENT OF AMERICAN PETROLEUM PIPELINES: A STUDY IN PRIVATE ENTERPRISE AND PUBLIC POLICY, 1862-1906, Ithaca, NY, Cornell University Press, (1956), xiii, 307pp.**
Here is the definitive work on early pipelines. While researching the subject, the author was able to document the critical part played by

pipelines in the political and economic history of the oil industry. His stated purpose was to trace the relationships of private enterprise to public policy prior to the establishment of the "Common Carrier" amendment to the Hepburn Act. He traces early oil history as the pipeline was developed by independents to out-flank Rockefeller's domination of oil transportation from wells to refineries through the railroads. The technology and logistical problems were overcome with great difficulty, particularly with Standard Oil fighting them legally and illegally. The pipeline concept was so efficient and logical that Rockefeller began building and acquiring pipeline systems, later dominating that form of transportation. Anti monopoly sentiment was building across the industry and country, finally catching up with Rockefeller in the Supreme Court ruling of 1911, which dissolved the Standard Oil Trust. Several maps and photos are included which punctuate the key events of pipeline development. A useful bibliography is included with notes and index. Cited in Swanson on page 162. The hardbound first edition is becoming scarce.

**A160. (TEXAS) Johnson, J.K., BORGER: THE NATURAL HISTORY OF AN OIL-BOOM TOWN, Dallas, SMU Press, 1940, pp3-10, S.M.U. Journal, Volume IV, No. 1&2.**
As per the title, the brief article looks at the effect of disorganized and out-of-control boomtown mentality on the participants. Borger was perfect choice for this sociological study, as it was one of the wildest boomtowns in Texas history. The softbound first edition is common.

**A161. (Texas) Johnston, Marguerite, HOUSTON: THE UNKNOWN CITY, 1836-1946, College Station, Texas A&M University Press, (1991), xii, 446pp.**
City history with emphasis on the wealthy and philanthropic citizens. The importance of oil in the oil capital of the world is stressed at times of The Depression, WW I and WW II. Many oil personalities are covered in good detail. From the cover: "..a history marked by murder, mutiny, and the ironies of war, by comedy and high jinks, by heroism and a remarkable generosity". The hardbound first edition is common.

**A162. (GENERAL) Jones, Howard Mumford:, THE AGE OF ENERGY: VARIETIES OF AMERICAN EXPERIENCE, 1865-1915, NY, The Viking Press, 1971, xix, 545pp.**
A general history of the U.S. with references to Rockefeller and others as they contributed to the development of industry in general. Largely a social history of the turn of the century. The hardbound first edition is common.

**A163. (OKLAHOMA) Jones, Richard Lloyd, et al (editors), OKLAHOMA AND THE MID-CONTINENT OIL FIELD, Tulsa, Oklahoma Biographical Association, 1930, 500pp.**
From the title page: "History of a great new state and man's quest for oil...An epoch in American history and progress...told in terms of individual achievement and biography...A work for the newspaper and reference library". History of major Oklahoma towns, industry and oil fields written by important men of the day including Overholser, Skelly, Gould, and Halliburton. Oklahoma counties are documented in detail, from their organization to their current assets. The book contains over 300 photos plus biographies of oilmen of Oklahoma. Not mentioned in Swanson. The hardbound first edition is very scarce.

**A164. (ROMANIA) Jordan, Constantin N., THE ROMANIAN OIL INDUSTRY, NY, NY University Press, 1955, 357pp.**
In 1955 the Romanian oil fields were under Russian control and at the time was the only important European oil area. Romania lost control of its oil fields first to Germany in the early 1940's, then after a brief round of attempts to rebuild, from 1944 to 1947, lost them to Soviet Block control. Quoting from Swanson "Rumanian oil industry in 1947: communization of oil industry, oil field equipment and supplies, labor force, exploration, drilling, production, refining, transport, marketing, consumption; statistics, 1928-1947". The book is somewhat political but a big part of European oil history. Cited in Swanson on page 55. The hardbound first edition is common.

**A165. (GENERAL) Josephson, Matthew, THE ROBBER BARONS: THE GREAT AMERICAN CAPITALISTS 1861-1901, NY, Harcourt, Brace and Co., (1934), viii, 474pp.**
Along with Rockefeller and his Standard Oil, Josephson looks at Carnegie, Gould, Frick, Morgan, and Vanderbilt in their late 1800's

heyday and after the turn of the century when their monopolistic power began to diminish. All were powerful monopolists, but Rockefeller stands out as a little more ruthless, cunning and thorough than the others. Josephson was already an established author, but this appears to be the first of his attacks on capitalists. He takes after them all, particularly John D., and provides a good bibliography for backup. Cited in Swanson on page 162. The hardbound first edition is scarce.

**A166. (OKLAHOMA) Kay County Gas Company (compilers), KAY COUNTY OKLAHOMA, Ponca City, OK, The Company, 1919, 75pp.**
Excellent history of the prolific oil area adjacent to the Osage reservation. Much on Indians and oil and of course E.W. Marland, founder of Marland Oil which became the core of Conoco. The book is profusely illustrated with black and white photos not published elsewhere. Many are panorama fold-outs. History of the 101 ranch is also found throughout the book. Not cited in Swanson. The hardbound first edition is scarce.

**A167. (ALASKA) Keeble, John , OUT OF THE CHANNEL: THE EXXON VALDEZ OIL SPILL IN PRINCE WILLIAM SOUND, NY, Harper Collins, 1991, xii, 290pp.**
Keeble previously wrote fiction, but was compelled to write this book when he became emotionally involved in the historic debacle. He attacks the Valdez spill with vigor, reviewing much of the history and circumstances which led to the inevitable accident. He criticizes Exxon thoroughly as he describes the devastation as he observed it, first-hand. Includes notes and a bibliography. The hardbound first edition is common.

**A168. (TEXAS) Kelly, Louise (editor), WICHITA COUNTY BEGINNINGS, Burnet, TX, Eakin Press, (1982), xvi, 286pp.**
The editor takes a hard look at the early history, largely pre- 1910. The Wagner, Burnet and Hart ranches dominated the county and it was populated by 332 people at the 1880 census. Much on early oil independents and majors, along with Red River boundary dispute. Discussion of Burkburnett Oil Field and North Texas Oil Field as they are interwoven with the history of the area. The hardbound first edition is becoming scarce.

**A169. (CANADA) Kennedy, Tom, QUEST: CANADA'S SEARCH FOR ARCTIC OIL, Edmonton, Reidmore Books, (1988), x, 222pp.**
The author followed the Arctic oil play from its inception. He is a journalist, reporter and commentator with fine credentials and many awards. The story begins in 1969 with Panarctic's Drake wildcat blowout. The well spewed 40 million cubic feet of gas and 40,000 barrels of salt water per day, a great disappointment to Dome, the drilling contractor and Panarctic who had been seeking oil. The story continues as Jack Galagher of Dome and the Panarctic people keep drilling for riches in Canada's Arctic over the next fifteen years. This story is a roller coaster ride of oil exploration adventure, a lesson in logistics and politics, a horrible plane crash and finally an economic disaster. Although fair quantities of oil and gas have been recovered from the area, politics and economics doomed the project to failure. The hardbound first edition is common.

**A170. (CANADA) Kerr, Aubrey, ATLANTIC 1948 NO. 3, Calgary, S.A. Kerr, (1986), x, 226pp.**
Kerr was a district geologist for Imperial Oil (Standard) during the Atlantic No. 3 blowout. After his retirement he decided to chronicle the event which he regards as a landmark of Canada oil history. Rapid development of the Leduc oil field was well underway and had been taking place without the benefit of blowout preventers. A shallow, gas bearing reservoir was being ignored as the deeper, highly-porous Devonian reservoir was being exploited. It was an accident waiting to happen. This is the exciting well- told story of Atlantic 1948 No. 3 which blew wild and produced 1.4 million barrels before finally being successfully plugged. The saga extended over several months and included such phenomena as a huge crater, a major fire, the presence of Red Adair, and finally the successful drilling of relief wells to flood the formation. The hardbound first edition is becoming scarce.

**A171. (CANADA) Kerr, Aubrey, CORRIDORS OF TIME, Calgary, S.A. Kerr, (1988), x, 331pp.**
Kerr gives his version of the history of Canadian oil through a series of short stories that are designed to educate as well as entertain. From the dust jacket: "From Petrolia to Prudhoe Bay, this collection of narratives spans the miles and the years, melding facts with folklore". It does that by studying some of the important personalities involved in the discoveries and significant dry wildcats. Much information on Canada's

important oil fields such as the Turner Valley anticline. Much of the narrative concerns stories of Prudhoe Bay, its discovery and the subsequent lease sale drama. Kerr does a good job of humanizing oil history. The hardbound first edition is common.

**A172. (CANADA) Kerr, Aubrey, LEDUC, Calgary, S.A. Kerr, (1991), xiii, 312pp.**
Kerr writes of that which he has considerable knowledge. He states that Canadian oil was nothing of consequence until the 1947 discovery at Leduc. His recollections are impressive and he fortunately had access to the best of resource material. An Imperial employee named Doug Layer compiled a three-volume set of memos, letters and maps documenting the 1947 discovery and development of the famous oil field. The Leduc discovery began the successful search for more similar fields and put Alberta on the oil country map. Again Kerr provides human interest as he explores the characters involved. The hardbound first edition is common.

**A173. (TEXAS) Kittle, Kit, ROUGHNECKS: OIL PATCH U.S.A., Dallas, Taylor Publishing, (1985), 189pp.**
This is a coffee table book with an incredible assemblage of color photos covering modern Texas characters from front office executives to the roughest of roughnecks and their friends. Little narrative...little needed. Unusual. The hardbound first edition is common.

**A174. (GENERAL) Knowles, Ruth Sheldon, THE FIRST PICTORIAL HISTORY OF THE AMERICAN OIL AND GAS INDUSTRY 1859-1983, Athens, OH, Ohio University Press, (1983), ix, 171pp.**
Historic black and white photos cover events from the Drake well forward. Knowles has been a prolific and popular author, providing history and insight of the oil industry for laymen and industry people alike. This is her first picture book effort and it seems to provide something for everyone, touching on all phases of the business. Also included are maps and drawings. The hardbound first edition is common.

**A175. (GENERAL) Knowles, Ruth Sheldon, THE GREATEST GAMBLERS: THE EPIC OF AMERICAN OIL EXPLORATION, NY, McGraw-Hill Book Co., (1959), xii, 346pp.**
The author was involved in the oil business, largely as a government consultant for several years. She also participated in drilling ventures with her oilman husband. Knowles gives her brand of oil history through biographical sketches of prominent successful independent oilmen. She spends little time on early oil history. Her thesis is that oil history began after the Spindletop discovery. Her subjects are Mike Benedum, E.W. Marland, Tom Slick, Charles Canfield, and Edward Doheny, although Rockefeller and Standard still play a large part in her story. Brilliant geologists, Gould, DeGolyer, Pratt, Levorsen and Weeks are discussed in relation to some of their contributions to oil finding. The second 1978 edition has 376pp and includes important new fields discovered, new tax laws, environmental activists and the Arab embargo. This is one of the top oil history books by an author who knows her stuff. Cited in Swanson on page 154. The hardbound first edition is becoming scarce.

**A176. (OKLAHOMA) Lambert, Paul F. and Kenny A. Franks (editors), VOICES FROM THE OIL FIELDS, Norman, University of Oklahoma Press, (1984), xii, 260pp.**
During the late 1930's the WPA's Federal Writers Project financed the recording of interviews with early Oklahoma oil people. Interviewers, Ned DeWitt, Welborn Hope and Dan Garrison transcribed the recordings and they are presented here with editorial notes preceding each. The stories are word-for-word recollections and not from the professionals, but the workers. Their general dissatisfaction with the dangerous and primitive working conditions in the oil fields is vivid. Twenty real stories worth reading. Several photos are also included. The hardbound first edition is common.

**A177. (CALIFORNIA) Latta, F.F., BLACK GOLD IN THE JOAQUIN, Caldwell, Idaho, The Caxton Printers, 1949, 344pp.**
The author was a totally dedicated California historian. Born in 1892, he interviewed 10,000 early Californians and published much history over the years. First hand reminiscences of old-timers comprises much of this book. It is limited to San Joaquin oil history from 1830 to 1900. Of particular interest is Latta's photo of a 110 year old Yokuts Indian he interviewed in 1929. The Indian told of hauling asphalt from the oil seeps

at Coalinga in 1830. Oil was sold from pits dug in that same area during the 1860's. An early independent tells of his 1890's drilling rig powered by real horse-power and complains of unfair competition by the larger oil companies. History of the 1928 Kettleman Hills discovery is well told as it produced rampant northward exploration and development. Rare photos of early oil people and places are included. Cited in Swanson on page 49. The hardbound first edition is scarce.

**A178. (PENNSYLVANIA) Lawrence, Albert A., PETROLEUM COMES OF AGE, Tulsa, OK, Scott-Rice, (1938), x, 227pp.**
Published the same year as Giddens Birth of the Oil Industry, Lawrence presents a well documented history of early oil. He cites much oil reference material throughout the book. He was inspired to write this book by a committee formed to preserve an oil seep at Cuba, New York. It had first been described in 1627 by a Franciscan Monk. Lawrence documents the history of the area and complements it with the now famous Allegheny Valley oil fields at Oil Creek, Pennsylvania. He covers the early indications of oil in New York and Pennsylvania and the slowly developing uses of Seneca Oil in some detail. The Drake discovery and the ensuing booms are well covered. The author also concentrates on the large amount of effort which was put into the development of technology for producing, transporting and marketing this new product. A memorial sketch of John D. Rockefeller is included. Cited in Swanson on page 154. The hardbound first edition is becoming scarce.

**A179. (STANDARD) Lawson, Thomas W., FRENZIED FINANCE, NY, Ridgway-Thayer Co., 1906, xix, 559pp.**
This book was not available for review by Neff, however it is regarded as important. Quoting from Swanson: "Landmark in muckraking works aimed at Standard Oil; author was financier and market operator, and for this reason his expose carried extra weight". Cited in Swanson on page 163. The hardbound first edition is very scarce.

**A180. (GENERAL) Leeston, Alfred M., et al, THE DYNAMIC NATURAL GAS INDUSTRY, Norman, University of Oklahoma Press, (1963), xviii, 464pp.**
Subtitle: "The description of an American industry from the historical, technical, legal, financial, and economic standpoints". Leeston, Crichton and Jacobs have all been associated with the natural gas industry for years

and present a well organized picture of its past, present and future. The industry began when the first U.S. gas well was completed at Fredonia, New York at a depth of twenty-seven feet. This was in 1821 and the small diameter well had enough volume to fire the thirty burners which were used for street lighting. The authors favor free markets for natural gas and make a good case for it. Extensive bibliographic references are included with each chapter. Also included are maps, graphs and drawings. The hardbound first edition is common.

**A181. (PENNSYLVANIA) Leonard, Charles C. (Crocus), THE HISTORY OF PITHOLE, Pithole, PA, Morton, Longwell & Co., 1867, xvii, 106pp.**

Leonard, as a journalist was one of the first to see the humor in the crazed situation as it developed at the oil boom town of Pithole. The town later became world-famous for its rapid rise and almost equally rapid fall. Boom town to ghost town in two years. It was almost gone when Leonard wrote this book and today there is little evidence that it ever existed except in literature. It is a delightfully humorous and factual account of the town. When Dr. Miller of West Penn reprinted this historic work in 1945, he added photographs and bibliographical notes. Cited in Swanson on page 155; Hardwicke page 35; Giddens page 89. The soft or hardbound first edition is rare and the 1946 reprint by West Penn Oil is very scarce.

**A182. (TEXAS) Logan, Chauncey, BURKBURNETT, San Antonio, TX, The Naylor Company, (1973), 253pp.**

This exciting story of wildcatting in North Texas just prior to WW I is a fast moving account which has been slightly fictionalized. The author describes what happens to a town in terms of the people and their sudden wealth. Jack Merrell, a Magnolia Petroleum Company engineer, and his family live the story of booming Burkburnett from elation to disaster and back again. Dialogue added along with some interpretation of motives of characters. Good reading in this popular book. The hardbound first edition is common.

**A183. (MIDDLE EAST) Longrigg, Stephen Hemsley, OIL IN THE MIDDLE EAST: ITS DISCOVERY AND DEVELOPMENT, London, Oxford University Press, (1954), xiii, 401pp.**

Longrigg was an oil company officer in the area for years. He traces the history of oil developments in the Arab countries and in Persia from the earliest days to 1953. Largely exploration and development history along with much detail on the individual oil operations. The second 1961 edition has five additional chapters updating developments to 1961. Useful maps are included. Cited in Swanson on page 59. The hardbound first edition is common.

**A184. (GENERAL) Look Magazine (editors), OIL FOR VICTORY: THE STORY OF PETROLEUM IN WAR AND PEACE, NY, McGraw-Hill Book Co., (1946), 287pp.**

A picture story about people doing their jobs to keep the critical supply of oil moving to the armed services during WW II. From the dust jacket: "Whatever their part, without their effort no plane could have flown, no ship could have sailed, no tank could have rolled, no gun could have fired. Without their effort, we might well have foundered on the road to victory". This fine book was assembled by the editors of Look Magazine who were assisted by the American Petroleum Institute, Petroleum Administration for War, and Petroleum Industry War Council. All aspects of the logistical problems and accomplishments are reported from around the world, such as the production, transportation, refining, and special products manufacturing. Great photo-history and propaganda piece. Cited in Swanson on page 155. The hardbound first edition is common.

**A185. (GENERAL) Lundberg, Ferdinand, THE RICH AND THE SUPER- RICH: A STUDY IN THE POWER OF MONEY TODAY, NY, Lyle Stuart, Inc., (1968), 812pp.**

Lundberg does his usual job of muckraking the capitalists. He attacks oilmen and oil wealth, whether inherited or self generated. Oil financed philanthropies are also a target. This is a major work, including much information with appendices and notes. Lundberg answers the questions "Who really owns America? and How do they keep their wealth and their power"? plus many more. From the dust jacket "...until no aspect of American life and culture goes unexamined in relation to the 500 families who rule our lives today". The hardbound first edition is common.

**A186. (TEXAS) Lynch, Gerald, ROUGHNECKS DRILLERS & TOOLPUSHERS: THIRTY-THREE YEARS IN THE OIL FIELDS, Austin, University of Texas Press, (1987), xv, 262pp.**

This history is largely an autobiographical account of the author working his way through the West Texas oil fields of the 1920's and into the

1950's. From the dust jacket: "...a much needed insider's view of the oil industry, describing life in various oil fields in and around Texas. He also chronicles changes in drilling methods and oil field technology and how these changes affected him and his fellow oil field workers". This time period saw most of the Permian Basin major discoveries and the rapid development programs which followed. Enjoyable reading. The hard-bound first edition is common.

**A187. (TEXAS) Mabry, Mrs. Grant, MARTIN COUNTY CENTENNIAL, Big Spring, TX, Martin County, 1984, 118pp unnumbered.**
Brief county history with many photos. The effort was supported mainly by oil and gas companies as the county lies within the giant Spraberry oil trend of West Texas which has been producing since the early 1950's. Many small communities are covered along with the people who are largely in farming and business with oil interests. The softbound first edition is common.

**A188. (GENERAL) Marcosson, Isaac F., THE BLACK GOLCONDA: THE ROMANCE OF PETROLEUM, NY, Harper & Brothers, 1924, xi, 369pp.**
The author did a credible job in what has been described as a very short time. Although Marcosson's story is somewhat rambling and he often seems to be in over his head, the history of world oil is fairly well covered from both historical and political perspectives. Quoting Swanson: "Pot Boiler...by a writer of industrial biographies". Photos are included. Cited in Swanson on page 187; Hardwicke page 26. The hardbound first edition is becoming scarce.

**A189. (VENEZUELA) Martinez, Anibal R., CHRONOLOGY OF VENEZUELAN OIL, London, George Allen, Ltd, (1969), 207pp.**
Martinez is a Venezuelan oilman who compiled considerable data on his subject while at Stanford. He later worked for Creole. His detailed chronology and description of historical events in Venezuelan oil are impressive from a historic and research point of view. The hardbound first edition is common.

**A190. (PENNSYLVANIA) Mather, John A., MATHER'S HISTORIC OIL REGIONS OF WESTERN PENNSYLVANIA, Titusville, PA, Mather's Historical Photographs, 1895.**

This short, but sweet work contains twelve historic photographs and descriptive text of early Pennsylvania oil fields, including Mather's famous shot of Drake at his well. According to Swanson, Mather set up his photographic studio in Titusville in 1860. Without this famous photographer's work, much history of early oil would not have been preserved. Although often cited in bibliographies and literature as just another early oil book, the uniqueness of this title is often underestimated. The scenes depicted in the twelve photos varied from book to book. They were individually selected, either by Mather or the customer and hand tied into an album type book. Reprinted in 1994 by M.A. Mong. Cited in Swanson on page 154; Hardwicke page 29; Giddens page 92. The hardbound first edition is rare.

**A191. (PENNSYLVANIA) May, Lioy, MISPLACED GLORY: THE LOST SOLDIERS OF 1847-1865 IN THE PENNSYLVANIA OIL REGION, Philadelphia, The author, 1946, 353pp.**

This book was not available for review by Neff, but is regarded as important. Quoting from Swanson: 'Early oil development in Pennsylvania and role of Civil War soldiers in its origin and growth". Cited in Swanson on page 155. The hardbound first edition is very scarce.

**A192. (GENERAL) McBeth, Reid Sayers, OIL: THE NEW MONARCH OF MOTION, NY, Markets Publishing, 1919, x, 210pp.**

Swanson simply describes this book as: "General description of oil industry". A more interesting quote comes from the title page: "An unbiased presentation of the whole oil industry. Written especially for the investor who desires the facts". A popular book of the time which not only contains much oil history, but covers all U.S. oil areas and the operational phases of the business. Investing in 1919 was rampant and the oil markets fluctuated wildly. In the last chapter, the author comes to his point and provides a treatise on the art of selecting oil securities. Photos and maps are included. Cited in Swanson on page 187; Hardwicke page 13; Giddens page 92. The hardbound first edition is common.

**A193. (TEXAS) McComb, David G., HOUSTON: THE BAYOU CITY, Austin, University of Texas Press, (1969), 315pp.**

Well written history of the city from its founding to 1969. The book is an expanded version of the author's doctoral dissertation. It was widely circulated and has become one of the most popular Houston books. It also contains a bibliography along with photos and drawings. The second, 1980 edition is entitled "Houston: A HISTORY." and includes much on the effects of the modern oil boom which was reaching its peak. The hardbound first edition is common.

**A194. (MEXICO) McConnell, Burt M., MEXICO AT THE BAR OF PUBLIC OPINION: A SURVEY OF EDITORIAL OPINION IN NEWSPAPERS OF THE WESTERN HEMISPHERE, NY, Mail and Express, 1939, viii, 320pp.**

An extensive compilation of editorial comment from across the U.S. regarding the Mexican confiscation of all foreign oil properties. McConnell had the reputation of being a free-thinking and fair editor as he had worked for years with the Literary Digest. His book is very well organized into eleven categories ranging from Mexican history to what to do about the problem. The author provides pertinent comment along with some forty political cartoons which clearly indicate the U.S. sentiment. Much on the break-down of the "Good Neighbor Policy" and universal belief that Mexico would not indemnify the oil companies. Not mentioned in Swanson. The hardbound first edition is scarce.

**A195. (TEXAS) McCoy, Dorothy Abbott, OIL, MUD AND GUTS, Mission, TX, Privately Published, (1977), xii, 188pp.**

This is not your usual dry oil town history. McCoy provides an incredible number of semi-chronological anecdotes about the local citizenry, which in effect tell the history of Freer, Texas. Freer is located in the infamous county of Duval, about fifty miles from Laredo, in rough country. The author grew up there. Freer sprung from nothing as the sleepy Norton Ranch came alive in 1928 with a wildcat oil discovery, the first well in what would become Government Wells Field (named for a federal water well). Much drilling occurred there in 1929 as oil people swarmed to the area to help develop the new oil discovery. They came mostly from the oil fields of Kansas and Oklahoma, as well as other areas of Texas. Housing accommodations were pathetic for the 8,000 people working in the oil field and other assorted businesses. Freer survived the Depression crash, but the East Texas field production which flooded the market almost closed it down. Economic recovery was finally assured by 1932. The famous dust or mud streets were paved in 1938 and by 1962, potable water was finally available to all residents. The title is apt in that it took lots of guts to live in this early oil town which was a constant mud-mire. Excellent local history. The hardbound first edition is scarce.

**A196. (TEXAS) McDaniel, Ruel, SOME RAN HOT, Dallas, Regional Press, 1939, 252pp.**

Many Texas historians believe this to be the most readable story of the East Texas Field and oil boom. McDaniel takes his major subjects from start to finish rather than doing a straight chronology. He concentrates on the effect that the hot oil runners had on shaping the modern oil conservation and proration legislation. He is also sensitive to the effect on royalty owners who largely did not get paid for this hot (stolen?) oil. The hot oil was usually sold to fly-by-night independent refiners after being siphoned from storage, pipelines, or well streams. Much on Governor Sterling's military shut-down of the field as well as the special state and federal investigators who vied for control of the economic, political and criminal oil boom. Quoting from Swanson: "What transpired in East Texas from date of discovery to frenzied days of trading, fighting, hot-oil running and martial law; of days of chaos and five-cent oil and coming of regulation in world's greatest oil field". Cited in Swanson on page 39. The limited hardbound edition is rare.

**A197. (MISSISSIPPI) McGowen, Norris Cochran, NATURAL GAS - THE GULF SOUTH'S SYMBOL OF PROGRESS, NY, Newcomen Society of NA, 1951, 32pp.**

McGowen was president of United Gas Company, the world's largest handler of natural gas. He reviews the history of gas from ancient Greece seeps, through the giant merger of five gas companies which formed United Gas. Much history of gas lighting of cities from the first in Soho, England in 1804 by "coal gas". The last one third of the book concentrates on the history of United Gas, before and after its big merger. Typical informative and propagandistic Newcomen pamphlet, promoting business. Contains drawings. Not mentioned in Swanson. The softbound first edition is becoming scarce.

**A198. (OKLAHOMA) McGuire, Paul F., GENESIS OF OKLAHOMA OIL BANKING, Muskogee, OK, Western Heritage Books, (1986), 160pp.**

McGuire was a banker in Oklahoma for forty-seven years. He was drawn into oil banking early in his career and over the years came in contact with many of the important oil operators in the state. He spends brief time on his biography and the history of banking, so that he can discuss the oilmen he knew. He comments on at least fifty companies and individuals on a personal basis. The book reads very well and is an important part of oil history. Includes photos. The first edition is becoming scarce.

**A199. (TEXAS) McKay, Seth S. and Odie B. Faulk, TEXAS AFTER SPINDLETOP: THE SAGA OF TEXAS 1901-1965, Austin, TX, Steck- Vaughn Co., (1965), vii, 247pp.**

This is part six of the Saga of Texas series. It is a cursory account of the years after the 1901 Spindletop discovery. Largely a review of the politics and politicians who directed Texas history. The hardbound first edition is common.

**A200. (GENERAL) McLaurin, John J., SKETCHES IN CRUDE-OIL: SOME ACCIDENTS AND INCIDENTS OF THE PETROLEUM DEVELOPMENT IN ALL PARTS OF THE GLOBE, Harrisburg, PA, The Author, 1896, xii, 452pp.**

Swanson says that this book is "Romanticized and unreliable in many respects, but a source of much primary information not found elsewhere." However, it remains one of the most popular of early oil histories and was widely circulated in three similar editions. With 500 pages of fine print (probably 500,000 words), it includes a surprising amount of information. From McLaurin's introduction: "The purpose is to give the busy outside world, by anecdote and incident and brief narration, a glimpse of the grandest industry of the ages and of the men chiefly responsible for its origin and growth". Over half of the book is early Pennsylvania oil. The remainder covers oil around the U.S. and the world with particular emphasis on the Russian Baku fields. The third 1902 edition appendix includes much information about the new Spindletop discovery which changed the rules of the game in almost every way. Profusely illustrated with 272 photos of the characters, places, events, wild wells, etc. Many drawings and maps of the interesting areas. This is not a boring scholarly history book, but a constant run of stories and people integrated with facts and humor. After reading McLaurin's account, the reader will know almost everything there was to know about oil up to the turn of the century. Second edition in 1889, 452pp; Third in 1902, 470pp. Cited in Swanson on page 155. The hardbound first edition is scarce and the others are becoming scarce.

**A201. (ALASKA) Mead, Robert Douglas, JOURNEYS DOWN THE LINE: BUILDING THE TRANS-ALASKA PIPELINE, Garden City, NY, Doubleday & Co., 1978, 609pp.**
From the dust jacket: "Late in July 1977 the first oil from Prudhoe Bay reached the end of the Trans-Alaska pipeline at Valdez and poured into the tanker that would carry it to the West Coast". Mead presents a complete look into the history, politics, alternatives, and what he believes are the motives of the companies and people involved. Readable and fair assessment of this gigantic project. The hardbound first edition is common.

**A202. (MEXICO) Meyer, Lorenzo, MEXICO AND THE UNITED STATES IN THE OIL CONTROVERSY, 1917-1942, Austin, University of Texas Press, (1972), xix, 367.**
The relationship between the two countries is explored with special attention to the Mexican nationalization of the oil industry. Mexican archival material not previously analyzed puts the problem in reasonable historic perspective. Much about the on-going historic dependency of Mexico and other Latin-American countries on the U.S. Half of the book is source notes, comprising an exceptional reference. Translated by Muriel Vasconcellos. The hardbound first edition is common.

**A203. (MIDDLE EAST) Mikesell, Raymond F. & Hollis B. Chenery, ARABIAN OIL: AMERICA'S STAKE IN THE MIDDLE EAST, Chapel Hill, University of North Carolina Press, (1949), xi, 201pp.**
This is a historic look at the methods and politics which have been involved in establishing the American position in the Middle East. Mikesell also documents the later strategic, economic and political developments as oil became more abundant and valuable. He was an economics professor and served as government economist in the Middle East during WW II. He discusses attempts by the U.S. government itself, to become directly involved in Middle East oil production. It is obvious that the author's interest lies in the creation and development of foreign

policy. Cited in Swanson on page 92. The hardbound first edition is common.

**A204. (PENNSYLVANIA) Miller, Ernest C., AN INVESTIGATION OF NORTH AMERICA'S FIRST OIL WELL: WHO DRILLED IT?, Rutland, VT, Charles E. Tuttle Co., (1964), 66pp.**
Dr. Miller is the historian-detective who finally deduces that the Canadians were actually producing and selling oil from oil wells prior to Drake producing oil from his discovery well. It turns out that the Canadian crude was produced from hand-dug wells rather than drilled holes, making the critical and semantic difference. Swanson cites Dr. Miller's sixteen page article of the same title which was published, softbound in 1959. The limited hardbound edition is scarce.

**A205. (PENNSYLVANIA) Miller, Ernest C., EARLY DAZE IN OIL, Philadelphia, Dorrance and Company, (1974), 85pp.**
Entertaining work with good sources and notes. Dr. Miller spent years reading and writing heavy history and decided to lighten up a little with this book. This is not by any means a capricious work, as there is considerable amount of history here. A chapter on the honor of early day oilmen reveals that not much has changed in 135 years. Thirteen pages are dedicated to the great early oil photographer, John Mather, without whom much oil history would not be available. Also covered is sin among the derricks and a series of sketches on some who struck oil and some who did not. The last chapter is dedicated to the famous Nellie Bly (Elizabeth Cochrane) who had an interesting influence on the early oil business. The hardbound first edition is common.

**A206. (PENNSYLVANIA). Miller, Ernest C., OIL MANIA: SKETCHES FROM THE EARLY PENNSYLVANIA OIL FIELDS, Philadelphia, Dorrance and Company, (1941), 140pp.**
This early history by Dr. Miller contains a chronological bibliography of Pennsylvania oil books from 1861 to 1938. He does a fine job in this short, compact book as he covers many of the early oil events, places and characters. Drake, Hogan, Booth, Coal Oil Johnny, Henry Rouse and many others are described. Titusville, Pithole and Oil City are covered. Much is included on the oil scouts in their role of helping to stabilize the frantic oil speculation as they reported actual events as they occurred, squelching rumors. Interesting details on the Oil Exchange as well as the

teamster versus pipeline conflict and saga. Miller's success in this his first oil history book was impetus for another seven fine Pennsylvania oil histories over the next forty-two years. Good bibliography. Cited in Swanson on page 155. The hardbound first edition is scarce.

**A207. (PENNSYLVANIA) Miller, Ernest C., PENNSYLVANIA'S OIL INDUSTRY, Gettysburg, Pennsylvania Historical Association, (1954), iii, 69pp.**
This pamphlet covers major aspects of Pennsylvania oil history from the Drake discovery. It is brief and concise with excellent photos of the derricks and early towns. The incidents and characters are also covered. Included are an excellent bibliography, a listing of monuments to Pennsylvania's oil industry, along with some data on the modern industry for the mid-1960's. Cited in Swanson on page 155. The softbound first edition is becoming scarce.

**A208. (PENNSYLVANIA) Miller, Ernest C., THE HISTORY AND DEVELOPMENT OF THE PETROLEUM INDUSTRY IN WARREN COUNTY, PENNSYLVANIA, Warren, Warren County Historical, 1983, xi, 129pp.**
From Dr. Miller's preface: "In the preparation of this history of the first century of the oil industry as related to Warren County, Pennsylvania, one hundred years of county newspapers were inspected, one by one. In addition, diaries, letters and county historical materials were examined in depth. The facts are as accurate as it is possible to have them at this writing". He provides much detail on the 646 Mystery Well which discovered Cherry Grove field, Warren County's most famous. The hardbound first edition is common.

**A209. (PENNSYLVANIA) Miller, Ernest C., THIS WAS EARLY OIL: CONTEMPORARY ACCOUNTS OF THE GROWING PETROLEUM INDUSTRY, 1848- 1885, Harrisburg, Pennsylvania Historical Commission, (1968), ix, 211pp.**
From Dr. Miller's preface "The first twenty-five years of growth and development were most important. During those formative years, the industry developed means, methods, and rules, most of which are still in current use". Includes a key map showing historical wells, areas and farms. Dr. Miller fills in a few gaps and adds detail to some of Giddens' earlier works. This is probably the best of Dr. Miller's numerous books

and pamphlets on the early oil industry. The hardbound first edition is scarce.

**A210. (PENNSYLVANIA) Miller, Ernest C., TINTYPES IN OIL, Rutland, VT, Charles E. Tuttle Co., (1961), 186pp.**
Tintypes, in this title, refers not to photos, but to biographical sketches. Here Dr. Miller reviews eight famous, or infamous, early Pennsylvania oil farms, or the characters related to them. One of the best is the story of Cherry Run, on which was created an above average boomtown called Cherry Grove. In 1882 the "Mystery Well" was drilled by Grace & Dimmrich on Tract No. 646. The stir it created is told here, eventually developing a field with 40,000 barrels per day capacity. Another sketch is on Henry Rouse who tragically died in one of the first large oil field fires created at a new discovery well on his leases. Another interesting story is about John Wilkes Booth, the actor who had an up and down oil speculation career. He owned small interests which were drilled in 1864, unprofitably. Another story is of John Washington Steele, "Coal Oil Johnny" who lost a fortune in the oil business, merely by inattention. Gib Morgan and his legendary story-telling career is also documented. The hardbound first edition is becoming scarce.

**A211. (GENERAL) Miller, Max, SPEAK TO THE EARTH, NY, Appleton-Century-Crofts, (1955), vii, 310pp.**
From the title page: "The story of the pioneers pushing into America's last frontier - straight down. A vividly human cross- section of America's most dynamic industry...oil:. From Swanson: "An oil tour through the U.S., California to Williston Basin to Appalachian oil region, back through Ohio to Southwest, including offshore operations; popular, generalized treatment". This widely circulated general history includes drawings. Cited in Swanson on page 31. The hardbound first edition is common.

**A212. (MIDDLE EAST) Mineau, Wayne, THE GO DEVILS, London, Cassell, (1958), xi, 243pp.**
Based on the author's tour of the oil fields and numerous interviews he conducted with a wide variety of oil people. He documents the history starting with Knox D'Arcy in Persia and follows the development as it spreads across the Middle East. Cited in Swanson on page 59. The hardbound first edition is common.

**A213. (KANSAS) Miner, Craig, DISCOVERY: CYCLES OF CHANGE IN THE KANSAS OIL & GAS INDUSTRY 1860-1987, Wichita, Kansas Independent O&G Association, 1987, xvii, 239pp.**
This is a much expanded version of the author's earlier "Fire in the Rock". Miner is again sponsored by KIOGA and provides no sources, notes, bibliography, references or index. This is however, an excellent and exciting book. The narration flows as he presents more early history punctuated with abundant sepia photographs and current political oil cartoons. Much of the added history is from 1976 to 1987 during which Kansas enjoyed the boom and suffered the bust as it kept on discovering small shallow economical pools and kept on pumping its stripper wells. The hardbound first edition is common.

**A214. (KANSAS) Miner, H. Craig, THE FIRE IN THE ROCK: A HISTORY OF THE OIL AND GAS INDUSTRY IN KANSAS 1855-1976, Wichita, Kansas Independent O&G Association, (1976), 110pp.**
Miner relies on published articles and personal interviews, particularly the Kansas Historical Society although he includes no sources or notes. From the 1892 Neodesha discovery which began the era of mid-continent exploration to the fabulous 1915 El Dorado discovery, Miner presents a fresh look at the people and history. Western Kansas exploratory drilling by the "Lucky Seven", a group which included a real estate salesman, bank cashiers, a druggist and a rancher, opened the Nemaha Ridge and the west part of the state with their Carrie Oswald discovery well in 1923. Much on the thirteen trillion cubic foot Panhandle field in southwest Kansas. The last two chapters reflect the politics of Miner's sponsor, KIOGA. Also included are photos and humorous drawings. The hardbound first edition is common.

**A215. (GENERAL) Mintrop, Ludger, ON THE HISTORY OF THE SEISMIC METHOD FOR THE INVESTIGATION OF UNDERGROUND FORMATIONS AND MINERAL DEPOSITS, Hanover, Germany, Seismos G.m.b.H., 1930, 128pp.**
This book was not available for review by Neff, but is considered important. Quoting from Swanson: "Account of development of seismic method by its chief inventor and pioneer. Baron Mintrop". A fair statement regarding Mintrop's invention could be made to the effect that the art and science of oil prospecting would not even remotely resemble

its present state without his work. Cited in Swanson on page 6. The first edition is scarce.

**A216. (TEXAS) Moore, Richard R., WEST TEXAS AFTER THE DISCOVERY OF OIL: A MODERN FRONTIER, Austin, TX, Jenkins Publishing, 1971, xii, 185pp.**
The author divides West Texas oil exploration activity into three phases and reviews the progress and effect of each. The first started with Ranger in 1917, the second near the end of the depression and the third following WW II. He tries to place the impact of the oil industry in economic perspective. Not too hard to do. A useful bibliography and map are included. The hardbound first edition is becoming scarce.

**A217. (PERU) Moreno, Federico, PETROLEUM IN PERU: FROM AN INDUSTRIAL POINT OF VIEW, Lima, F. Masias & Co., 1891, 161pp.**
Moreno was the ex-prefect of Peru and dedicated this work to the Geographic Society of Lima. Part I is the history of oil in the area. Part II covers his version of the history of oil in the U.S. and Russia. Part III covers the origin and composition of petroleum along with where to find it and how to refine it. In Part IV, several conferences, seminars and reports are reproduced covering aspects of the oil industry in Peru and South America. Cited in Swanson on page 53; Hardwicke page 14. The hardbound first edition is rare.

**A218. (TEXAS) Morgan, Jonnie R., THE HISTORY OF WICHITA FALLS, Wichita Falls, Nortex, 1931, 221pp.**
Morgan was a teacher in Wichita Falls and wrote this history. It starts with the heirs of Tom Scott laying out the townsite in 1876 on lands he had won in a poker game some twenty years earlier. As in most West Texas townsites, a planned railroad through the area stimulated the action. Nearby oil development further stimulated growth as Electra field came in during 1911, the first of many in Wichita County. Several small discoveries around Burkburnett in 1913 eventually led to more and more production, flooding the market. The main Burkburnett field was found in 1918, fifteen miles north of Wichita Falls. The shallow depth (500-1000 feet), and high gravity oil caused a rampage. Refining and other operations spread in the major boom. Wichita Falls was far enough away to avoid some of the boomtown chaos, but grew rapidly in response

to the activity. Up to the 1970's the city maintained regional offices for many major and independent oil companies. It is still a preferred operating area for the small independent. Many original family histories are found in the last part of the book. Reprinted as a Collector Edition in 1971. Hardbound first edition is scarce and reprint is becoming scarce.

**A219. (PENNSYLVANIA) Morris, Edmund, DERRICK AND DRILL, NY, J. Miller, 1865, 277pp.**
Morris was the editor and author for this valuable piece of oil history. Another unnamed person had gathered every available written account of oil interest (largely newspaper reports) for the first five years after the Drake discovery. Sources for many of the articles are lacking and many were anonymous. From the preface: "...in this promiscuous collection of articles, there will be found a mass of facts and opinions which will possess a certain value with all who are seeking information on the subject. The development of the vast oil interest in Pennsylvania is of too recent occurrence to have its history written at this early day". Morris submitted his final text to several other knowledgeable oil editors before going to press. Even so, some conflicting and obviously incorrect information is included. He does a fine job of weaving the story of oil from the wild and primitive, but beautiful Titusville-Oil Creek area of Pennsylvania through the ups and downs of a fledgling industry which was learning everything by trial and error. A fold-out map of the oil regions as of 1865 is pasted in the front of the book and is useful for orientation. Much information is presented on the early towns and the changing conditions as the local citizens tried to cope with the oil progress. Considerable data is presented regarding individual wells and family farms as they were drilled. The major fires and incidents are covered from several sources, giving a fairly clear picture of the excitement. Oil fever carried from the first small original pumping wells in the creek beds near Titusville to two years later when the high capacity wells began flowing in the Oil City area. Morris covers all oil, from the origin of petroleum to worldwide distribution of oil springs and seeps to the marketing of U.S. oil in Europe. The unstable and speculative nature of the business is punctuated with several stories of fortunes made and lost. Great history with a journalistic flair. A 2nd edition was issued in 1872 with little change. Cited in Swanson on page 34; Hardwicke page 36; Giddens page 93. The hardbound first edition is very scarce.

**A220. (OKLAHOMA) Morris, John W. (editor), CITIES OF OKLAHOMA, Norman, University of Oklahoma Press, (1979), vii, 170pp.**
The Oklahoma Series covers all phases of Oklahoma history. This is Volume XI of the series and is specific to the history and development of the early towns, the oil centers as well as the major cities. Includes photos and maps. The hardbound first edition is common.

**A221. (OKLAHOMA) Morris, John W. (editor), DRILL BITS PICKS & SHOVELS: A HISTORY OF MINERAL RESOURCES IN OKLAHOMA, Norman, University of Oklahoma Press, (1982), ix, 201pp.**
Volume XVI of The Oklahoma Series. This volume is specific to the history and development of oil, natural gas, coal, lead and zinc, gold and other non-metallic resources. It is well done and prepared by contributing author experts in their fields (Kenny Franks does oil, Paul Lambert does Natural gas and Claren Kidd does Coal). Much detail on the Osage, Glenn Pool, Seminole, Cushing, Healdton and Oklahoma City fields. Much on the natural gas boom of the late 1970's and early 1980's where the frantic exploration in the Anadarko basin was fueled by demand and the price rose from sixteen cents to ten dollars per thousand cubic feet. Includes photos and maps. The hardbound first edition is common.

**A222. (TEXAS) Myres, Samuel D., THE PERMIAN BASIN: PETROLEUM EMPIRE OF THE SOUTHWEST, ERA OF DISCOVERY, FROM THE DISCOVERY TO THE DEPRESSION, El Paso, TX, Permian Press, (1973), xxviii, 708pp.**
One of the top oil history books. Mentioned on page 559 of Jenkins. BASIC TEXAS BOOKS, as being a comprehensive study of West Texas oil discoveries. Myres is a little dry, but his coverage is complete and is regarded as a reasonably accurate reference for oil and gas field development in West Texas and Southeast New Mexico. Much on the participants in this industry saga. After Corsicana, Spindletop, Ranger and Desdemona, West Texas opened up with the Lou Tex #1 Moore, discovery of Westbrook Field in Mitchell County in 1920 which was followed by Magnolia's discovery of Iatan-East Howard Field in 1922. This all led to the big show in 1923, at Big Lake, as Santa Rita #1 really

ushered in the West Texas oil era. In 1926 the fabulous Yates Field was discovered. Several early giant fields discovered in the Southeast New Mexico portion of the Permian Basin were Hobbs, Artesia and Vacuum. This first of two volumes provides much detailed history of the discovery and development of these important oil fields and the people involved up to the Depression. The limited leather edition is scarce and the hardbound first edition is becoming scarce.

**A223. (TEXAS) Myres, Samuel D., THE PERMIAN BASIN: PETROLEUM EMPIRE OF THE SOUTHWEST, ERA OF ADVANCEMENT, FROM THE DEPRESSION TO THE PRESENT, El Paso, TX, Permian Press, (1977), xii, 624pp.**
Volume two is the second half of one of the top oil histories. Myres did an incredible amount of research and compilation to put this excellent chronology together. West Texas and Southeast New Mexico are covered in detail up to the modern oil boom of the late 1970's. A different format for this book separates the various geological provinces and describes their developments chronologically and separately, being the: Central Basin Platform, Northwest Shelf, Midland Basin, Delaware and Val Verde Basins. Scholarly, definitive work. The limited leather edition is scarce and the hardbound first edition is becoming scarce.

**A224. (GENERAL) Neal, Harry Edward, OIL: RICHES OF THE EARTH, NY, Julian Messner, (1970), 192pp.**
Compact and precise history of most phases of the oil business from Drake to 1970. The work includes historic photos and a well written history. Much early history in this popular, widely circulated book. The hardbound first edition is common.

**A225. (GENERAL) Nevins, Allan, et al, ENERGY AND MAN: A SYMPOSIUM, NY, Appleton-Century-Crofts, Inc, (1960), xiii, 113pp.**
Included in this symposium are the following papers: Energy in the History of Western Man, by Nevins; The Petroleum Revolution, by Dunlop; Energy Patterns of the Future, by Teller; Low-cost and Abundant Energy, by Mason; and Energy and Public Affairs by Herbert Hoover, Jr. The hardbound first edition is common.

**A226. (PENNSYLVANIA) Newton, J.H. (editor), HISTORY OF VENANGO COUNTY, PENNSYLVANIA, Columbus, OH, J.A. Caldwell, 1879, 651.**
This book was not available for review by Neff, but is regarded as important. Quoting from Swanson: "History of county containing Drake well, chapter devoted to history of petroleum, including Silliman Report; excerpts relating to drilling, production, refining, with lists of wells and refineries". Cited in Swanson on page 34; Hardwicke page 37; Giddens page 93. The hardbound first edition is very scarce.

**A227. (OKLAHOMA) Newsom, D. Earl, DRUMRIGHT! THE GLORY DAYS OF A BOOM TOWN, Perkins, OK, Evans Publications, (1985), xiii, 231pp.**
This is volume one, of what later turned out to be a two volume set. From the dust jacket "In March 1912 the discovery of oil on the Wheeler farm in Creek County, Oklahoma, started the greatest oil field in the world, to that time. Thousands rushed into the barren hill-country and founded a dozen oil field towns. One of these, Drumright, was the heart of the great field and has survived for more than 70 years". Many stories, legends and characters are explored. Early photos, not published elsewhere are included. The hardbound first edition is becoming scarce.

**A228. (OKLAHOMA) Newsom, D. Earl, DRUMRIGHT II (AND SHAMROCK, PEMETA, OILTON, AND OLIVE): A THOUSAND MEMORIES, Perkins, OK, Evans Publications, (1987), xiii, 231pp.**
By popular demand, this second volume covers the surrounding towns as well as other facets of Drumright history. More than 200 rare unpublished photos are also included. A special collection by Ben Russell, Oklahoma's the premier oil field photographer, is the highlight. The hardbound first edition is becoming scarce.

**A229. (CALIFORNIA) Nichols, Allen G. (editor), OIL: AN HISTORICAL EDITION OF THE OIL INDUSTRY, No Place, Curran Printing Co., 1909, 122pp.**
Brief history of oil, particularly California told with some of the best and most interesting photographs available. Cover price was $15.00. Nichols was the editor of The Oil Industry, a first- class California magazine which went out of business in 1910. Subtitle: BEING A BRIEF HISTORY OF OIL DEVELOPMENT IN THE UNITED STATES AND

MORE ESPECIALLY IN THE SOUTHWEST. Nichols lightly touches on oil around the world but concentrates on what he knows best, California oil. Aside from the 122 pages, an article on Standard Oil versus the People is included (this was before the breakup). Also included is a tabulation of annual U.S. oil production and another of annual world oil production by country. Another added attraction is a listing of oil companies incorporated under California laws. Many historic ads in the back. An unusual hand-painted color oil scene is found on the front cover. Not mentioned in Swanson. The limited leather edition is rare.

**A230. (WYOMING) Noggle, Burl, TEAPOT DOME: OIL AND POLITICS IN THE 1920'S, LA, Louisiana State University, 1962, ix, 234pp.**
From the dust jacket: "Teapot Dome, a sandstone formation in Wyoming that overlooks a U.S. Naval Oil Reserve, furnished a name for one of the most sensational scandals in the history of American politics. Professor Noggle unravels this complex tale of corruption, tracing Teapot Dome from its origins in the conservation feuds of 1920-1921 through the bitter partisanship that evolved between the two major political parties". Albert Fall, as Secretary of the Interior, seemed to have the government's interest at heart and his explanation for leasing the proven oil properties to Sinclair and Doheny was rational and reasonable, he just should not have taken the bribe. It appears that this was the one serious mistake in a political career that was otherwise extremely productive. The saga ends with Fall's 1944 death, in a state of poverty some twelve years after serving a year in the Santa Fe prison. Reprinted by Greenwood press in 1980. Extensive bibliography. The hardbound first edition is scarce.

**A231. (GENERAL) O'Connor, Harvey, THE EMPIRE OF OIL, NY, Monthly Review Press, 1955, xii, 372pp.**
The prolific, Socialist-leaning author of MELLON'S MILLIONS and HISTORY OF OIL-WORKERS UNION continues in his usual vein with this book. It is a bit of a tirade, but interesting and very readable. Quoting Swanson: "Sensational, left-wing view of international oil industry and its control of State Department and foreign ministries abroad; need for international control of oil through cooperative alliance; holds up nationalized Mexican industry as beacon of hope". Cited in Swanson on page 82. The hardbound first edition is common.

**A232. (GENERAL) O'Connor, Richard, THE OIL BARONS: MEN OF GREED AND GRANDEUR, Boston, Little, Brown & Co., (1971), x, 502pp.**
O'Connor writes of the oil moguls as if he was there and knew them all and how they were thinking. He leaves no oil mogul or major company unchecked. Fun reading although this is another one which falls a little left of center. A considerable amount of biased history. What other kind is there? The hardbound first edition is common.

**A233. (PENNSYLVANIA) Oil City Derrick, DIRECTORY OF OIL MEN, Oil City, Derrick Publishing Co., 1902, 791pp.**
This comprehensive directory gives the names and addresses of producers, pumpers, refiners, rig builders, drillers, tool dressers also dealers in and manufacturers of oil well supplies, drilling and fishing tools, cordage, pumping powers, gas engines, torpedoes, dynamite, etc. Includes the entire U.S., by area with photos and biographies of many, particularly Pennsylvania and New York. Also includes comprehensive directory of Beaumont and Corsicana, Texas. Also Indian Territory and Russia. Cross-referenced alphabetically. Oil Well Supply Company ads on front and rear covers. The ads are historic to say the least. Not mentioned in Swanson. The hardbound first edition is rare.

**A234. (PENNSYLVANIA) Oil City Derrick, THE DERRICK'S HAND-BOOK OF PETROLEUM, Oil City, Derrick Publishing, 1884, 110pp.**
This is a concise compilation of early Pennsylvania oil information on a daily basis which is presented in a "you are there" monologue. Fires, dry holes, gushers, big leases, accidents, incidents and characters are all covered. A major work edited by F.H. Taylor, this was the predecessor to many excellent reference books on the industry published by Oil City Derrick. Cited in Swanson on page 156; Giddens page 89. The hardbound first edition is rare.

**A235. (PENNSYLVANIA) Oil City Derrick, HAND BOOK OF PETROLEUM, Oil City, Derrick Publishing Co., 1898, 1062pp.**
Volume 1. Swanson describes this major work as: "Extremely detailed chronological and statistical review of petroleum developments from 1859 to 1898, with daily market quotations, tables of runs, shipments, and stocks, oil exports, field operations and other subjects of interest and

importance to oil trade; biographical sketches of, and memoirs written by many notable persons engaged in early oil business. Indispensable sourcebook for any historian interested in early oil industry". This most impressive book includes 121 pages of historic ads. Cited in Swanson on page 156; Giddens page 89. The hardbound first edition is very scarce.

**A236. (GENERAL) Olien, Roger M. and Diana Davids Olien, EASY MONEY: OIL PROMOTERS AND INVESTORS IN THE JAZZ AGE, Chapel Hill, University of North Carolina Press, (1990), xi, 215pp.**
The Oliens examine the go-go decade of the 1920's and the rampant oil speculation and promotion that took place there. The main characters are S.E.J. Cox and Frederick Cook. Also included are C.C. Julian and many others. This is both a social history and a business history. The authors examine the behavior of perpetrators and victims alike as they try to understand the mentality of the time. Included are many photos. The hardbound first edition is common.

**A237. (TEXAS) Olien, Roger M. and Diana Davids Olien, LIFE IN THE OIL FIELDS, Austin, TX, Texas Monthly Press, (1986), 263pp.**
This book consists largely of verbatim personal interviews with oil field characters. The Oliens have done history a big favor by recording and publishing these interviews before the subjects passed from the scene. The book looks at everyday life of the oil workers from Texas fields along with their families. Eighty good photos are included, many of which are not seen elsewhere. The hardbound first edition is common.

**A238. (TEXAS) Olien, Roger M. and Diana Davids Olien, OIL BOOMS: SOCIAL CHANGE IN FIVE TEXAS TOWNS, Lincoln, University of Nebraska Press, (1982), xiv, 220pp.**
The authors examine five communities in the Permian Basin of West Texas: Odessa, Midland, Snyder, Wink and McCamey, to present a new and full picture of oil booms and their effects on individuals and communities. The book is a result of 300 individual interviews. Chapters explore population, housing, public services and education, women and family, Black and Mexican Americans, crime and vice with respect to each town. The hardbound first edition is common.

**A239. (TEXAS) Olien, Roger M. and Diana Davids Olien, WILDCATTERS: TEXAS INDEPENDENT OILMEN, Austin, Texas Monthly Press, (1984), xi, 234pp.**
From the dust jacket: "There are more than 18,000 independent oilmen in Texas and the U.S. today. Wildcatters are the entrepreneurial heart and soul of the oil business. They are the prospectors who find and deliver oil where the major companies have traditionally feared to tread...Some, like oilmen H.L. Hunt, Hugh Roy Cullen, and Sid Richardson, have gambled and won. many more have lost their shirts". The old wildcatters and the new are studied from personal interviews and historical research. The Oliens do much to dispel the myth of independent oil men being wild and crazy. With a few exceptions they have proved to be shrewd, smart, hard working gamblers who have historically provided most of the nation's oil. Probably the best of the Olien's several varied oil history works to date. Includes photos. The hardbound first edition is common.

**A240. (GENERAL) Owen, Edgar Wesley, TREK OF THE OIL FINDERS; A HISTORY OF EXPLORATION FOR PETROLEUM, Tulsa, American Association of Petroleum Geologists, 1975, xv, 1647pp.**
Owen's book is dedicated to Everette DeGolyer and the introduction is made by Wallace Pratt, two of the worlds finest geologists and oil finders. The massive work covers the entire history of oil geological and geophysical exploration from day one, and is not just limited to the U.S. Important world wide exploratory history is documented in this epic volume which was published in commemoration of the fiftieth anniversary of the AAPG. Definitive. The hardbound first edition is scarce.

**A241. (GENERAL) Peckham, S.F., REPORT ON THE PRODUCTION, TECHNOLOGY, AND USES OF PETROLEUM AND ITS PRODUCTS, Washington, Government Printing Office, 1884, viii, 319pp.**
This great work is usually found bound-in with the Tenth Census of the U.S. Swanson's description is useful: "History of petroleum, production, transportation, sale, statistics of production, technology of petroleum and uses. Extensive bibliography. A pioneer document". High quality photo plates and fold-out maps accompany the book. DeGolyer called it one of the top American monographs on petroleum. Peckham was a well

respected oil chemist and refiner from the early 1860's. He put this major work together for the feds after retiring from their service. Cited in Swanson on page 156; Hardwicke page 38; Giddens page 101. The hardbound first edition is very scarce.

**A242. (GENERAL) Peeples, Malcolm W.H., EVOLUTION OF THE GAS INDUSTRY, NY, New York University Press, (1980), xvi, 235pp.**
Comprehensive history of the natural gas industry from its first commercial enterprise, the street lighting in London in 1812, to modern LNG and transportation. Includes many of the ups and downs in between. The hardbound first edition is common.

**A243. (MEXICO) Person, Harlow S., MEXICAN OIL: SYMBOL OF RECENT TRENDS IN INTERNATIONAL RELATIONS, NY, Harper & Brothers, 1942, 83pp.**
This book provides historical background of events which led to the March 18, 1938 Mexican expropriation of oil properties owned by various U.S. and foreign oil companies. Subsequent developments are covered in detail up to the settlement of U.S. company claims. Much political negotiation took place between the U.S. and Mexican governments. There were also legal maneuverings and attacks by the U.S. oil companies (largely the Standard companies). The two governments kept it pretty cool, finally settling for less than twenty-five million dollars in early 1942. The author was a business economics and management consultant who headed the negotiating group for the U.S. companies. Interesting story, but more interesting were later settlements with British and Dutch interests for much higher figures. Cited in Swanson on page 88. The hardbound first edition is common.

**A244. (General) Peto, Sir S. Morton, THE RESOURCES AND PROSPECTS OF AMERICA: ACERTAINED DURING A VISIT TO THE STATES IN THE AUTUMN OF 1865, London, Alexander Strahan, 1866, xv, 428pp.**
In Peto's oil chapter, he gives ancient history and early uses of petroleum, but praises America for finding the uses and discovering an efficient method for recovering the substance. He quotes from several sources as he reviews the important wells and event along with the economic and logistical problems encountered in producing and refining crude oil. He

recognizes the importance of the products to U.S. and world commerce. Good discussion of other American resources as well. Cited in Swanson on page 32. The hardbound first edition is very scarce.

**A245. (MONTANA) Petroleum County Public Library (compilers), PAGES OF TIME: A HISTORY OF PETROLEUM COUNTY, MONTANA, MT, Petroleum County Library, 1989, xvi, 856pp.**
Montana's centennial celebration edition of Pages of Time. The county is famous for STOCK RAISING, OIL & GAS, AND AGRICULTURE. Everything you always wanted to know about Petroleum County from a historical and cultural standpoint is in this massive book. Even some oil. Map endpapers and photos. The hardbound first edition is becoming scarce.

**A246. (TEXAS) Philips, Shine, BIG SPRING, NY, Prentice-Hall, Inc., 1942, vi, 231pp.**
Pre-oil history of "The biggest little town in the U.S.A.". The author gives us reminiscences and his philosophy of life, all done in the first person. Big Spring has within its city limits one of the largest oil fields in West Texas, Howard-Glasscock. Also, the massive Cosden Refinery is located on I-20, right in town. The hardbound first edition is common.

**A247. (OKLAHOMA) Ponca City D.A.R. (editors), THE LAST RUN: KAY COUNTY, OKLAHOMA, 1893, Ponca City, OK, The courier Printing Company, 1939, 352pp.**
This is a major compilation of at least a hundred individual stories which were related by area pioneers in 1939. They told of their life in the county and participating in the last of the five Oklahoma land runs. Kay is one of five counties which were formed in north central Oklahoma from confiscated Indian lands, called the Cherokee Strip. The run for Strip land and homesteads took place in 1893. Ponca City was formed and subscribed prior to the run. Cattlemen were removed from the county along with fences and the area was laid out in 160 acre tracts for homesteads prior to the run. Many of the original farmer-rancher pioneers later became involved in the oil business as the Osage boom spread westward during the early 1900's. Much history is given in regards to Kay county's two claims to fame. First, E.W. Marland, one of the world's greatest oil finders and Continental Oil Company which is the direct result of his success. Second, the Miller brother's 101 Ranch which at its

zenith was known across the country more for its agricultural research and Wild West Shows than for its oil production. Highlights of Marland's career include the establishment of oil and gas production in the county in 1910, the Marland Refinery in 1915, development of the giant Burbank field in 1920, losing it all in the crash of 1929 due to overextension (to J.P. Morgan). Marland was a congressman in 1932 and governor for four years starting in 1934. At the time this book was published he was back in the wildcatting business again. Major work. The hardbound first edition is scarce.

**A248. (MEXICO) Powell, J. Richard, THE MEXICAN PETROLEUM INDUSTRY: 1938-1950, Berkeley, University of California Press, 1956, 269pp.**
Mexico had long felt that it had been at the mercy of more powerful foreign countries as they exploited the Mexican resources and used economic pressure to influence political decisions. The large oil companies put little back in Mexico compared to what they took out. In fact, the Mexican government and private enterprise had effectively been barred from participating in their lucrative industry. Much antagonism and labor relations problems led to the 1938 expropriation. Most evidence points to the cause being the major oil company's short sighted policy of not taking care of business. They were doing no exploratory drilling and not maintaining production levels in the field. Activity had almost ground to a halt. The book points out results of expropriation were mixed as some objectives were met and others not. The government monopoly continued to have labor problems and the private Mexican citizens did not get their right to participate in the oil industry as they had been promised. The hardbound first edition is common.

**A249. (GENERAL) Pratt, Wallace E., THE VALUE OF BUSINESS HISTORY IN OIL FINDING, Houston, Texas Gulf Coast Historical Association, 1961, 14pp. Volume V, Number 2, August.**
Speech by the famous geologist given May 9, 1961 at the annual meeting. Parts of this delivery appeared in Harvard's OIL'S FIRST CENTURY. Pratt reviews the early business history of Standard Oil and the modern business problems with global oil markets and production. Brilliant man. Reprint limited to 500 copies. The softbound first edition is becoming scarce.

**A250. (TEXAS) Presley, James, A SAGA OF WEALTH: THE RISE OF THE TEXAS OILMEN, NY, G.P. Putnam's Sons, (1978), 464pp.**
Presley is a fine writer and knows his oil people. His personal interviews have resulted in an interesting study of oilfield characters, both famous and obscure. From the dust jacket: "..men like Dad Joiner, the Hamill brothers, James Hogg, John Gates, Senator J.W. Bailey, Doc Lloyd, H.L. Hunt, the Murchisons, John Mecom, and Glenn McCarthy...are the story of this book...also the story of the oil companies themselves, ..Sun Oil, Gulf, Shell, Humble, Standard". A very good bibliography is included. The hardbound first edition is common.

**A251. (PENNSYLVANIA) Quaker State Oil Refining Corporation, A HISTORICAL SKETCH OF OIL CITY 1871-1971, Oil City, PA, The Corporation, 1971, 32pp.**
This pamphlet is a brief sketch of the city which was prepared for its centennial celebration by one of the city's largest employers. Very brief coverage from pre-oil history to the 1860's. Interesting discussion of how the barrel became 42 gallons instead of 40 or 45. The first pipelines, the famous oil wells, the oil exchange, personalities, fires and flood are all touched on. The softbound first edition is becoming scarce.

**A252. (VENEZUELA) Rabe, Stephen G., THE ROAD TO OPEC: UNITED STATES RELATIONS WITH VENEZUELA, 1919-1976, Austin, University of Texas Press, (1982), ix, 262pp.**
Venezuela was one of the original five who organized OPEC in 1960 and probably the instigator. OPEC was largely ignored by the U.S. and other oil importing countries, until 1973. Much of this is the story of early conflicts between U.S. and Venezuelan governments and oil companies which led to nationalization of many U.S. owned properties. Much history and political intrigue. A useful bibliography and map are included. The hardbound first edition is common.

**A253. (WYOMING) Ravage, M.E., THE STORY OF TEAPOT DOME, NY, Republic Publishing, 1924, v, 198pp.**
This book was written as the scandal was breaking and the major players had been identified as Albert Fall, Edward Doheny and Harry Sinclair.

The book obviously stirred public and political opinion against Fall and was published long before he was convicted and jailed. In fact the investigations and trials went on for another seven years before being resolved. A reference table of dates and events is included. Cited in Swanson on page 157; Hardwicke page 39. The softbound first edition is very scarce.

**A254. (CALIFORNIA) Redpath, Lionel V., PETROLEUM IN CALIFORNIA, Los Angeles, The author, 1900, 158pp.**
This book was not available for review by Neff, but is regarded as important. Quoting from Swanson: "California oil industry at close of 19th century; biographical sketches of California oilmen; illustrations. Regarded as authoritative". The hardbound first edition is rare.

**A255. (GENERAL) Redwood, Sir. Boverton, PETROLEUM AND ITS PRODUCTS, London, Charles Griffin & Co., Ltd., 1896, xxv, 900pp.**
In Two volumes, this definitive work is commonly called "A Treatise On Petroleum" from the subtitle. It covers: "The geographical distribution and geological occurrence of petroleum and natural gas; the physical and chemical properties, production, and refining of petroleum and ozokerite; the characters and uses, testing, transport, and storage of petroleum products; and the legislative enactments relating thereto; together with a description of the shale oil and allied industries; and a full bibliography". Quoting from Swanson: "One of the landmark oil reference books by a distinguished British petroleum consultant and chemist. Author knew his subject from firsthand information acquired through a lifetime of travel and consulting work". Detailed history of oil by countries. Frontispiece is a map of the world showing oil productive areas in red. Major worldwide oil bibliography. Fold out map plates of Europe and U.S. oil fields as well as many others around the world. The 2nd 1906 edition in two volumes has 1064pp. The 3rd 1913 edition in three volumes has 1167pp. The 4th 1923 edition in three volumes has 1353pp. The 5th 1926 edition in three volumes has 1353pp (probably a reprint of 4th). Cited in Swanson on page 180: Hardwicke page 16. The hardbound first edition is very scarce.

**A256. (GENERAL) Reinholt, Oscar H., OILDOM: ITS TREASURES AND TRAGEDIES, Philadelphia, David McKay Co., 1924, 384pp.**
Quoting Swanson: "Information on all aspects of oil industry, based on author's contacts and experiences". From title page: "A profusely illustrated book of late and basic facts about petroleum and the dependent oil and automotive industries, popularly presented for the benefit of investors, motorists and operators". Pictorial front cover has gushing oil well with red flames issuing from burning oil tanks in foreground. Includes photos, charts and maps. Cited in Swanson on page 188; Hardwicke page 39. The hardbound first edition is very scarce.

**A257. (CALIFORNIA) Rintoul, William, DRILLING AHEAD: TAPPING CALIFORNIA'S RICHEST OIL FIELDS, Santa Cruz, Valley Publishers, (1981), ix, 291pp.**
The third Rintoul pictorial oil history. This one starts after WW II and ends after the 1969 Santa Barbara oil spill and debacle. There is moderate oil growth in California during this period, but the workers rights, environmental control and regulation era begin there, with a vengeance. Much of the book is on the Pico Canyon area which had tempted explorers from the earliest days and was not blessed with economical-commercial oil production until the late 1940's. Somehow the operators managed to drill 145 wells on 70 acres in the hysterical Pico boom. Much on innovative projects, such as that of Occidental going back into old field areas and discovering significant new reserves. Over one hundred photos are included. The hardbound first edition is common.

**A258. (CALIFORNIA) Rintoul, William, OILDORADO: BOOM TIMES ON THE WEST SIDE, Santa Cruz, Valley Publishers, (1978), ix, 241pp.**
The second of Rintoul's photo histories of the California oil industry. This one is different as it concentrates in the Taft- Bakersfield area from the early 1900's to the mid-1930's. Rintoul works oil history into the telling of "The great Washington's birthday automobile road race", which utilized oil field roads as it coursed through the oil fields. A Stutz won. This is a very readable book and hard to put down once started. Coverage of the 1920 Oil Workers Union strike which set the precedent for the rest of the country. The book ends with Standard Oil drilling their #1 Mascot (started in 1928), the worlds deepest well at the time...a saga in itself. The softbound first edition is common.

**A259. (CALIFORNIA) Rintoul, William, SPUDDING IN: RECOLLECTIONS OF PIONEER DAYS IN THE CALIFORNIA OIL FIELDS, San Francisco, California Historical Society., (1976), xii, 240pp.**

The title is an oil field term for starting of oil drilling operations and this book is the start of Rintoul's three volumes on California oil history. It was the first widely circulated oil history of the area to be published in thirty years. Rintoul used thirty-five oil pioneer interviews which had been recorded in 1963 to compile the book, thus it is a narrative of personal experiences which captures the tempo and the spirit of the times from 1890 to 1940. Old city and oil field maps are shown. Over 200 photos and illustrations covering La Brea, Ventura, Signal Hill, and the Lakeview gusher plus many others make this a valuable history. The hardbound first edition is becoming scarce.

**A260. (TEXAS) Rister, Carl Coke, OIL TITAN OF THE SOUTHWEST, Norman, University of Oklahoma Press, (1949), xxiii, 467pp.**

Another of the top oil history books. This is #175 and one of only three oil books given bibliographical status in Jenkins' BASIC TEXAS BOOKS. From Jenkins: "This volume is the best general history of the Texas oil industry that has so far been written". From the dust jacket: "...the rush for oil riches in the great Mid-Continent and Gulf producing area, from the era of the Indians' oil springs through the blustery years of wildcatting to the recent more orderly, but equally dramatic, period of exploration and development". As Spindletop before it, East Texas became the vehicle which made small oil companies large and which made individual speculators and investors wealthy enough to start their own oil companies. H.L. Hunt was the biggest name to come out of that field as he built an empire on Dad Joiner's success. H.L. does not come off too well in the telling of this story. Joiner and Lloyd essentially did it all by themselves but were quickly swarmed by others who would do much better, financially than they. Rister was a research professor of history at Oklahoma University and produced this major work after three years of research which was funded by a grant from Standard Oil. Swanson says the book is considered authoritative. It is readable and organized for further research. Cited in Swanson on page 157. The hardbound first edition is becoming scarce.

**A261. (OKLAHOMA) Roberson, Courtney Ann Vaughan and Glenn Roberson, CITY IN THE OSAGE HILLS, Boulder, CO, Pruett Publishing, (1984), vii, 196pp.**
A fresh look at Tulsa's growth and development. The story begins with early history of "Creek Tulsa" and moves through the cattle era. The influence and history of oil is followed by the Depression and finally to the modern city. Chapter three is entitled Oil Capitol, a self-conferred title the city held for nearly forty years. Oil had been known in the area from the early 1800's in seeps and as an unwanted contaminant from salt wells. Jesse Heydrick, a Pennsylvania oil man and John Wick, a local well driller made a significant strike at Red Fork, near Tulsa in 1901, but the area did not come alive until Galbreath and Chesley discovered Glenn Pool, just south of Tulsa in 1905. Good detail on the early oil. The hardbound first edition is common.

**A262. (WYOMING) Roberts, Harold D., SALT CREEK, WYOMING: THE STORY OF A GREAT OIL FIELD, Denver, CO, Midwest Oil Corp., (1956), 211pp.**
Almost a half billion barrels of oil were produced from the field from 1889 to the mid-1950's. Much history on the participation of Stanolind and Midwest oil as they developed this large field. The two companies financed Roberts' book and made their records freely available. As an attorney, the author was involved in much Salt Creek litigation and negotiation over the years, both for government and private interests. A large chart is included which unravels the complex relationships of the many companies involved with Midwest-Stanolind in the Salt Creek field. Quoting from Swanson: "History of Salt Creek from early developments in Territorial days through formation of Salt Creek Unit and its final approval on September 1, 1939, with epilogue covering subsequent 16 years". Good photos are included. Cited in Swanson on page 46. The hardbound first edition is common.

**A263. (TEXAS) Robinson, Spencer W. (editor), SPINDLETOP: WHERE OIL BECAME AN INDUSTRY 1901-1951, Beaumont, Spindletop Commission, (1951), vii, 201pp.**
This is the record of proceedings of a program conducted by the Spindletop 50th Anniversary Commission. The celebration took place when some of the original participants were still alive and able to attend. Although many old timers came, the stars of the show were Al Hamill

and Patillo Higgins, both looking rather old. Hamill was the driller on the wildcat and Patillo Higgins was the persistent idea man. Most of the Houston Geological Society rode a train in for the event. Three photos in the back demonstrate changes in the field: first, the crowded derricks of 1902; second, the crowded, but spaced derricks of 1926 as flank production was being developed; last, in 1951 showing the modern derricks which had developed even deeper production in the field. Many historic photos plus speeches given by Everette DeGolyer, David Rockefeller, John Pew, Glenn McCarthy and others. Cited in Swanson on page 38. The hardbound first edition is becoming scarce.

**A264. (TEXAS) Rogers, John William, THE LUSTY TEXANS OF DALLAS, NY, E.P. Dutton and Co., Inc., 1951, 384pp.**
Author and playwright Rogers recounts the lurid past from Dallas on the frontier to the modern day social and political scene. This is essentially a social history of Dallas and covers the interesting and important characters over the years in journalism, the arts, entertainment, oil, railroads, business and industry. Chapter 21 is dedicated to Dallas oil people. Second revised and updated 1965 edition has 444pp. The hardbound first edition is becoming scarce.

**A265. (MIDDLE EAST) Roosevelt, Kermit, ARABS, OIL AND HISTORY: THE STORY OF THE MIDDLE EAST, NY, Harper & Brothers, (1949), 271pp.**
The author was a professor-historian and grandson of Teddy Roosevelt. From the dust jacket: "...simple, anecdotal, and authoritative account of these people and their way of life. It is packed with information...some of it curious, all of it indispensable to anyone who hopes to understand, travel in, or do business with these countries". Not too much history is covered as Roosevelt gives a cultural analysis of how oil has affected the people of the area. The conclusion is: substantially. Cited in Swanson on page 92. The hardbound first edition is becoming scarce.

**A266. (ALASKA) Roscow, James P., 800 MILES TO VALDEZ: THE BUILDING OF THE ALASKA PIPELINE, NJ, Prentice-Hall, Inc., (1977), vii, 227pp.**
The first oil moved through the Alaska pipeline in 1977 after a long fought battle and major triumph over technical and human obstacles. The actual construction was rapid after political and environmental problems

had been overcome. From the dust jacket: "..story of human accomplishment against awesome odds, it is richly entertaining and moving". Includes a map and photos. The hardbound first edition is common.

**A267. (GENERAL) Ross, Victor, THE EVOLUTION OF THE OIL INDUSTRY, Garden City, NY, Doubleday, Page & Co., 1920, xvi, 178pp.**

This is one of the most popular general oil books and was widely circulated. Quoting Swanson: "Review of oil industry, exploration, drilling, gathering and transporting, refining, shipping, oil in First World War, and future outlook; recommends diplomatic support of American oil companies". Unusual photos, not seen elsewhere, are from around the world and of earthen oil reservoirs being temporarily used to store wild well production. Cited in Swanson on page 188; Hardwicke page 39; Giddens page 94. The hardbound first edition is common.

**A268. (WYOMING) Rountree, Russ, WESTERN OIL REPORTER'S ROCKY MOUNTAIN OIL HISTORY, Denver, CO, Hart Publications, (1984), 210pp.**

The first half of the book is oil history, broken down into five time periods within the 1862-1984 range, and each of those are further broken down into events of the individual states or Rocky Mountain geological sub-provinces. Interesting review of drilling near oil seeps at Florence in the 1860's and near Salt Creek in the 1880's. The remainder of the book is titled Company Histories, and includes majors, independents and seismic companies. Much good promotional material for them. Many photos and drawings are included. The hardbound first edition is becoming scarce.

**A269. (TEXAS) Rundell, Walter, Jr., EARLY TEXAS OIL: A PHOTOGRAPHIC HISTORY, 1866-1936, College Station, Texas A&M University Press, (1977), 260pp.**

Brief historical comments of the author are enhanced by historic photos, first of the early Nacogdoches area where Barret brought in the 1866 producer, then of Corsicana for the first real boom, and all of that is just for starters. This is one of the better "Picture Books" in its organization and completeness. The hardbound first edition is common.

**A270. (TEXAS) Rundell, Walter, Jr., OIL IN WEST TEXAS AND NEW MEXICO: A PICTORIAL HISTORY OF THE PERMIAN BASIN, College Station, Texas A&M University Press, (1982), xii, 183pp.**

This pictorial study complements Myres' Permian Basin histories very well. The 271 Photos cover major events from Santa Rita #1 in 1923 to modern day Permian Basin oil industry. There is also much on the characters that made it all happen. The hardbound first edition is common.

**A271. (PENNSYLVANIA) Sargent, M.P., PIONEER SKETCHES: SCENES AND INCIDENTS OF FORMER DAYS, Erie, PA, Herald Printing & Publishing, 1891, 512pp.**

Sargent was a brilliant and talented man. The text reveals that he was an entrepreneur and was involved directly in many phases of early Pennsylvania history. Aside from coal and lumber, he speculated in oil and ran boat commerce on the Allegheny river. In the preface he states his object is to give the reader life sketches of the early settlers, the scenes and incidents and current topics of the day. He relates stories of Drake, Pithole, Oil City, Titusville, Coal Oil Johnny, the Bennehoff family and much more. He punctuates his many short stories with his own poetry. A very classy book. Cited in Swanson on page 34. The hardbound first edition is scarce.

**A272. (PENNSYLVANIA) Schooley, John S., THE PETROLEUM REGION OF AMERICA, NY, Harpers Magazine, 1865, 13pp unnumbered.**

Classic early history with seven beautiful engravings reproduced. It is an accurate account of oil field activity in Venango County and surrounding region and contributes significantly to the history of the area and industry. Reprinted as a pamphlet by Oil Region Books. Cited in Giddens on page 168. The softbound reprint is common.

**A273. (KANSAS) Schruben, Francis W., WEA CREEK TO EL DORADO: OIL IN KANSAS, 1860-1920, Columbia, University of Missouri Press, 1972, x, 176pp.**

From the dust jacket: "...traces the first 60 years of the development of the Kansas oil industry. It describes the men and the events involved in that development and the ensuing conflict between small, independent companies and Standard Oil over the oil bonanza". Excellent early history is told here as Kansas came into its own during WW I. The state has never had the oil exploration romance for historians like that of the Pennsylvania, California, Oklahoma, and Texas oil sagas, although it is comparable in many ways. The first exploratory drilling efforts were at WEA Creek in 1860, giving only slight encouragement. By 1865, scattered shallow production was being found and Standard Oil began to move in, first buying a small refinery. #1 Norman at Neodesha in 1892 really opened the Mid-Continent Field (Kansas & Oklahoma fields). The great wildcatters from Pittsburgh, Guffey and Galey gave Kansas a good effort, drilling numerous wells. Unimpressed by their meager results, they ended up selling out to Standard. By far the biggest happening in this period is Henry Daugherty's Empire Oil & Gas #1 Stapleton, discovery well of the El Dorado field at 2465 feet in 1915. The wildcat was drilled on recommendation of Charles Gould, famous early geologist. This one field had much to do in the war effort, producing twenty-nine million barrels during the year of 1918. The hardbound first edition is common.

**A274. (GENERAL) Schweitzer, P., A LECTURE ON PETROLEUM, Columbia, MO, Statesman Book, 1879, 64pp.**
This book was not available for review by Neff, but it is regarded as important. Quoting from Swanson: "Lecture delivered early in 1879, before Missouri House of Representatives, Jefferson City, and followed by enactment of coal-oil bill. Historical review, industrial aspect, chemical composition, refining and testing; statistics; draft of proposed legislation. Contains early petroleum bibliography". Cited in Swanson on page 157. The softbound first edition is very scarce.

**A275. (GENERAL) Schweizer, Peter, VICTORY: THE REAGAN ADMINISTRATION'S SECRET STRATEGY THAT HASTENED THE COLLAPSE OF THE SOVIET UNION, NY, Atlantic Monthly Press, 1994, xx, 284pp.**
This sensational book provides a missing piece of oil history which most people in the industry had been wondering about. What was the real cause of the January 1986 crude price collapse? This book answers the question and indicts Ronald Reagan. Evidently Reagan with his inner circle (not including the secretary of energy) and the Saudi Arabian government agreed that the one way to defeat the superpower U.S.S.R. was through economic war. That war would be designed to deprive them of their

primary source of hard European cash which was desperately needed by Gorbachev to rebuild his infrastructure and achieve some economic stability. War was engaged when the Saudis flooded the oil markets and forced the price of crude from $30.00 to $10.00 per barrel. This was way below their cost of producing and they suffered tremendous losses. The plan worked, but unfortunately it took the U.S. domestic oil industry with it, which probably caused the real estate market to crash, which probably caused the savings and loan debacle. It was expensive, but probably worth it. The hardbound first edition is common.

**A276. (TEXAS) Schwettmann, Martin W., SANTA RITA: THE UNIVERSITY OF TEXAS OIL DISCOVERY, Austin, TX, Texas Historical Association, 1943, xix, 43pp.**
This short, but sweet booklet covers most of the players and history of the discovery which really brought the oil industry to West Texas in 1923. Santa Rita...Saint of the Impossible, a perfect name for a remote and hopeless wildcat well. Frank Pickrell was the prime mover; Bill Latham, the logistics man; R.S. McDonald the derrick builder; and Carl Cromwell, the driller. These men were the key actors and Big Lake field was the stage. A major discovery on University of Texas lands in the middle of nowhere started the Permian Basin oil play with a bang. The book is printed on fine paper and illustrated by Tom Lea with pen and ink drawings. A 1958 reprint is common. Cited in Swanson on page 40. The hardbound first edition is scarce.

**A277. (CALIFORNIA) Scott, Charles C. (editor), PETROLEUM INDUSTRY COMMITTEES IN WORLD WAR II, DISTRICT V, 1941-1946., San Francisco, CA, The Committee, (1947), xx, 351pp.**
This large volume contains the history, function, and makeup of the West Coast division of PAW. With Harold Ickes as overall PAW administrator, several districts were formed to coordinate all phases of the oil industry in geographic areas. District V covered the West Coast plus Alaska and Hawaii. The first hurdle was to circumvent the pending government anti trust litigation against the oil companies which claimed lack of competition. The thrust of PAW was to reduce competition to a minimum, thus eliminating duplicated efforts and promoting efficiency. The work was enormous but performed well by oil company and government experts. Major committees were: Production; which controlled allocation; Refining which controlled what products were to

be produced and where; Supply and Transportation, which administered pipelines and dealt with the severe tanker loss problem; Distribution and Marketing, which controlled use of oil products and dealt with shortages in everything from asphalt to aviation gas. An exceptional job of coordination by the industry and by Mr. Ickes. Not mentioned in Swanson. The hardbound first edition is becoming scarce.

**A278. (TEXAS) Shelton, Hooper, FROM BUFFALO...TO OIL: HISTORY OF SCURRY CO., TEXAS, Noplace, Feather Press, (1973), 272pp.**

Voted best local or regional Texas history of 1973, this is a very comprehensive book. Only the last five pages are dedicated to oil. The author was not too impressed with that phase of the history although his county embraces one of the very few billion barrel oil fields in the U.S. The hardbound first edition is common.

**A279. (TEXAS) Siegel, Stanley E., HOUSTON: A CHRONICLE OF THE SUPERCITY ON BUFFALO BAYOU, Woodland Hills, CA, Windsor Publications, (1983), 296pp.**

More than 350 illustrations and 32 color plates. From the dust jacket: "In this richly illustrated, panoramic history of Houston....stunning assemblage of images to produce a significant and beautiful chronicle of America's quintessential Supercity". Well done history of the oil capital of the world. Includes a bibliography. The hardbound first edition is common.

**A280. (CALIFORNIA) Silliman, Benjamin, Jr., A DESCRIPTION OF THE RECENTLY DISCOVERED PETROLEUM REGION IN CALIFORNIA, WITH A REPORT ON THE SAME, NY, Francis & Loutrel, 1865, 24pp.**

This is a very important oil book (pamphlet) in that Silliman had previously gained immense fame by predicting many uses for petroleum and essentially causing the Drake well to be drilled in 1859. Unfortunately he was not proven correct in his glowing predictions for the California oil area until some twenty years later. The professor was attacked as a quack promoter by his colleagues and suffered great humiliation during this time. Quoting from Swanson: "Report of Professor Silliman upon outcrops of mineral oil, or petroleum, which appear on the Ranch of Ojai, near Buenaventura, Santa Barbara County, in areas of present Ventura Field. Property owned by California Petroleum Co., organized by Thomas Scott of the Pennsylvania Railroad. Historical review of petroleum in California, with discussion of transportation and markets. Report dated San Francisco, Sept. 1, 1964". Cited in Swanson on page 50; Hardwicke page 40; Giddens page 167. The first edition is rare.

**A281. (PENNSYLVANIA) Silliman, Benjamin, Jr., REPORT ON THE ROCK OIL, OR PETROLEUM, FROM VENANGO COUNTY, PENNSYLVANIA, New Haven, 1855, 20pp.**
This is without doubt, the most important book (pamphlet) relating to the oil industry. Most historians regard this early work by the famous chemist and Yale professor, Silliman, to be the impetus that launched the oil industry. He was the son of a pioneering chemist and his astute opinions were much sought by industry and academia and he was well respected. He determined that crude oil had commercial value and reviewed its properties with potential products which could be exploited. Its illuminating properties were stressed and compared with other illuminants in use at the time. The report was solicited and funded by Eveleth, Bissell, & Reed, who were promoters of the Pennsylvania Rock Oil Co. Cited in Swanson on page 158; Hardwicke page 40; Giddens page 167. The pamphlet has been reprinted many times in Pennsylvania oil history books, but no stand alone reprints are known at this time. The softbound first edition is very rare.

**A282. (TEXAS) Sinise, Jerry, BLACK GOLD & RED LIGHTS, Burnet, TX, Eakin Press, (1982), xi, 152pp.**
From the dust jacket: "Some towns are born casual and calm-like. Borger had a street walker for a midwife and an oil boom for a birth announcement. Oil, blood, and money flowed freely when this Texas Panhandle town burst upon the scene in the 1920's like a stampede of rawhide Texas Longhorns hell bent for water". Sinise survived the town and writes here about it. More fun reading about the wild boom towns. The hardbound first edition is common.

**A283. (GENERAL) Solberg, Carl, OIL POWER, NY, Mason/Charter, 1976, xi, 308pp.**
Solberg takes a political, economic, social and military look at the oil industry from its beginning to the aftermath of the Arab embargo. The

author is adverse to: monopoly, tax breaks, Teapot Dome, offshore drilling, etc. He believes that for continued growth and stability the U.S. needs alternate energy sources. Extensive bibliography plus a bibliographical essay. The hardbound first edition is common.

**A284. (TEXAS) Southwest Texas Oil Scouts Association, OIL AND GAS DEVELOPMENT IN SOUTHWEST TEXAS, Bulletin Number One, San Antonio, The Association, 1930, 84pp.**
All information valuable to oil scouts including maps and production data by areas and fields. Quoting Swanson: "Describes producing fields, pipe lines, refineries, geophysical activities, geology, production and development activities in Southwest Texas, with special articles". Cited in Swanson on page 40. The softbound first edition is scarce.

**A285. (TEXAS) Spratt, John S., THE ROAD TO SPINDLETOP: ECONOMIC CHANGE IN TEXAS, 1875-1901, Dallas, TX, SMU Press, (1955), xxix, 337pp.**
Much history and economic development trends are explored as they lead up to the famous Spindletop discovery in this comprehensive work. What Spindletop did for Texas and the nation can hardly be measured as it not only brought the new oil industry to Texas from Pennsylvania, but created innumerable related industries and businesses. Only the last chapter is dedicated to the burgeoning oil industry as Spindletop increased the statewide oil flow from meager to over twenty million barrels per year, almost instantly. Cited in Swanson on page 38. The hardbound first edition is common.

**A286. (CALIFORNIA) Stalder, Walter, A CONTRIBUTION TO CALIFORNIA OIL AND GAS HISTORY, CA, California Oil World, 1941, pp31-72.**
The Author was a long time consulting geologist as his father had been. He had access to diaries, scrapbooks, and other files to prepare this excellent work. Maps and photographs are included along with some bibliographical material on early California oil. Lithographs in sepia and a map of California oil fields in color are exceptional. This publication is part of the November 12, 1941 edition of California Oil World and Petroleum Industry magazine. Cited in Swanson on page 158. The softbound first edition is scarce.

**A287. (GENERAL) Standard Oil Development Co., THE FUTURE OF INDUSTRIAL RESEARCH: PAPERS AND DISCUSSIONS, NJ, The Company, 1945, viii, 173pp.**

Research and development were running rampant in the country during and after WW II with little coordination. Standard was trying to get government and industry to work together by presenting these interesting papers by various authorities in their fields. In three parts: 1, What should be the principles and objectives for industrial research and development; 2, What role should small business play in this; 3, How much effort should be in the national security area. The obvious overtones of the seminar were military in nature. Although WW II was over, the general feeling within this group was that another war would soon follow. That sentiment was not held by the general public, who thought they had just fought the war to end all wars. The hardbound first edition is common.

**A288. (TEXAS) Stanley, F., THE ISOM, TEXAS STORY, Nazareth, TX, Privately Published, 1973, 23pp.**

Stanley pamphlet. Isom lives today as a community within the collective city of Borger. It began long before oil came. The Weatherly family first settled at Isom in 1898. There were probably a total of forty people in all of Hutchison County at the time. The family was still there when the Panhandle field boom caused great interest and speculation in town lots at Borger and Isom. The two towns were close enough together to almost be considered one, on the new Santa Fe rail line. Later the two merged with another town, Dixon Creek and Borger was the surviving name. Nearby Whittenburg and Phillips also merged but did not join Borger. The softbound limited edition is common.

**A289. (TEXAS) Stanley, F., THE LEFORS TEXAS STORY, TX, Privately Published, 1975, 20pp.**

Stanley pamphlet. Lefors, the townsite was established in 1852, although not settled. The small ranching, farming town gained a post office in 1892. It was the county seat of Gray County, but Pampa wanted the honor and by 1928 got it. Lefors grew from the Panhandle oil field boom, but not like Pampa. During 1928, gushers were drilled within the townsite, which was precarious in that it is almost an island between branches of the North fork of the Red River. In 1975 a tornado demolished most of the town but it continues on, oil boom or not. The softbound limited edition is common.

**A290. (TEXAS) Stanley, F., THE PHILLIPS TEXAS STORY, Nazareth, TX, Privately Published, 1975, 20pp.**

Stanley pamphlet. In 1926 the oil boom changed the character of 28 small towns in the Texas Panhandle. Phillips (named for the oil company) and Borger had the advantage of good local management as well as proximity to the drilling activity. The railroads also helped. At Phillips, one hundred new houses were constructed for company personnel during 1934 and rented out at the rate of two dollars per room per month with utilities furnished. Many of the employees had been living in shacks and the company was embarrassed. Phillips grew and absorbed the town of Whittenburg, south of Pampa and survives today. The softbound limited edition is common.

**A291. (TEXAS) Stanley, F., THE PLEMONS (TEXAS) STORY, Nazareth, TX, Privately Published, 1973, 22pp.**

Stanley pamphlet. The 1926 oil boom had a temporary favorable effect on the sleepy, ranching and farming community in the Texas panhandle named Plemons. It was the county seat of Hutchison County. The Whittenburg family was also becoming pervasive. Competition between countless small towns for the economic boost of a frenzied oil boom was exacerbated by the necessary railroads and their routes. Plemons lost the county seat in 1926 to Stinnet and began to fade, losing its post office in 1952. Borger, Amarillo and Canyon grew and absorbed the people which even a sustained oil boom could not maintain. The softbound limited edition is common.

**A292. (TEXAS) Stanley, F., THE SIGNAL HILL (TEXAS) STORY, Nazareth, TX, Privately Published, 1973, 23pp.**

Another Stanley pamphlet. Today Signal Hill has ceased to exist. There is still oil and gas activity at the abandoned townsite but that is all. Signal Hill was entirely a promotion and on May 14, 1926 the townsite was opened in the center of the gigantic Panhandle oil field. Speculators anticipated the Santa Fe would build a line through the townsite. The nearest rail then was forty miles north and another was located forty miles south. The Promoter (C.C. Anderson) was also anticipating a refinery and gasoline plant. The deeded town lots included oil royalty which was a big drawing card and by 1927 possibly as many as 100 businesses were located there (largely places of ill-fame). Signal Hill developed a bad reputation and most of the resident moved to nearby Stinnett. The town became an outlaw hangout and slowly faded away. The softbound limited edition is common.

**A293. (TEXAS) Stanley, F., THE TEXAS PANHANDLE: FROM CATTLEMEN TO FEED LOTS, Borger, TX, The Author, (1971), xx, 656pp.**

This is essentially a handbook of the Texas Panhandle by the most prolific Texas author. He wrote countless city and area histories of Texas and New Mexico. Chapters cover: Cattle Herds, Panhandle Counties, Fort Elliott, Mobeetie, Oil in the Panhandle (pp 461- 500), Panhandle Railroads, Panhandle Newspapers, and Feed Lots. His story of oil is first of the ranchers and farmers who needed drilling for water much more than oil and gas. While doing summer mapping with a small geological field party in the summers of 1904 and 1905, Charles N. Gould created and later published maps which led to the identification and subsequent drilling of the giant Panhandle Field. In 1918 the first gas discovery was drilled by Amarillo Oil Co. as their #1 Hapgood on the Masterson ranch. Oil also was soon located, and after countless boomtowns like Borger and Pampa, production rose to over forty million barrels in 1927. The majors and independents flooded in during the early 1920's to develop the shallow oil reserves. Along with Amarillo Oil, Phillips, Texaco, Cosden, Gulf and numerous others participated. Stanley gives brief histories of the major players and events. The hardbound first edition is becoming scarce.

**A294. (MIDDLE EAST) Stegner, Wallace, DISCOVERY: THE SEARCH FOR ARABIAN OIL, Doubleday & Company, Garden City, NY, (1971), xii, 190pp.**

Wallace Stegner is one of the finest fiction and non-fiction writers in America. This books is actually historical fiction at its very best. So little of it is fiction that inclusion in this book is necessary. Most of the places and events are accurately described along with the characters who began the Arabian oil search in the late 1920's and early 1930's. The book is full of fast action and was a favorable Aramco propaganda tool. The softbound first edition is scarce.

**A295. (CALIFORNIA) Steinhart, Carol E. and John Steinhart, BLOWOUT: A CASE STUDY OF THE SANTA BARBARA OIL SPILL, Duxbury Press, (1972), xvi, 138pp.**

The Steinhart's give a very detailed account of the now-famous January 28, 1969 blowout of Union Oil's A-21 well, offshore Santa Barbara. The chronology clearly shows drilling crew negligence coupled with a sequence of adverse geological-engineering events which took a year to bring under control. The authors are environmental activists but manage to present a fair picture of events. Political opinions are given by the authors only in their analysis and future of society chapters. Good notes and references. The softbound first edition is common.

**A296. (OKLAHOMA) Stewart, Roy P., BORN GROWN: AN OKLAHOMA CITY HISTORY, Oklahoma City, Fidelity Bank, 1974, xvi, 352pp.**

This is one of the most popular and comprehensive Oklahoma City histories. The city was created in the first land run into Indian Territory, on April 22, 1889. From noon to night that day, many unassigned Indian lands were up for grabs (homesteading) throughout central Oklahoma. The city has had amazing growth, not the least of which was due to the giant Oklahoma City oil field discovered in 1929. Much on the development and importance of oil and gas to the city. Contains photos, a useful bibliography and list of important early Oklahoma Citians. The hardbound first edition is common.

**A297. (GENERAL) Stillwagon, C.K., THE ROPE CHOKERS: A COLLECTION OF HUMAN INTEREST STORIES, ANECDOTES, HISTORICAL FRAGMENTS AND PICTURES OF THE OIL FIELDS, Houston, TX, Well Equipment Manufacturing, (1945), xi, 227pp.**

Stillwagon worked in all phases of the business, first in Pennsylvania, then at Ranger and East Texas. He was also the tool dresser on Santa Rita #1, the Big Lake field discovery well which started the West Texas boom. Quoting Swanson: "Title is nickname by which cable tool drillers are known. Stories are author's experiences in oil fields of Pennsylvania from around 1915". Excellent, historic photos. Cited in Swanson on page 188. The hardbound first edition is common.

**A298. (GENERAL) Stotz, Louis, with Alexander Jamison, HISTORY OF THE GAS INDUSTRY, NY, The Author, (1938), 534pp.**

Author was highly qualified gas industry pioneer. Quoting Swanson: "One hundred and twenty years of natural gas history; uses of gas; management problems; pipe line systems; regulation of public utilities.

Written when Natural Gas Act of 1938 was under consideration in Congress". Includes photos, maps and drawings. Cited in Swanson on page 68. The hardbound first edition is common.

**A299. (TEXAS) Straton, Florence, THE STORY OF BEAUMONT, Beaumont, TX, The Author, ca 1924, 196pp.**
This is a delightful city history which was prepared by the author so that the school children would be aware of the special history of Beaumont. By 1849, several log cabins had been built on the banks of the Neches and inside the 1837 chartered Beaumont townsite. A short eleven pages and two photos are given to the Spindletop oil boom which exploded this peaceful city. From the text: "The merchant left his desk, the clerk his counter, the lawyer his books, and the laborer laid down his tools. Every man saw the opportunity of making a million without the slow, tedious efforts of ordinary calling". Not mentioned in Swanson. The hardbound first edition is scarce.

**A300. (ALASKA) Strohmeyer, John, EXTREME CONDITIONS: BIG OIL AND THE TRANSFORMATION OF ALASKA, NY, Simon and Schuster, (1993), 287pp.**
This is a pretty harsh review of the effect of oil on Alaska over the last thirty years. The author reviews Richfield's initial bold move into Alaska oil exploration during the 1950's which was finally rewarded by the monster discovery at Prudhoe Bay in 1968. Richfield had found oil with moderate success, but at great expense prior to Prudhoe, however the initial tests there indicated significant reserves. From this point, the author takes a critical look at effects of the discovery on all aspects of Alaska, through the help of numerous individuals who were caught up in it. Government bungling and corruption was pervasive, from the local politicos to top U.S. officials. The unscrupulous and often incompetent oil company dealings and operations are discussed in scenes that are not pretty, and unfortunately on-going. The hardbound first edition is common.

**A301. (GENERAL) Sullivan, Mark, OUR TIMES: THE UNITED STATES, 1900-1925, NY, Charles Scribner's Sons, Vols II and VI, 1927 and 1935, xxxvii, 1342pp.**
These books were not available for review by Neff, but are regarded as important. Quoting from Swanson: "Vol. II: History of petroleum,

activities of Rockefeller, Standard Oil Company, dissolution of Standard Oil Trust. Vol. VI: The Twenties, reviews leasing of Teapot Dome and Elk Hills naval petroleum reserves and legal actions which followed. Probably best account extant of Teapot Dome Scandal". Cited in Swanson on page 158. The hardbound first edition of these books is very scarce.

**A302. (GENERAL) Swanson, E.B., A CENTURY OF OIL AND GAS IN BOOKS: A DESCRIPTIVE BIBLIOGRAPHY, NY, Appleton-Century-Crofts, (1960), xiii, 214pp.**
This fine book is regarded as the bible for oil booksellers and collectors. Swanson's impressive work replaced earlier efforts by DeGolyer and Hardwicke to become the most valuable and useful oil industry book reference. It is well organized by subject (everything from exploration to marketing) and is indexed by author. Swanson personally examined every item in the bibliography after amassing a gigantic collection. The API largely funded the effort which was edited by Kendall Beaton and aided to a great degree by Dr. Ernest Miller. The hardbound first edition is becoming scarce.

**A303. (HUNGARY) Sweetman, John, PLOESTI: OIL STRIKE, NY, Ballantine Books, (1974), 159pp.**
Part of a series of about one hundred Ballantine books on all aspects of WW II, this one concerns the allied raids on Ploesti, the Romanian refinery and oil complex which was feeding Hitler's war machine. Germany had overrun Romania in late 1939 and essentially taken over the large oil facility. Germany had virtually no oil reserves of its own, except for synthetic oil manufacturing plants (distilled from coal), which were becoming important. Logistics of an Allied air raid on Ploesti were exceedingly complex and a preliminary attack was made from bases in Iraq in June 1942, to test feasibility. Big mistake. Little damage was done, but the Germans were warned and took preventive measures which later proved costly. The big raid called Tidalwave was flown from airfields in Benghazi on August 1943. Countless errors and unfortunate circumstances decreased the effectiveness of the raid and led to tremendous plane and personnel loss. Of the 178 aircraft involved, over 50 B-24's and 500 men were lost. Permanent damage to the facility after the Germans got back to operation was estimated at only thirty percent of capacity. The softbound first edition is scarce.

**A304. (PENNSYLVANIA) Szalewicz, Steve S., OIL MOON OVER PITHOLE: A STORY OF A PHENOMENON IN AN OIL TOWN, Noplace, The Author, (1958), vi, 92pp.**

This is a somewhat embellished, non fiction historical work which contains excellent photos of Pithole, the first and most famous of Pennsylvania boomtowns. The pamphlet concerns a fire which burned the Tremont Hotel and the unusual manner in which it was somehow related to an unusual moon phase which caused previously pumping wells to begin flowing and creating all kinds of havoc. Cited in Swanson on page 159. The softbound first edition is becoming scarce.

**A305. (GENERAL) Tait, Samuel W., Jr., THE WILDCATTERS: AN INFORMAL HISTORY OF OIL-HUNTING IN AMERICA, Princeton, Princeton University Press, 1946, xvi, 218pp.**

Quoting from Swanson: "Scattered accounts of oil exploration history in substantially all major oil producing areas of the U.S., and early Canadian industry. While not a comprehensive survey, each vignette is carefully done and contains good material not found elsewhere". According to the dust jacket, although Tait was an attorney of some prominence, he concentrated his interest in oil and had a lifetime of experience in scouting and oil producing. His father was brought up in the early oil fields of Pennsylvania. Tate covers development of many of the large pre-1930, fields across the country and adds stories of the people and events to make them all interesting. Excellent historic photos are included. Cited in Swanson on page 158. The hardbound first edition is common.

**A306. (GENERAL) Talbot, Frederic A., THE OIL CONQUEST OF THE WORLD, London, William Heinemann, (1914), x, 301pp.**

Talbot covers everything from the earliest drilling in the U.S. through the spread of oil fever throughout the world and up to the start of WW I. His photos are among the best of early scenes around the world. Many of which are not published elsewhere. Many of the photos are of wild wells and clearly show the primitive equipment in use at the time. He relates the early oil events in several countries and how they are progressing and what their future possibilities are. Cited in Swanson on page 159; Hardwicke page 16. The hardbound first edition is becoming scarce.

**A307. (PENNSYLVANIA) Tennent, James C., THE OIL SCOUTS: REMINISCENCES OF THE NIGHT RIDERS OF THE HEMLOCKS, PA, Derrick Publishing, 1915, 85pp.**
Tennent was an old man when he dictated this book. The job of early oil scout was to keep oil operators honest, by any means possible. Speculation on the Oil Exchange hinged on correct information from the wildcat and development wells. Rumors caused havoc. His stories are real and give a good history of the heyday of scouting from the early 1870's to the early 1890's. Much biographical and historical information included. Reprinted in 1931 by Tiona Petroleum and again in 1986 by Venango County Historical. Cited in Swanson on page 159; Hardwicke page 41; Giddens page 94. The hardbound first edition is scarce.

**A308. (GENERAL) Terry, Pamela, FIFTY-FIVE GALLONS: THE HISTORY OF STEEL DRUM RECONDITIONING, Landover, MD, NABADA, 1991, xv, 448pp.**
Scholarly work which follows the development of the wooden barrel industry to the large barrel reconditioning industry of today. The author worked for the barrel & drum association for several years and wrote this book for their account. Includes photos. Hardbound first edition is common.

**A309. (TEXAS) Texas Parade Magazine, SPINDLETOP...WHERE OIL BECAME AN INDUSTRY: AN HISTORIC EDITION, Dallas, TX, Texas Parade, (1951), Volume XI, Number 7, 36pp.**
This issue commemorates the Spindletop discovery and real beginnings of the Texas oil industry. Article by Wallace Pratt on where oil is found. Another by Everette DeGolyer on the art of oil prospecting. Another by Ernest Thompson on the history of oil conservation. Many more articles are included in this impressive magazine. Not mentioned in Swanson. The softbound first edition is becoming scarce.

**A310. (TEXAS) Texas Permian Historical Society (editors), WATER, OIL, SAND AND SKY: A HISTORY OF WARD COUNTY, TEXAS, Monahans, TX, Monahans Jaycees, 1962, 64pp.**
Ward County is the center of some of the best oil and gas production in the Permian Basin of West Texas, but only the last chapter of the booklet is dedicated to oil. It had been found in the surrounding counties earlier, but not until 1928 did Shipley Oil Company test their No. 1 S.F. Hazlett

wildcat near the town of Royalty, naturally flowing 105 barrels per day from the Queen sand at 2480 feet. In 1929, the county's largest oil field, Ward-Estes was opened by Gulf Oil with 1,350 barrels per day from the Yates sand at 2,600 feet. Several other large fields followed as exploration proceeded westward across the county. Much anecdotal material on the independent and major companies as they developed the vast shallow oil reserves of the county. The softbound first edition is common.

**A311. (TEXAS) The Oil and Gas Journal (editors), GOLDEN ANNIVERSARY: 50 YEARS OF OIL IN THE SOUTHWEST, Tulsa, OK, The Oil and Gas Journal, 1951, 484pp, Volume 50, No. 4.**
Twenty-six feature stories and articles, by-lined by the highly competent magazine staff are presented in this fine work. Aside from the excellent oil history review from exploration to refining, separate sections are provided to give sketch-biographies of the well known or important individuals in their respective fields. Numerous early photos are presented, taking the reader back to the Spindletop discovery and following industry progress to 1951. Some of the old ads are almost as interesting as the history. The softbound first edition is becoming scarce.

**A312. (GENERAL) The Oil and Gas Journal (editors), THE PETROLEUM INDUSTRY, 1859-1934, Tulsa, OK, Petroleum Publishing Company, 1934, 268pp.**
This book was not available for review by Neff, but it is regarded as important. Quoting from Swanson: "Special historical supplement to Oil and Gas Journal, August 27, 1934". Cited in Swanson on page 156. The softbound first edition scarce.

**A313. (GENERAL) The Oil and Gas Journal (editors), PETROLEUM 2000: SEVENTY-FIFTH ANNIVERSARY ISSUE, Tulsa, OK, The Oil and Gas Journal, 1977, pp57-531.**
Almost all of the issue is dedicated to the future of the oil industry and covers exploration, drilling, production, pipelining, refining, petrochemicals, gas processing, international, alternate fuels, plus the story of The Oil and Gas Journal, itself. The Journal started at Beaumont in 1902, where it was originally called The Oil Investors Journal. Much history and speculation in this well-compiled series of articles by industry experts. The softbound first edition is becoming scarce.

**A314. (GENERAL) The Oil and Gas Journal (editors), PETROLEUM PANORAMA: AN ISSUE TO COMMEMORATE OIL'S FIRST ONE HUNDRED YEARS, Tulsa, OK, The Oil and Gas Journal, 1959, iv, 623pp. Volume 57, No. 5.**

Exceptional work by the magazine staff. Each of the six main subjects are self contained, detailing oil history and progress through oil's first century. Section A is 162 pages of oil industry history in light of economics, politics, technology and competition. Section B is 47 pages of exploration history. Section C is 108 pages of drilling history. Section D is 92 pages of producing history. Section E. is 82 pages of transportation history. Section F is 132 pages of refining and petrochemical history. Numerous dated ads are included. This is the most comprehensive and impressive of commemorative issues by the Journal. Cited in Swanson on page 156. The softbound first edition is scarce.

**A315. (OKLAHOMA) Tinkle, Lon, THE STORY OF OKLAHOMA., NY, Random House, (1962), 185pp.**

This is a juvenile state history probably aimed at sixth or seventh grades. Tinkle covers the occupation of Indian Territory, the land runs, and statehood. In chapter ten he documents the early oil discoveries and Everette DeGolyer, a leading Oklahoma geologist (The author later wrote DeGolyer's excellent biography). There is also mention of famous Oklahomans, Will Rogers and Jim Thorpe. Drawings are included. The hardbound first edition is common.

**A316. (GENERAL) Tower, Walter S., THE STORY OF OIL, NY, D. Appleton and Co., 1909, xi, 270pp.**

The author was knowledgeable and well respected in the oil industry. His book covers oil history from day one, to 1909. A chapter is dedicated to Standard Oil, which was still intact at the time. Much on Baku, the only real rival of the U.S. for oil supremacy at the time. Quoting Swanson: "Oil developments during preceding fifty years, magnitude and range of operations, and American participation in growth". Includes photos. Cited in Swanson on page 188; Hardwicke page 17; Giddens page 95. The hardbound first edition is scarce.

**A317. (GENERAL) Tugendhat, Christopher, OIL: THE BIGGEST BUSINESS, NY, G.P. Putnam's Sons, (1968), xvi, 318pp.**
The bulk of the book describes how the oil industry grew up, from Drake to OPEC and the 1967 Middle East crisis. In the last part, Tugendhat describes today's industry operations and how companies and political factors work together. Includes photos, maps and charts. The hardbound first edition is common.

**A318. (MIDDLE EAST) Twitchell, K.S., SAUDI ARABIA: WITH AN ACCOUNT OF THE DEVELOPMENT OF ITS NATURAL RESOURCES, Princeton, NJ, Princeton University Press, 1947, xiii, 192pp.**
Fascinating early work on the country by an American mining engineer who had been asked to investigate the natural resources for King ibn-Saud in the 1930's. He was instrumental in selling the oil concession idea to the oil companies and tells the oil story the way he saw it in the first person. The hardbound first edition is common.

**A319. (WYOMING) United States Senate, LEASES UPON NAVAL OIL RESERVES (CONTINENTAL TRADING COMPANY, LTD., OF CANADA), Washington, Government Printing Office, 1929, xx, 1369pp.**
Full Senate hearings in the last throes of the Teapot Dome, Wyoming and Elk Hills, California oil leasing scandal. Doheny and Sinclair had bribed Albert Fall, Secretary of the Interior under Harding, with about $400,000 in return for federal oil and gas leases on the two Naval Petroleum Reserves. They were both found not guilty on those charges, however Fall eventually spent a year in jail on related charges and Sinclair served six months for contempt of court. These hearings uncovered detailed workings of Continental Trading Company of Canada which was the source of the bribe money. Continental was nothing more than a slush fund for political payoffs, its cash being generated by skimming 25 cents per barrel from oil trades between four large oil companies, Sinclair, Prairie, Standard of Indiana and Midwest Refining. The saga ranges across the U.S. and Canada. On reading the testimony as Senator Welch interrogates Sinclair, Stewart and others, it is obvious they are lying. Fascinating piece of history. Not mentioned in Swanson. The leather bound first edition is rare.

**A320. (OKLAHOMA) United States Senate, OSAGE INDIAN LEASES IN OKLAHOMA: LETTER FROM THE SECRETARY OF THE INTERIOR, TRANSMITTING......, Washington, Government Printing Office, 1917, 31pp.**

Senate Document 63, 65th Congress. In response to a Senate resolution of April 16, 1917, information is given relative to the number of leases of lands of Osage Indians in Oklahoma issued or authorized since the first day of January 1916. The Foster lease had been broken and no more subleases were approved after 1915. Several companies holding over 100,000 acres are listed along with new leases approved and pending applications. The paper gives details of leasing procedure which is very nearly the same as that being practiced in 1994. Not mentioned in Swanson. The softbound first edition is becoming scarce.

**A321. (OKLAHOMA) Wade, Henry F., SHIP OF STATE ON A SEA OF OIL., OK, The State, (1975), xiii, 67pp.**

Wade is the guide for tourists visiting the Oklahoma Capitol building. He spent eleven years interviewing people as he collected and compiled information included in this book. The title refers to the capitol being located on top of one of the largest oil fields in the U.S. Most of the book is history of the capitol from its original location in Guthrie to the actual construction of the building in Oklahoma City. Various monuments and related buildings are also discussed along with the important people who worked there over the years. The producing oil well named Petunia #1 (drilled in a petunia bed) is legally deviated underneath the capitol and in 1942 was completed at 6,618 feet. By 1973 the well had made over 1,000,000 barrels while generating considerable royalty and tax revenue for the state. Numerous other wells are drilled on capitol grounds, up and down Lincoln boulevard. The softbound first edition is common.

**A322. (OKLAHOMA) Walker, James P., A HISTORY OF THE INTERNATIONAL PETROLEUM EXPOSITION AND CONGRESS 1923-1979, Oklahoma City, Oklahoma Heritage Association, 1984, xiv, 129pp.**

From the dust jacket: "In 1923 the civic and business leaders of Tulsa staged the first International Petroleum Exposition in the city increasingly being called The Oil Capital of the World". Also called the Tulsa Oil Show, it survived the depression and during the 1950's became the world's leading industrial exhibit, all petroleum related. The great exhibit closed in 1979, due largely to competition from imitators and specialized shows. This important piece of oil history includes many photos. The hardbound first edition is common.

**A323. (GENERAL) Ward, Edward, OIL IS WHERE THEY FIND IT, London, George G. Harrap & Co., (1959), 255pp.**
The author had much personal experience with the oil industry and traveled extensively to view oil fields and operations. he was a BBC commentator and gives his account of the status and history of oil in various countries including Venezuela, New Guinea, Ecuador, Nigeria, Holland and many others. His work is largely based on conversations with the staff members of Royal Dutch- Shell. Many photos. Cited in Swanson on page 189. The hardbound first edition is common.

**A324. (TEXAS) Warner, C.A., TEXAS OIL & GAS SINCE 1543, Houston, Gulf Publishing Co., (1939), vii, 487pp.**
One of three Texas oil books to get full treatment in Jenkins. BASIC TEXAS BOOKS. From Jenkins: "The most extensive and detailed statistical treatment of the oil industry in Texas, this monograph was written by a prominent petroleum engineer and geologist who had worked in the oil industry in Texas since 1917". Swanson calls it authoritative and says it contains much information not found elsewhere. Date in the title refers to survivors of the De Soto expedition who, in 1543, used pitch like substance to repair vessels". A review of the general history and growth of the industry in Texas is followed by a review of the important oil and gas provinces of: East Texas, Gulf Coast, North Texas, Panhandle, Southwest Texas and West Texas. Reprinted in 1951 for the Texas Centennial. Cited in Swanson on page 160. The hardbound first edition and reprint are very scarce.

**A325. (MEXICO) Watkins, Lew, LANDMAN UNDER OLD SPANISH LAW, Houston, TX, The Author, (1962), 62pp.**
This is a fine summary of the Mexican oil history and land problems encountered there by the U.S. and foreign companies which first developed the oil resources. The author worked for Doheny's Pan American Petroleum company and was deeply involved with unraveling the private and government land ownership as well as fighting the subsequent title suits and finally expropriation. An unusual book. The hardbound first edition is scarce.

**A326. (TEXAS) Weaver, Bobby D. (editor), PANHANDLE PETROLEUM, Amarillo, TX, Panhandle-Plains Historical, (1982), ix, 132pp.**
This is a compilation of articles regarding the Panhandle field, its history and development. A fine article by Charles N. Gould on the beginning of the field and another by Ernest Thompson regarding conservation regulations are particularly informative. The book is in three parts: 1) the raw materials which includes geology and origins of the oil gas and helium in the field, 2) the events leading to discovery of the field and its first boomtown, Borger, 3) the field development for gas, carbon black and helium. Great photos of boomtown, Borger. The softbound first edition is common.

**A327. (OKLAHOMA) Wells, Laura Lou, YOUNG CUSHING IN OKLAHOMA TERRITORY, Perkins, OK, Evans Publishing, (1985), v, 221pp.**
Cushing was a tiny farming area which had been acquired by homesteaders in the 1891 land run of formerly Sac and Fox Indian lands. It was a sleepy little town until opening of Cushing oil field in 1912. An incredible boom followed. Within two years the field was producing twenty percent of the U.S. oil and the town grew from 700 to 7000 people. Tom Slick and B. B. Jones drilled the discovery well on Anna Jones property. They had leased much of the surrounding countryside but the field became much bigger than anyone anticipated, soon turning into the independent operator's paradise. Well written early history with historic photos and many newspaper quotes to back it up. Maps are included. The hardbound first edition is common.

**A328. (OKLAHOMA) Welsh, Louise, et al, A HISTORY OF THE GREATER SEMINOLE OIL FIELD, Oklahoma City, Oklahoma Heritage Association, (1981), xii, 128pp.**
All three authors live in the Seminole area. One is a librarian and the other two are teachers. The greater Seminole area contains six giant oil fields. From 1923 to 1935 it was the world's largest producer of crude oil. The area was known to have oil as early as 1913 with small production being established, however the 1923 Wewoka field discovery began a steady exploratory growth pattern. Robert Garland was the lease owner

and driller for Independent Oil & Gas Company when he drilled the #1 Frixico discovery well. The oil-charged Wilcox sand was encountered at 4,073 feet and the well was completed flowing 10,000 barrels per day. About one third of the book is historic photos. Maps are also included in this fine history. The hardbound first edition is common.

**A329. (WYOMING) Werner, M.R. and John Starr, TEAPOT DOME, NY, The Viking Press, 1959, x, 306pp.**
Swanson calls this a journalistic account of the Teapot Dome and Elk Hills scandals. It was definitely the most popular and widely circulated version. From the dust jacket: "A mammoth conspiracy, how it was exposed by the most remarkable feat of detective work in the annals of government, the complete story told for the first time". Authors claimed to be the first to use Secret Service documents to verify their account of the scandal. Political cartoons of the day are used to good effect. Further from the dust jacket: "..crooked oil barons, thieving politicians, apathetic congressmen, jury-fixing henchmen, and a gullible president.." What a cast of characters. Cited in Swanson on page 160. The hardbound first edition is becoming scarce.

**A330. (CALIFORNIA) White, Gerald T., SCIENTISTS IN CONFLICT: THE BEGINNINGS OF THE OIL INDUSTRY IN CALIFORNIA, San Marino, CA, The Huntington Library, 1968, xiii, 272pp.**
Great story of early California oil exploration efforts which were stimulated by Dr. Silliman's report of potential riches there. The quality of crude sampled from live seeps at Ojai and Pico indicated a potential bonanza which was verified by chemist-refiner, Stephen Peckham. The saga and mystery gets complicated as certain oil samples evidently had been salted. The titled conflict lay in the person of Josiah Whitney of the California Geological Survey who discounted Silliman's optimistic reports and accused him of aiding swindlers and promoters. Whitney appeared to have a case as the early digging and drilling efforts were unsuccessful and many operators and investors went under. However, Silliman's reputation was later saved and even enhanced when prolific producers were completed. Vindication after several years of ridicule. The hardbound first edition is becoming scarce.

**A331. (TEXAS) White, John H., BORGER, TEXAS: A HISTORY OF THE REAL FACTS ABOUT THE MOST TALKED OF TOWN IN TEXAS AND THE SOUTHWEST, Waco, TX, Texian Press, 1929, 139pp unnumbered.**
The city was founded amidst the Panhandle oil field boom in 1926 and was instantly crowded by unsavory characters...lawlessness prevailed. A.P. Borger, a promoter, had earlier paid $12,000 for speculative land in the area, and within a week of the oil discovery was selling town lots for up to $5,000 each. 40,000 people crowded into the Borger, Isom, Whittenburg and Dixon Creek area. The governor sent the Texas Rangers into Borger to calm things down. It was effective and they left. Author, John White was a local lawyer and State Representative who strongly felt that a second coming of the Texas Rangers to secure the unruly town, was unjustified and that the governor had overstepped his authority. His book was written shortly after that occurred. Photos of early Borger are impressive. Actual reports and proclamations are reprinted in the appendix. The book has an oil field panorama photo dust jacket. Reprinted as a collectors edition in 1973. Not mentioned in Swanson. The hardbound first edition is very scarce and the reprint is scarce.

**A332. (OKLAHOMA) White, Wayne E., FROM SALT KETTLES TO NUCLEAR FUELS: CHEMISTRY IN OKLAHOMA, Tulsa, OK, The Author, 1976, 135pp.**
History of chemistry in Oklahoma. Much on the work of oil companies, although agriculture and other fields are covered in good detail. Chapter five gives history of the work and individual accomplishments of several companies. Amoco Research with a staff of 600 did much on corrosion control, formation fracturing and enhanced recovery. Dowell developed acidizing and cementing chemicals and processes. Halliburton pioneered many different chemical treatments for oil field problems. Cities Service did much early petrochemical work. Texaco, ARCO and Skelly's contributions are also listed. Chapter six documents the accomplishments of the two large Oklahoma based majors, Conoco and Phillips. Only highlights of the work of these companies are given as books could be written on each. Conoco started early work on benzene from the 1930's. At the same time, Phillips started with work on paraffins. Photos and biographies of the early oil chemists are included. The softbound first edition is common.

**A333. (PENNSYLVANIA) Whiteshot, Charles A., THE OIL-WELL DRILLER: A HISTORY OF THE WORLD'S GREATEST ENTERPRISE, THE OIL INDUSTRY, Mannington, WV, Whiteshot, 1905, 895pp.**

This second, and greatly expanded edition of his book covers most every area, company and important character in the oil business at the time. It is largely concentrated in Pennsylvania, but covers the entire U.S. and most of the foreign oil areas. "The universally recognized standard authority of the world's history of petroleum, crude and refined oil, natural gas, carbon black and its by-products". Quoting from Swanson: "Large part of 2nd edition covers the beginning and rise of the Standard Oil Company. Author was publisher of The Oil Region News, established 1900". Many photos and drawings. Definitive work. Cited in Swanson on page 167; Hardwicke page 42. The 1st edition has 127pp. and is rare. The hardbound second edition is very scarce.

**A334. (PENNSYLVANIA) Wilkie, Franc Bangs, PETROLIA: or THE OIL REGIONS OF THE UNITED STATES, Chicago, John R. Walsh & Co., 1865, 82pp.**

This book was not available for review by Neff, but is regarded as an important work. Quoting from Swanson: "Guide book to Oil Regions of Pennsylvania, Ohio, and West Virginia; includes six letters written by author between Feb 22, 1865 and March 6, 1865, from various Oil Region communities; drilling, travel routes, costs, hotels; discussion of reliable companies". Cited in Swanson on page 160. The first edition is rare.

**A335. (GENERAL) Williamson, Harold F. and Arnold R. Daum, THE AMERICAN PETROLEUM INDUSTRY: THE AGE OF ILLUMINATION 1859-1899, Evanston, IL, Northwestern University Press, (1959), xvi, 864pp.**

This is Volume I of an authoritative history of American oil industry. The authors prepared it under grant from API. It was to consolidate all of the previous research. Swanson believes that it is of high quality and "is excellent for researchers and students seeking definitive information in a single volume". It is a widely circulated major work with photos, maps and drawings. Cited in Swanson on page 160. The hardbound first edition is common.

**A336. (GENERAL) Williamson, Harold F., et al, THE AMERICAN PETROLEUM INDUSTRY: THE AGE OF ENERGY 1899-1959, Evanston, IL, Northwestern University Press, (1963), xx, 928pp.**
This is Volume II of the comprehensive work by several authors covering the time period indicated. Research compilation much in the same manner as Volume I. This was published after Swanson's bibliography so we do not have his opinion. Another major oil history with photos, maps and drawings. The hardbound first edition is common.

**A337. (OKLAHOMA) Wilson, Terry P., THE UNDERGROUND RESERVATION: OSAGE OIL, Lincoln, University of Nebraska Press, (1985), xiv, 263pp.**
Wilson particularly explores the effect of oil wealth on the Osage Indians. From the first oil found on their lands in 1897, the royalty payments increased until the late 1920's where each headright received several thousand dollars annually. To great benefit, the Tribe had retained mineral rights to their 1.5 million acres before surface rights had been allotted to individuals in 160 acre tracts. The tribal council controlled all oil royalty and lease bonus income and distributed it to those who retained head rights. Wilson studies their complex and difficult relationships with the federal government as well as the relationships between the mixed-bloods and whites of the area. He finds that many of the turn of the century conflicts are still in existence in one form or another today. The lawlessness, corruption and violence which were commonplace back then have now subsided. The Tribe and its unique place in American culture (and in the framework of American Indian culture) is placed in historical and social perspective. The author teaches at University of California and previously published "Bibliography of the Osage". As might be expected, this work contains a useful bibliography. The hardbound first edition is common.

**A338. (GENERAL) Winslow, Ron, HARD AGROUND: THE STORY OF THE ARGO MERCHANT OIL SPILL, NY, W.W. Norton & Co., (1978), 286pp.**
In the early morning darkness of December 15, 1976, a nightmare became real. The Argo Merchant, an aging Liberian tanker, went aground on a sandy shoal southeast of Nantucket with 7.6 million gallons of industrial heating oil aboard. Winslow's unbiased reporting is accentuated with spectacular photos. The hardbound first edition is common.

**A339. (PENNSYLVANIA) Witteman, A, THE OIL COUNTRY OF PENNSYLVANIA AND NEW YORK, NY, The Author, 1893, 37pp.**
From the cover, stamped in gilt: "Oil Country Illustrated". This popular little photo album was printed by Albertype Company of New York two years before Mather's historic album. It contains fourteen photo plates showing: The Drake Well, A Flowing Well, Jo Jo, A Typical Oil Find Town, Derrick With Side Flow, City of Tanks (Below Olean, NY) Oil City, PA, Burning Oil Well, Burning Oil Tank, Bradford PA, The Great Kijnzua Bridge, A Torpedo Wagon With Load of Nitro-Glycerine, Filling the Torpedo Shell, A Pump, Loading an Oil Train, Oil City Crude Tar Stills, Natural Gas Scenes. The photos are not of great quality, but are not seen in other publications. Not cited in bibliographies. The hardbound first edition is scarce.

**A340. (GREAT BRITAIN) Woodward, Guy H. and Grace Steele Woodward, THE SECRET OF SHERWOOD FOREST: OIL PRODUCTION IN ENGLAND DURING WORLD WAR II, Norman, University of Oklahoma Press, (1973), xviii, 266pp.**
From the dust jacket: "...tells for the first time the story of how Britain's Ministry of Fuels, enlisting the aid of forty-four American oilfield roughnecks, met the crisis by developing shallow pools of oil in Sherwood Forest". During the heat of WW II, Anglo-Iranian, British Petroleum's predecessor, solicited Noble Drilling and Fain-Porter Drilling companies of Oklahoma for their expertise and crews to come to England and help develop their shallow oil field. This story is told as a running dialogue and commentary which makes the true events very exciting. The project went a long way towards relieving the British wartime fuel shortage. The hardbound first edition is common.

**A341. (OKLAHOMA) Work Projects Administration Writers, THE WPA GUIDE TO 1930'S OKLAHOMA, Lawrence, University Press of Kansas, (1941), xxxviii, 442pp.**
In four parts: 1. The general background on Oklahoma, including an essay on its history by Angie Debo. A section on the oil industry, Oklahoma's largest. 2. History and development of the twelve largest cities, individually. 3. Tours of points of interest around the state. 4. A Chronology of Oklahoma history and a selected reading list. The original 1941 University of Oklahoma title was, "Oklahoma: A guide to the

Sooner State". Maps are included.. The softbound first edition is common.

**A342. (PENNSYLVANIA) Wright, William, THE OIL REGIONS OF PENNSYLVANIA: SHOWING WHERE PETROLEUM IS FOUND; HOW IT IS OBTAINED AND AT WHAT COST, NY, Harper & Brothers, 1865, 275pp.**

Highly readable account by the author who was a correspondent for the New York Times. He made an extensive tour of the Oil Regions, visited all of the important sites personally (on foot), and conducted numerous interviews prior to completing this excellent early work. From the preface: "I have described the processes of boring the wells, of repairing them after getting out of order, and of refining the oil". He was largely concerned about the amount of misinformation which was being spread from the farmers, drillers, operators, investors, promoters, etc. He tries to bring some order to the chaos and give a true picture of how much oil, how much cost and how much gain was being made on which farm and in which area. Also from the preface: "Petrolia needed a searching examination and a scathing exposure; it has got both. Yet let me not be misunderstood. Underneath a system of falsehood and fraud, that might almost be termed magnificent, there is a great basis of fact, which needs to be presented in its true light". He does that in fine fashion. Cited in Swanson on page 35; Hardwicke page 43; Giddens page 95. The hardbound first edition is scarce.

**A343. (GENERAL) Yergin, Daniel, THE PRIZE: THE EPIC QUEST FOR OIL, MONEY, AND POWER, NY, Simon & Schuster, (1991), 877, xxxiipp.**

Yergin is a prolific oil and energy writer. He spent ten years researching this definitive work on the history of the oil industry. Most historians thought this book was long overdue. Yergin took the highlights and watershed events of the 132 year oil industry history and through expert narrative molded it into a classic saga. The story begins with Dr. Benjamin Silliman, Jr's famous report which showed that rock oil could be processed into an effective illuminant, and ends just before the Iraq invasion of Kuwait. The book is technical and a documentary masterpiece which has been criticized only in that it is dry reading. Not so for those really interested in oil history. Chapters are: The Founders, The Global Struggle, War and Strategy, The Hydrocarbon Age, The Battle for World

Mastery. Included are an epilogue, chronology, oil prices, and notes. The book also contains a comprehensive, eighteen-page bibliography. This is the finest overall history of oil, to date. The hardbound first edition is common.

**A344. (GENERAL) Young, Bob and Jan Young, GUSHER: THE SEARCH FOR OIL IN AMERICA, NY, Julian Messner, 1971, 190pp.**
The authors provide a popular history of oil from "Drake's Folly" to the Prudhoe Bay discovery in an easy reading format. The book fairly covers important men and events with a generally uncritical eye, designed to please most everyone. It has a useful glossary and bibliography. The hardbound first edition is common.

# PART II - BIOGRAPHIES

**B1. (JOHN D. ROCKEFELLER) Abels, Jules, ROCKEFELLER BILLIONS: THE STORY OF THE WORLD'S MOST STUPENDOUS FORTUNE, NY, The MacMillan Co., (1965), xiii, 386pp.**
One of the top Rockefeller family biographies, this book also covers Standard Oil history. It is well-balanced and highly recommended as a first-read for aspiring historians or anyone else interested in the subject. Abels is an attorney, editor, economic analyst, and published author. He presents an organized, insightful look at the unique Rockefeller saga. He discusses the good and the bad, but not the libel. From the introduction: "This writer believes that the lives of the Rockefellers have served the public good. The reader will be able to judge for himself". British edition retitled ROCKEFELLER MILLIONS. The hardbound first edition is common.

**B2. (LOUIS ABRAMSON) Abramson, Louis, Jr., INSIDE THE PETROLEUM INDUSTRY, NY, The Carlton Press, Inc., (1977), 96pp.**
Louisiana boy makes good. For nearly fifty years (1920's to 1970's) Abramson was involved in almost all phases of petroleum marketing and distribution. His autobiography contains anecdotal accounts of doing business in Louisiana, particularly the years before 1950. Stories of fuel oil marketing in the 1920's to LPG marketing in the 1930's and then petrochemicals in the 1940's. He also served as an assistant to Secretary of the Interior, Harold Ickes, on the Petroleum Administration for War (WW II) council. The hardbound first edition is common.

**B3. (EDWARD ACHESON) Acheson, Edward Goodrich, A PATHFINDER: DISCOVER, INVENTION AND INDUSTRY, HOW THE WORLD CAME TO HAVE AQUADAG AND OILDAG, NY, Press Scrap Book, 1910, 143pp.**
This is a remarkable story of a bright and curious young man who became a successful inventor and industrialist. Born in Pennsylvania in 1856, he received some schooling in math and engineering. He left school at sixteen to go to work. He found a menial job at an iron furnace, but soon became restless and began inventing things. He patented a boring

machine which was designed for work in coal mines. He worked a while for Allegheny Valley Railroad which serviced the booming Oil Regions, and where he was quickly elevated to engineer. For the next few years he worked for Standard Oil's United Pipelines Company in Bradford as a tank gauger. Intellectually frustrated, he barged his way into Thomas Edison's Menlo Park complex, which then employed 200 people and asked for work...he got it. He left Edison at age thirty and proceeded to invent Carborundum, which he patented. He received the first patent to manufacture synthetic graphite. from which he developed an exceptional graphite-oil lubricant which he called Oildag. As this book was written, Acheson was only forty-four and on a roll. High achiever. The hardbound first edition is scarce.

**B4. (HENRY FLAGLER) Akin, Edward N., FLAGLER: ROCKEFELLER PARTNER AND FLORIDA BARON, Gainesville, University Press of Florida, 1992, xiii, 305pp.**
Flagler's life is presented in two parts, the Standard Years and the Florida Years. The author did considerable research on the subject and has the sources and notes to back it up. He begins with a detailed account of the Rockefeller, Anderson and Flagler partnership and closes with a discussion of Flagler's incredible accomplishments in developing the East Coast of Florida. Part of Akin's thrust seems to be comparing Rockefeller and Flagler in history to see how they stack up. As far as he is concerned Flagler comes out on top since he essentially put the state of Florida on the map. The author was also motivated to get an academic-scholarly type record of Flagler's accomplishments into the state archives. The softbound first edition is common.

**B5. (ROBERT HILL) Alexander, Nancy, FATHER OF TEXAS GEOLOGY: ROBERT T. HILL, Dallas, TX, S.M.U. Press, (1976), xii, 317pp.**
Hill was born in 1858 in Nashville, Tennessee into a wealthy merchant family. Bad luck was quickly encountered as the family fortune was lost in the Civil War and both of his parents died early. Raised by his grandmother, he went to work for a brother on a Texas newspaper at sixteen. He attended Cornell at age twenty-three, studying geology. He worked for Major Powell at the U.S.G.S. and by age thirty was doing brilliant innovative geological work over the country and publishing important papers. His pioneering geological work in Texas eventually

gained him the title of Father of Texas Geology. Although he never had success in oil and gas exploration, his maps and articles were much sought after by geologists and drillers alike, which they used to great benefit in locating prospective drillsites. He left the survey in 1903 and worked as a consultant over a good part of the world, but mainly Texas where he was the first to unravel surface stratigraphy, structure and physiography. He was devastated in the depression, but continued writing history and geology mainly for the Dallas Morning News until his death in 1941. A very well researched book on an important early geologist. The hardbound first edition is common.

**B6. (JOHN D. ROCKEFELLER) Allen, William Harvey, ROCKEFELLER: GIANT, DWARF, SYMBOL, NY, Institute for Public Service, 1930, xviii, 619pp.**
This book was not available for review by Neff, but it is regarded as important. Quoting from Swanson: "Review and appraisal; comments concerning others who wrote about Rockefeller". Cited in Swanson on page 160. The hardbound first edition is very scarce.

**B7. (JOHN D. ROCKEFELLER) Allstorm, Oliver, OLD ROCKY: THE WORLD'S FIRST BILLIONAIRE, Chicago, IL, M.A. Donohue & Co., 1939, 186pp.**
A most unusual book, it was prepared as a tribute to Rockefeller and to be delivered on his 100th birthday. Even though the target date was missed by two years and two months, the book was published on time. There is no information included about the author, but he was a true-blue John D. fan. From the dust jacket: "This is the story of John D. Rockefeller, the world's first billionaire. It tells of his rise from near-poverty to the spectacular heights of wealth, greater perhaps than the combined fortunes of the Nabob of the Arabian Nights, the Nizan of Hyderabad, the Gehwar of Baroda, or of any other Midas or Croesus that ever lived in legendary lore...how he made his riches and of how he employed his wealth to work for the uplift of humanity. These pages also include the names of the friends, the books, and the songs that he loved, and mention is made of many of the greater historical events that took place during his life". The entire book is written in a form of non-rhyming poetic verse and is surprisingly easy and interesting to read. A good deal of history can be gleaned as the author recaps 1839 to 1937, touching on

people and events. According to the author, John D. wrote the following poem which is included.

> I was early taught to work
> As well as play.
> My life has been one long
> Happy Holiday,
> Full of work and full of play
> I dropped the worry by the way,
> And God was good to me every day.

The book should probably be classified as fiction, however, it has a large amount of factual biographical material. Offbeat. The hardbound first edition is scarce.

**B8. (PETER LESLEY) Ames, Mary Lesley, LIFE AND LETTERS OF PETER AND SUSAN LESLEY, NY, G.P. Putnam's Sons, 1909, 1088pp.**
This two-volume set was not available for review by Neff, but is regarded as important. Quoting from Swanson: "Lesley was employed on the 1st Geological Survey of Pennsylvania and was Director of 2nd Survey. Wrote on coal oil, Department of Agriculture report, 1862. Biographical notice by his nephew, Benjamin Smith Lyman, gives full account of his work". Cited in Swanson on page 173. The hardbound first edition is rare.

**B9. (STELLA DYSART) Armitage, Merle, STELLA DYSART OF AMBROSIA LAKE: COURAGE FORTITUDE AND URANIUM IN NEW MEXICO, NY, Duell, Sloan and Pearce, (1959), 160pp.**
This is a most entertaining book about a very tough lady. She was an obsessed explorationist. She was unique, and her biography stands somewhere between that of Sue Sanders, whose tenacity is legendary, and that of C. C. Julian, whose ability to promote high risk exploratory ventures is legendary. Stella essentially spent her life promoting oil drilling on her vast land holdings at Ambrosia Lake, about an hour's drive west of Albuquerque, New Mexico. She never found the oil, and at this 1958 writing had still not given up. However she was deeply involved in a billion dollar uranium discovery. The discovery which lay partly on her land consists of 68% of the U.S. uranium reserves. Trials and tribulations and legal problems galore. Stella was born in Missouri in the late 1800's

and moved to New Mexico. By the 1920's she was a successful businesswoman and began trading in land. She was fascinated by the oil business and believed she could find oil. Her story of perseverance in the face of adversity is impressive. It paid off too. The hardbound first edition is becoming scarce.

**B10. (HOWARD HUGHES) Bartlett, Donald L. and James B. Steele, EMPIRE: THE LIFE, LEGEND, AND MADNESS OF HOWARD HUGHES, NY, W. W. Norton and Co., (1979). xviii, 766pp.**

By far the most professional and comprehensive biography of Hughes, this work has detailed chronology, appendixes, notes and index in the back. There is a good description of Hughes Sr. as a child, a college man, a lawyer, then oilman. He was one of the first to go to Beaumont and capitalize on the 1901 Spindletop discovery activity. Hughes Sr. was a restless genius. After following the oil boom for several years he started wildcatting across south Texas and Louisiana under the Sharp and Hughes partnership name. Hughes bought a rock bit design idea from a man named Humanson and refined and patented it in 1908. By 1909 it was proved to be superior to any drill bit in cutting hard formations such as limestone. This rock drill bit is fundamentally unchanged today and revolutionized the drilling business. By 1918 he had bought out his partners and owned the company 100% He was a flamboyant salesman and evidently had a tremendous impact on Hughes, Jr., then thirteen. The authors go on to cover every phase of Hughes, Jr.'s incredible career to his death in 1976. Required reading for Hughes or mogul fans. The hardbound first edition is common.

**B11. (KIM BARTLETT) Bartlett, Kim, GULF STAR 45, NY, W.W. Norton & Co., (1979), 224pp.**

Bartlett is a good writer and an oil worker. Here he relates experiences during one event-filled week on an offshore gulf coast drilling platform. He changed the names of people and companies in his story, probably to avoid lawsuits. He gives the reader a real-life feel for the people and the operation through use of dialogue in the vernacular. He begins with a harrowing, bad weather, crew-boat ride to the platform by the forty-man crew and ends with the anxiety of a helicopter ride back to shore a week later. He makes you feel like you are there. Anyone thinking about working on an offshore rig should read this before trying it. One minute there is a hair-raising emergency and the next thirty-six hours are total boredom...then another emergency. Fortunately the work pays very well and the workers get every- other week off. Enlightening. The hardbound first edition is common.

**B12. (RICK BASS) Bass, Rick, OIL NOTES, Boston, Houghton Mifflin, 1989, 172pp.**

Bass was an oil production geologist for several years. Currently (1995) he is a much sought after writer, since his way with words has been recognized by the critics and public. Seven top selling books so far. This particular journal covers a few years of oil and gas development drilling while in the employ of the highly successful and respected Dudley Hughes. Bass makes the tedium of development well drilling seem as exciting as wildcatting although he shows no interest in the latter. He is a pragmatic philosopher and makes oil geology interesting to about anyone who will listen, much as John McPhee does with hardrock geology. Thought provoking book. The hardbound first edition is becoming scarce.

**B13. (SAMUEL BAYNE) Bayne, Samuel Gamble, DERRICKS OF DESTINY, NY: Bretano's, (1924), xv, 259pp.**

Bayne was born in Ireland about 1840 and emigrated to the U. S. in 1869. He was well born and educated. He had made and lost a fortune in Irish linen speculation prior to his U.S. expedition. For starters in the States, he purchased a nearly depleted oil well in Titusville, right next to the Drake discovery. He then proceeded to learn the oil business the hard way, from the ground up. Eventually he drilled hundreds of oil producers and founded an oil well supply business. A successful oil career was later enhanced by his creation of a large banking empire. Good reading with lots of reminiscences, oil and otherwise. Good reading. Cited in Swanson on page 173; Hardwicke page 44; Giddens page 88. The hardbound first edition is scarce.

**B14. (ESSAD BEY) Bey, Essad, BLOOD AND OIL IN THE ORIENT, NY, Simon and Schuster, (1932), 315pp.**

The author's father was a powerful oil-capitalist with holdings in Baku until the Russian Revolution came and "those interests were swept away". This is the story of the frantic years from 1912 to 1920, in the middle of which, young Bey had to flee his home on an odyssey covering many nations, finally ending in Germany. It is a fascinating and very exciting part of oil history viewed through a personal biography. Bey was thirteen when the revolution took place and his adventure began. Although at the time this was written, he was already an established author with a tendency to sensationalize. His description of the oil fields and oil industry intrigue, as it worked before the revolution, are very informative. Baku was probably the greatest oil producing area in the world at the time of the revolution. It never recovered its former glory. No dates, sources or notes are provided in this excellent book. Cited in Swanson on page 93. The hardbound first edition is scarce.

**B15. (ARMAND HAMMER) Blumay, Carl with Henry Edwards, THE DARK SIDE OF POWER: THE REAL ARMAND HAMMER, NY, Simon & Schuster, (1992), 494pp.**
Definitely the best characterization of Hammer, this 1992 book was written by his public relations manager of twenty-five years. Blumay gets away with writing the book by being one of the few Hammer employees who had not been required to sign a vow of silence. There are many intimate details about family and deals throughout the book as it carries the Hammer story on from the last biography (which was done by Weinberg in 1989), to his death in 1990. There are forty photos not seen in the other biographies. Blumay describes Hammer as driven to make money only for the power it could buy him. Verbatim quotes used in describing Hammer's one-man handling of the Libyan exploration program and subsequent monster discovery is particularly compelling when put in the context of its almost certain failure. Lots of details are included on Oxy's part in the Love Canal and North Sea Piper Alpha disasters. Blumay would certainly be defending his book in court if Hammer was still alive, as Weinberg did. His description of Hammer getting out of the wheelchair and dancing a jig after receiving a non-jail sentence for the Nixon campaign contribution conviction alone would have gotten Blumay in a heap of trouble. Excellent reading about an incredible man. The hardbound first edition is common.

**B16. (GIB MORGAN) Boatright, Mody C., GIB MORGAN: MINSTREL OF THE OIL FIELDS, E. Paso, Texas Folklore Society, 1945, xi, 104pp.**
When Boatright began research to do a folktale story on Gib Morgan, he quickly realized he was dealing with a real man who had become a

remarkable teller of tales rather than just a tale himself. From page 2: "What Paul Bunyan is to lumbering, what Pecos Bill is to the range cattle industry, what Mike Fink is to keelboating, Gib Morgan is to oil....he is more like David Crockett about whom has survived a large body of authenticated fact, another large body of popularly believed legend, and a third body of tall tales of the backwoods in which he reaches supernatural proportions". Morgan was born in Pennsylvania in 1842. His father was a successful businessman, indirectly involved in the incipient oil industry which began about forty miles south of his home in 1859. Gib Morgan witnessed the beginnings of the industry but decided to enlist in the Union army in 1861. His company fought in many of the major battles. During the war years, he gained fame as a story teller. Morgan's tall stories, many with moral themes, mainly satirized people and things he disliked. He eventually returned to the oil business, beginning as a tool dresser and later becoming a driller. He worked the oil fields for the next thirty years, telling his tall tales as he went. In 1894 he was admitted to a disabled soldiers home in Indiana, suffering from "chronic malarial poisoning". He kept the old soldier's home entertained with his storytelling until his death in 1909. Rumors circulated for years that he had published a book of tales, however Boatright has not been able to verify this. The last half of the book contains a compilation of over fifty Morgan tales. Overwhelming! Reprinted in 1965 by SMU Press. Cited in Swanson on page 189. The hardbound first edition is scarce.

**B17. (IDA TARBELL) Brady, Kathleen, IDA TARBELL: PORTRAIT OF A MUCKRAKER, NY, Seaview/Putnam, (1984), 286pp.**

Brady was fascinated by Tarbell as a youngster, and has now become the authority on her life. Through considerable research plus the recent discovery of numerous Tarbell letters, she is able to present a very interesting picture of this complex and extraordinary woman. Brady's view is much more objective than that of Tarbell herself as presented in her 1939 autobiography "ALL IN THE DAY'S WORK. Tarbell is given credit for swaying public opinion against Rockefeller which subsequently influenced the government action to break up the Standard Oil Trust, which finally occurred in 1911. Much detail is given of Ida's career and worldwide accomplishments in the literary field, outside of the Standard Oil history and Rockefeller confrontation. Besides her muckraking investigative reporting for McClure's Magazine, she published widely acclaimed biographies of Lincoln and Napoleon. The hardbound first edition is common.

**B18. (DEE BROWN) Brown, Dee, WHEN THE CENTURY WAS YOUNG: A WRITER'S NOTEBOOK, Little Rock, AR, August House Publishers, (1993), 223pp.**
Brown is without question one of America's favorite authors, both in fiction, such as "Creek Mary's Blood" and non-fiction such as "Bury My Heart At Wounded Knee". This is his autobiography of sorts, relating parts of his most interesting life from childhood to being a best selling author. He was born in 1908 and in the second chapter relates his early exposure to oil. His mother became postmistress of the oil boom town of Stephens, Arkansas. Dee was twelve years old when the post office became swamped with mail to and from the growing oil population. He was placed in charge of special deliveries. He gained knowledge of wildcatters, lease hounds and particularly the flim-flammers who were obviously using the mails for all sorts of schemes. He got big tips from the promoters when he delivered good news. Cox and Ladd were two of the biggest promoters of the day and Brown describes their operation as he saw it. Noting their apparent success, he even tried a mail scheme of his own. This is a delightful book, done with humor in Brown's usual captivating style. The hardbound first edition is common.

**B19. (JOHN D. ROCKEFELLER) Brown, Marcus M., A STUDY OF JOHN D. ROCKEFELLER, THE WEALTHIEST MAN IN THE WORLD: WITH HIS NAME LEFT OUT, THE HISTORY OF EDUCATION AND RELIGION COULD NOT BE WRITTEN, Cleveland, The author, 1905, 150pp.**
This book was not available for review by Neff, but it is regarded as important. Quoting from Swanson: "Title illustrates author's belief that Mr. Rockefeller is a good man". Cited in Swanson on page 160. The hardbound first edition is very scarce.

**B20. (H. L. HUNT) Brown, Stanley, H. L. HUNT, Chicago, IL: Playboy Press, (1976), 217pp.**
This fine biography was the first to take on Hunt and really delve into his lifestyle and oil dealings. Prior to Brown's work, Hunt had flooded the market with his own right-wing propaganda books while establishing himself as a brilliant and successful good-guy oilman. Brown was

probably the first to make the general public aware of Hunt's bizarre childhood, his three families, and his rough business dealings (particularly with Dad Joiner, discoverer of the East Texas oil field). Brown is very biased in his writing and finds absolutely no redeeming value in the person of H.L. Hunt. While he touches briefly on most of the undesirable aspects of Hunt's character, the attack is only pressed beyond reasonable limits on one subject. The author is convinced that Hunt drove his first son, Hassie into mental instability through psychological badgering while doing little to help his son's situation. Very good reading for anti-Hunt fans. The hardbound first edition is common.

**B21. (HOWARD BRUNSON) Brunson, Howard, THE OILMAN WHO DIDN'T WANT TO BE A MILLIONAIRE, NY, Exposition Press, (1955), 84pp.**

Howard Brunson was born in 1884, the last of thirteen children, to a family of hard-working, church-going cotton farmers. He tried farming the family acreage for a few years but crop failures discouraged him. He headed for the California oil fields at Kern River in 1907. His first job was digging ditches for Imperial Oil, then a few years later he hired on with Kern Trading and Oil Co. He worked on some of the first rotary drilling rigs at Taft and Maricopa. He returned to Louisiana in 1915 then to Oklahoma. He worked for Champlin, Gulf and Magnolia during the Desdemona boom through 1920. He tells of Jake L. Hamon building a town south of Desdemona called Jakehamon where he supposedly made a fortune selling town lots before it turned into a ghost town. He tells of federal agent two-gun Gonzaullas scouting out the boomtown of Pioneer disguised as a teamster and later returning with other agents to clean up the rampant illegal activities. He was a drilling superintendent for Humble at Corsicana 1923-24 and then back to Oklahoma where he spent ten years with W.C. McBride Company. Brunson went into business for himself in 1935 and endured five years of dry holes and debt. Finally he had some shallow oil success in Kansas and continued drilling there. As the book ends in the early 1950's, he had established some good oil production and his older son had joined him in the business. Good guys don't always finish last. Easy reading about the life of a fairly normal person. Cited in Swanson on page 174. The hardbound first edition is becoming scarce.

**B22. (ALFALFA BILL MURRAY) Bryant, Keith L., Jr., ALFALFA BILL MURRAY, Norman, University of Oklahoma Press, (1968), xiii, 287pp.**

This book largely covers Murray's political life and adds much information and analysis from the date of Hines' earlier biography. Bryant claims that: "...Murray is the most important figure in the political history of Oklahoma...no more colorful or controversial character ever strode into the state's political scene". He came on strong, politically before statehood and served as Speaker of the State House of Representatives. Later he became a U.S. Congressman, serving for two terms and then governor for four years. His handling of the oil crisis of the early 1930's by military intervention showed incredible courage....if not stubbornness. It was his usual way of handling difficult problems throughout his career, regardless of public outrage. He was for the average person and believed first of all that land was for farming and belonged in the hands of the little people, not the corporations or wealthy individuals. He contributed much to the formation of Oklahoma's political institutions. Besides colorful and stubborn, this legendary character is commonly described as flamboyant and unpredictable. Excellent reading about a popular hero. The hardbound first edition is common.

**B23. (ARMAND HAMMER) Bryson, John, THE WORLD OF ARMAND HAMMER, NY, Harry N. Abrams, Inc., (1985), 255pp.**

Two hundred fifty color photos by the author, taken of Hammer as he followed the 87 year old tycoon pursuing his social and business interests around the world. In the air and on the phone constantly, Hammer meets with heads of state and celebrities on his dizzying schedule. Introduction by Walter Cronkite. This is an impressive book prepared by an award winning photo-journalist. The hardbound first edition is becoming scarce.

**B24. (WILLIAM BUCKLEY, SR.) Buckley, Priscilla L. and William F., Jr. (editors), W.F.B., AN APPRECIATION, NY: Privately Printed, 1959, xii, 285pp.**

Many of William F. Buckley, Sr.'s friends and some of his ten children contributed to this combination history, biography and eulogy of him. Buckley was born in 1881 and died in 1958. The book is not very objective but shows that he was a wildcatter by philosophy, preferring the oil chase, even to the oil finding. A lawyer with a nose for oil, he was described as instinctive in his pursuit of it. He had many very successful oil ventures from Mexico, south to Venezuela and from the U.S. north to Canada. Risk-taking wildcatters with conservative politics is a more common phenomenon than one might expect. The hardbound first edition is common.

**B25. (H. L. HUNT) Burst, Ardis, THE THREE FAMILIES OF H.L. HUNT, NY, Weidenfeld & Nicholson, (1988), xiii, 256pp.**
At the publishing of this work in 1987, the crude oil market had crashed and the oil business was going down hill fast. The Hunt companies were in bankruptcy, although the hammer had not fallen and the extent of the losses could not even be guessed at. The book concentrates on H. L. Hunt's family relationships, first with wife Lyda Bunker and the seven children then with Frannia Tye (Lee) and the four children, and last Ruth Ray and the four children. From the dust jacket: "...a story that spans over sixty years of episodes and incidents that make the Ewings of Dallas look like pale imitations". The author recognizes Hunt's character flaws but does not chastise him. A sympathetic look at a pathetic family. The hardbound first edition is common.

**B26. (WILLIAM BURTON) Burton, William M., WILLIAM M. BURTON: A PIONEER IN MODERN PETROLEUM TECHNOLOGY, Cambridge, MA, Cambridge University Press, 1952, 97pp.**
Burton was born in 1865, and with a chemistry Ph.D. from Johns Hopkins, went to work for Standard Oil. He quickly became a gasoline expert. As automobiles became important so did his research in the field where he became known as "The Father of Modern Refining". The Burton Still was widely used by industry, although developed while he was at Standard. Quoting from Swanson: "A record of his accomplishments with Standard Oil Company of Indiana as pioneer in oil cracking and awards bestowed upon him". He received the API Gold Medal Award in 1947 for his numerous achievements and the citation is reproduced along with others in this booklet. Cited in Swanson on page 174. The softbound first edition is becoming scarce.

**B27. (DEWEY BYERS) Byers, D.E.L., THE LONE GAMBLER, North Dakota, Privately Published, (ca 1974), xix, 163pp.**
Byers was born in Kansas in 1896. He relates stories of oil and family which cover most of the next eighty years. He was involved in cable tool

drilling in 1913 in the Augusta and Eldorado areas of Kansas, also several early Oklahoma fields. He was a driller for wages and later became a lease broker, then a well promoter. He began buying mineral rights all over the U.S. based on speculation, and finally late in life became a successful player when one of his many mineral spreads became very valuable. The hardbound first edition is common.

**B28. (DAVID BYRD) Byrd, David Harold, I'M AN ENDANGERED SPECIES: THE AUTOBIOGRAPHY OF A FREE ENTERPRISER, Houston, Pacesetter Press, (1978), vii, 108pp.**

Born in Detroit, Texas in 1900 to a farming family, Byrd had finished two years of geology study at University of Texas when he decided to leave school for work in the oil fields. He caught on fast and after three years of working for oil companies, became a consulting geologist and drilling contractor in Brownwood, Texas. The nickname "Dry Hole Byrd" was gained for his being involved in eighty-six straight dry holes (seems like an impossible feat). However, he shook that name and had considerable success from 1928 on, primarily in west and central Texas. An oil pioneer in the East Texas Field, at one time he held 50,000 acres of leases in the Kilgore area under the Byrd-Frost, Inc. organization. Byrd proceeded to diversify and built a solid business career. The hardbound first edition is common.

**B29. (HERBERT DOW) Campbell, Murray, and Harrison Hatton, HERBERT H. DOW: PIONEER IN CREATIVE CHEMISTRY, NY, Appleton Century Crofts, 1951, xi, 168pp.**

Dow was born in 1866. His ancestors settled in Massachusetts in 1637. His dad was a master mechanic, also an inventor and engineer. It must have rubbed off. Young Dow's innovative work in extracting bromine from the salt wells of Michigan became the prelude to his giant worldwide diversified chemical company. The company which bears his name also became a major factor in oil exploration and development. Dow is now generally regarded as the creator of the modern American chemical industry. The hardbound first edition is becoming scarce.

**B30. (ELLA CARL) Carl, Ella Lane, LETTERS OF A TEXAS OIL DRILLER'S WIFE, NY, Comet Press Books, 1959, 222pp.**

The book consists of four years worth of letters from Mrs. Carl to her friend Stella, describing the hard times in the oilfield boomtowns as she and her kids and husband (a driller for Prairie Oil & Gas) bounce from town to town. The correspondence starts from Borger, Texas in 1927 and ends at Pampa, Texas in 1931. She gives good color descriptions of the barren West Texas and Panhandle areas, while reflecting back on the lush and beautiful Washington state home she dreams of returning to (Stella lives there). Many oilfield accidents and natural disasters are described as the family survives the difficult life. One of the top oil biographical works. Cited in Swanson on page 186. The hardbound first edition is scarce.

**B31. (JOHN D. ROCKEFELLER) Carr, Albert Z., ROCKEFELLER'S SECRET WEAPON, NY, McGraw-Hill Book Co., (1962), x, 383pp.**
Carr is a historical writer and had previously been an economic advisor on the White House staff. He expands on his thesis that the railroad oil tank car was the most devastating weapon in Rockefeller's arsenal. When the railroads gave up control of the tank cars, it was all over for the industry and Rockefeller was in charge. Carr gained access to the archives of the Rockefeller owned, Union Tank Car Company and found that through manipulation and organization they controlled an entire sphere of the oil industry. The interwoven history of Union Tank and Standard Oil is fascinating; and their problems, politics and accomplishments throughout the years are detailed here. Union Tank reserved the right to footnote Carr's book where interpretations disagreed. The hardbound first edition is common.

**B32. (CULLEN DAVIS) Cartwright, Gary, BLOOD WILL TELL: THE MURDER TRIALS OF T. CULLEN DAVIS, NY, Harcourt Brace Jovanovich, (1979), x, 401pp.**
Cartwright followed the story of Priscilla and Cullen Davis, writing several articles as the saga unfolded. T. Cullen Davis is the son of multimillionaire oil man, Ken Davis, who built a major conglomerate from nothing. During the height of his power, he owned Kendavis, Inc. which was a private corporation, controlling such entities as Great Western Drilling, a west Texas oil producer and drilling contractor. He also owned Loffland Brothers Drilling which was the largest international drilling contractor, along with many others. Ken Davis's two sons took over the business. Cullen, who was named for the famous oil finder, Hugh Roy Cullen, dominated the business. Against the wishes of almost

everyone he married a flashy blonde named Priscilla. She was from the wrong side of the tracks. A protracted divorce action was in progress when a masked man entered the Davis mansion, shooting Priscilla and murdering her daughter and boyfriend. Priscilla and almost all evidence pointed to Cullen as the perpetrator, but thanks to his brilliant legal defense headed by Racehorse Haynes, Cullen was acquitted. The sensational story of Texas society run amok is vividly portrayed here as Cullen was considered the richest man ever to be tried for murder. The hardbound first edition is common.

**B33. (DON CHAMPION) Champion, Don F., GROWING UP IN RANGER: BOOMTOWN U.S.A., Texas, Privately Published, 1988, 42pp.**
Champion was six years old when his father was the driller on the Ranger Oil Field discovery well in 1916. Champion describes his childhood through 1934 in anecdotal style as the oil boom developed and quickly began a rapid decline. Although he gives good color commentary on the times, the reality of the situation shows through and is somewhat depressing. The book leaves no doubt that Ranger was a classic flash-in-the-pan boomtown. The softbound first edition is common.

**B34. (DON CHAMPION) Champion, Don, RUMBLING IN THE EARTH: RECOLLECTIONS OF A LIFETIME IN THE OIL PATCH, Austin, TX, Nortex Press, (1992), v, 109pp.**
Champion wrote this book in a more polished style than his previous two. No less than ninety-two vignettes are presented, which cover his father's early drilling career and how the family coped with the hardships. Much on the Ranger field and his Dad's part as driller on #1 McKlesky, the discovery well there. Most stories are childhood remembrances from the 1920's and 1930's, growing up in the oilfields. Champion's writing shows a great feeling for the times and a sensitivity to the players. The title chapter was named from a supposed geologist called "Doc" who worked with Champion's dad and had a strange method of finding oil prospects. He would stand on a spot that looked likely and stare into space, north, south, east and then west. When he felt rumbling in the earth, he knew there was oil below and that he had located another perfect drillsite. The softbound first edition is common.

**B35. (DON CHAMPION) Champion, Don, WOOD DERRICKS, IRON MEN & GOLD WOMEN, Waco, TX, Privately Published, (1987), vii, 106pp.**
Champion gives his personal and factual account of surveying in the oil fields of Texas and Louisiana, beginning with childhood stories of the Ranger oil boom in 1917. His story ends with a claim surveying job in Alaska in 1964. The story in between is filled with an array of almost unbelievable characters doing interesting things as well as their oil industry jobs. Colorful narration with good photos. The hardbound first edition is common.

**B36. (HENRY FLAGLER) Chandler, David Leon, HENRY FLAGLER: THE ASTONISHING LIFE AND TIMES OF THE VISIONARY ROBBER BARON WHO FOUNDED FLORIDA, NY, Macmillan Publishing Co., (1986) x, 324pp.**
The first eighty pages are dedicated to Flagler's early years and the Rockefeller connection. After generating incalculable wealth with Standard Oil, Flagler took on the task of developing Florida. The remainder of the book covers thirty years of his Florida experience. He poured millions of dollars into the state and he was the driving force in development of St. Augustine, Miami, Palm Beach and most of the east coastal cities. His East Coast railroad and Key West railroad projects which connected the main cities were an awesome project and performed almost single handedly. Flagler built the cities, the schools, the libraries and whatever else they needed. Chandler presents the Flagler story in an easy reading novel-like format. There are plenty of fascinating characters and events in this well constructed book. The hardbound first edition is common.

**B37. (ABBY ROCKEFELLER) Chase, Mary Ellen, ABBY ALDRICH ROCKEFELLER, NY, The MacMillan Company, 1950, 159pp.**
Chase is a well-respected, much published biographer and novelist. This is her tribute to the wife of John D. Rockefeller, Jr. There is serious doubt that Junior would have survived under the burden and pressure of purposely disposing of his Dad's money, without the support of his wife. His job was to prudently give away a billion dollars of family money. She must have been very understanding. Abby was very bright and organized and became a stabilizing influence as the matriarch of a large family

dynasty. She was born to a wealthy family in 1874 and married John, Jr. in 1901. They had six children who came to dominate the family businesses as they spread themselves through U.S. business, politics and industry. Nice family. The hardbound first edition is common.

**B38. (MICHAEL CHENEY) Cheney, Michael Sheldon, BIG OIL MAN FROM ARABIA, NY, Ballentine Books, (1958), 282pp.**
Secondary titles are: FROM CAMEL BACK TO CADILLAC - OR THE AMAZING ADVENTURES OF ARAMCO, THE AMERICAN OVERSEAS OIL COMPANY THAT IS TRANSFORMING SAUDI ARABIA. The titles tell all, or most of it. Cheney spent ten years working in public relations for oil and construction companies in the Middle East. He describes early (1948-1958) exploration and development of Aramco's gigantic concession along with entertaining stories of the events he witnessed as tenth century culture met twentieth century culture in the form of oil men and their oil industry. Cited in Swanson on page 58. The hardbound first edition is common.

**B39. (ROBERT WELCH) Clark, James A., A BIOGRAPHY OF ROBERT ALONZO WELCH, Houston, Clark Book Co., Inc., (1963), ix, 153pp.**
Welch was born in South Carolina in 1872 and spent an unpleasant childhood growing up under reconstruction in the South. His parents owned a small plantation. Fortunately at age fourteen, Welch was allowed to attend a school in Houston which was run by one of his cousins. He loved the growing city of 15,000 people and the strict discipline of the school. He hired on as a bookkeeper for Bule Company, a seller of paints and insecticides and stayed with them for thirty-six years, retiring as Secretary- Treasurer and part owner. The 1901 Spindletop discovery at Beaumont intrigued him and he visited the site several times. A quick trade of one acre in the field within a week of acquiring it netted him $15,000 and he was hooked on oil. He studied geology and oil-finding for clues as to how to locate more. While fishing with a buddy at Goose Creek, he observed gas bubbles at the water surface and envisioned oil below. He bought a lease from Mr. Gaillard and brought in three partners to form Goose Creek Oil Co. Drilling started two years later, in 1907, after Goose Creek sold their operating rights to Farish, Sharp and Producers Oil for cash plus thirty-seven percent of the oil. The group drilled fifteen dry or marginal producers and turned the lease back to

Welch. Finally in 1916 American Production Company hit a 10,000 barrel a day well at 2,030 feet. Goose Creek turned out to be an unusually large field and became the proving ground for several innovations such as the rock bit, cementing, and oil field electrification. Welch died in 1952, leaving considerable amounts of money to his employees and several million more to his foundation which had been designed to further chemical research in Texas. Good reading about an impressive oil pioneer who had little public acclaim. The hardbound first edition is becoming scarce.

**B40. (WALTER LECHNER) Clark, James A., AN OILMAN'S OILMAN: A BIOGRAPHICAL TREATMENT OF WALTER LECHNER, Houston, TX, Gulf Publishing Company, (1979), vii, 143pp.**
Clark and his editor Judith King prepared a very interesting biography of a true oil pioneer. Lechner was born in 1890 in the East Texas town of Terrell. He worked his way through college and moved from job to job including a short term in the oil fields before serving in WW I. In 1919 he went to work in the oil fields for Currin and Kean at the booming town of Burkburnett. He dabbled in a few oil deals and broke about even. In 1923 with his partners in Loutex Company he drilled #1 J.J. Moore the wildcat discovery well for Sharon Ridge Field near Snyder, Texas. Lechner went on to direct participation in the #1 Lathrop, a significant confirmation well at the East Texas field. He went into politics and government service but continued to invest in oil after his East Texas field development. Lechner's personal diary helped Clark prepare much of the early Texas oil history, not only for this book, but for several other important works. Much good Texas history. The hardbound first edition is common.

**B41. (GENERAL) Clark, James A., FOUNDERS OF THE OIL INDUSTRY, Noplace, Schlumberger, 1967, 108pp.**
A weekly calendar for the year 1967 with fifty-two profiles of oil industry leaders from the earliest days in Pennsylvania to the oilmen of the early 1960's. Clark gives much historical detail through biographical work on the innovators and promoters as well as the researchers. The softbound first edition is becoming scarce.

**B42. (MARRS McLEAN) Clark, James A., MARRS McLEAN: A BIOGRAPHY, Houston, TX, Clark Book Co., 1969, xi, 194pp.**
Author, journalist, historian Clark calls Marrs Mclean the second prophet of Spindletop for good reason. Unlike Patillo Higgins, the first prophet, who had a vision of Spindletop oil, McLean developed a theory of oil being trapped in sands which flank the salt domes of the Gulf. His idea came from much experience in dealing with oil reservoirs and their varied nature. From the foreword: "He was one of the important pioneers in the modern petroleum industry. He had an idea, and no one could divert him from it". McLean was born in Sherman, Texas in 1883. His father was an attorney. Marrs, as a teenager was exposed to the Spindletop in 1901 as his father was helping in some of the multitude of litigation accompanying that boom. He decided to be an oilman on the spot. McLean bailed out of law school on the eve of his final exam to go into the world of business. He succeeded in most everything he tried and began investing in oil drilling around the edge of Spindletop with a man named Benckenstein. They established some flank producers at another pool called Vinton and Marrs decided to promote the idea back at Spindletop and High Island. The road to success was twelve years long and consisted of countless dry holes. Young finally drilled the Spindletop flank discovery on McLean leases in 1925, thus reviving the dead dome. New production exceed that produced from the dome previously. Two years later McLean established commercial oil production on the High Island dome. He controlled most of the leases and promptly drilled a number of 10,000 barrel per day producers, establishing another great field. He died in 1953, remembered as a top oilman and a powerful behind the scenes political figure. The slipcased hardbound first edition is scarce.

**B43. (ERNEST THOMPSON) Clark, James A., THREE STARS FOR THE COLONEL: THE BIOGRAPHY OF ERNEST O. THOMPSON, FATHER OF PETROLEUM CONSERVATION, NY, Random House, (1954), 265pp.**
The book opens with the East Texas Oil Field discovery scenario and the havoc it wrought within a short eighteen months, collapsing the oil economy. Ernest was born in north central Texas in 1892, but was raised in the small town (1500 people) of Amarillo. He was so industrious that he hardly had a childhood. He sold newspapers and later cars prior to going to the University of Texas in pursuit of law. He had been self

supporting since teenage years. At Austin he fortunately met Judge Gaines, one of the most influential men in Texas and became his personal assistant while still in school. Gaines became his mentor and greatly helped him mature in law and life. Thompson received his law diploma while he was training for WW I along with the rest of his law class. By the end of the war in 1919 he had gained the rank of Lt. Colonel from battlefield promotions. He successfully practiced law and business and was elected mayor of the now booming Amarillo in 1928. In 1932, Governor Ross Sterling called Thompson and asked him to take the place of Pat Neff who had resigned from the Railroad Commission. The state was desperately in need of someone who had great courage and the scruples to handle the job of controlling a petroleum industry which had gone wild. The state guard had already been called out to control the producers in the East Texas Field. Thompson accepted and the rest is history. Thompson turns out to be one of the most important oil men because of his expeditious handling of the new idea of conservation and prevention of waste in the oil fields. He was instrumental in forming the Interstate Oil Compact Commission and with that tool squelched the feds attempt to regulate or otherwise control U.S. oil output. This is an excellent biography by Texas' honored historian. Cited in Swanson on page 174. The hardbound first edition is becoming scarce.

**B44. (CHARLES PAGE) Clark, Opal Bennefield, A FOOL'S ENTERPRISE, Sand Springs, OK, Dexter Publishing, 1988, 288pp.**
Charles Page was a very successful oilman in the Tulsa, Oklahoma area. He was born in 1860 and had an up and down career of high risk ventures in real estate, mining and eventually oil. He was drilling as a contractor during the Red Fork and Glenn Pool booms, taking well interests for fees to great success. Having been orphaned at an early age, he became obsessed with the mission of helping orphans. He did an incredible amount of work in that and other civic and philanthropic areas. The author was seventy-six when she completed this labor of love. She was one of Page's orphans. The hardbound first edition is common.

**B45. (CULLEN DAVIS) Cochran, Mike, TEXAS VS. DAVIS, NY, The Bobbs-Merrill Co., (1980), 373pp.**
Touted as the only complete account of the bizarre Davis murder case. From the dust jacket: "On the night of August 2, 1976, a twelve year old girl was gunned to death in the basement of a six million dollar mansion

in Fort Worth. A short time later three other people were shot at the mansion: one died; one was paralyzed for life; one escaped with a critical chest wound. The survivors named Thomas Cullen Davis, heir to a billion dollar financial empire, as the gunman". Can you believe Davis got off and the real perpetrator was never found? The hardbound first edition is common.

**B46. (JOE BAILEY) Cocke, W.A., THE BAILEY CONTROVERSY IN TEXAS WITH LESSONS FROM THE POLITICAL LIFE-STORY OF A FALLEN IDOL, San Antonio, TX, The Cocke Co., 1908, xxv, 1012pp.**

This massive two volume work was written, compiled and published by William A. Cocke, arch enemy of U.S. Senator from Texas, Joe W. Bailey. Cocke was also a Senator. Bailey was a corrupt politician who had several criminal and ethical scrapes in Mississippi before moving to Texas in 1885. He continued to skirt the law and through his charisma alone was elected to the U.S. Senate in 1891. At the writing of this book, he was still in office. Evidently he was owned entirely by Standard Oil who had been banned from Texas by state anti trust law. Bailey was instrumental in getting Waters-Pierce Oil Co. (a Standard subsidiary) reinstated and allowed to do business in the state in 1900. He took considerable money in bribes and lied about it. What really set Cocke off was the fact that Bailey was able to maintain support across Texas, even with these revelations. Cocke and half of the senate were distressed that Bailey was gearing up for his re-election campaign. Countless letters, speeches and public accusations of Bailey are printed in these volumes along with Cocke's interpretation of them. Cocke left no stone unturned in his attack. He was a third generation legislator for Texas and mild-mannered attorney. Bailey almost sent him over the edge. It is comforting to see that nothing changes in Texas politics. The hardbound first edition is scarce.

**B47. (ROCKEFELLER FAMILY) Collier, Peter and David Horowitz, THE ROCKEFELLERS: AN AMERICAN DYNASTY, NY, Holt, Rinehart and Winston, (1976), 746pp.**

In preparing this major biography of the Rockefeller family, authors Collier and Horowitz were allowed unprecedented access to private Rockefeller papers. Covering four generations in this epic work, they had numerous interviews with the four living sons of John D., Jr., as well as their sons, inheritors of the legacy. This is an impartial balance between the previous books attacking the Rockefellers by Winkler and Flynn and those with excess sympathy for the Rockefellers by Nevins and Fosdick. A rehash of the lives of John D., Sr. and John D. Jr. shortly gives way to concentration on the five sons of John D., Jr. From the dust jacket: "Here then is Laurence, clever and charming as a youth, burnt out and cynical by middle age; Winthrop, shy, awkward, the black sheep who finally made a mark for himself in the eyes of everyone but his family; John D., III, introverted and anxious even after years of proving himself; David, a man on the move who took the nation's front-ranking bank and made it number three; and Nelson, ambitious, aggressive, the brother who broke the unwritten family code". The last hundred pages are devoted to John D. Sr.'s great grandchildren who appear to be in a state of disarray as the dynastic cohesion is breaking down. This is an excellent family biography. The hardbound first edition is common.

**B48. (WILLIAM CONNELLY) Connelly, William L., THE OIL BUSINESS AS I SAW IT: HALF A CENTURY WITH SINCLAIR, Oklahoma, University of Oklahoma Press, (1954), xiii, 177pp.**
Born in Cleveland, Ohio in 1873, Connelly's father owned a supply store in the oil boom town of Petrolia as the oil business was beginning. By 1896 Connelly had established his own oil production in Ohio, and by 1906 he had found oil in Osage County, Oklahoma. There, he met Harry Sinclair and they became close friends, shortly afterward establishing the predecessor to Sinclair Oil and Gas Company. Many interesting worldwide adventures are related in his career with Sinclair and its many subsidiaries. Connelly was in charge of Mammoth Oil, with which he spent a good part of six years in Casper, Wyoming developing and overseeing the Teapot Dome field. Sinclair had gained the federal lease under suspicious circumstances (not mentioned by Connelly). Connelly describes one of the producers as making 2,000 barrels per hour, the largest he had seen. Early on, the feds began pressing a suit to declare the lease illegal and in the midst of this Connelly had a nervous breakdown. He recovered and continued work in the area until the feds prevailed and the lease was declared illegal and invalid in 1927. A long and interesting career is well described. Cited in Swanson on page 174. The hardbound first edition is becoming scarce.

**B49. (VARIOUS OILMEN) Connor, Seymour V. (editor), BUILDERS OF THE SOUTHWEST, Lubbock, TX, Texas Technological College, 1959, xiii, 321pp.**

This book consists of seventy-two biographical sketches which were prepared by twelve authors. The subjects are a select group of West Texas businessmen, farmers, ranchers and educators. A screening committee in the late 1950's picked these people from the past and present. The surprisingly few oil men include: Amon Carter, Paul Cates, S.E. Cone, M.C. Lindsey, and Paul Moss. There is also a good company history of Pioneer Natural Gas. This is an excellent reference of important West Texas figures. A 1957 abbreviated version of the book including twenty of the profiles with 77pp. is also available. The hardbound first edition is becoming scarce.

**B50. (ARMAND HAMMER) Considine, Bob, THE REMARKABLE LIFE OF DR. ARMAND HAMMER, NY, Harper & Row, (1975), xii, 287pp.**

The title hardly does justice to the man. The author just breezes through the life of this incredible person. The work is generally believed to have been heavily edited by Dr. Hammer, if not dictated by him. There are good early photos, indexes and appendixes. Hammer was born in 1897 in New York to Julian and Rose Hammer. What followed is generally known by the public and is now an important part of worldwide history. A business genius who became a millionaire in pharmaceuticals while still in college, he let no opportunity for profit get by him. He graduated from Columbia Medical School, but plunged immediately into the business world. By age 58 (1956) he was a multimillionaire in whiskey, cattle, etc. and in severe need of tax relief. A friend recommended oil drilling as the best tax shelter (still is in 1995). In 1956 you could explore for oil with the government picking up 90% of the cost if you were in the top bracket. The friend introduced him to Occidental Petroleum Company, which had a net worth of less than $50,000 at the time. Hammer decided to make a loan to Occidental in the amount of $50,000 for which he would gain one half interest in the company. The money was provided specifically to drill two of their best wildcat prospects. Obviously he expected both to be dry and he would get a charge off against income. However, one of the two discovered a good field north of Los Angeles and Oxy's stock increased ten fold. Hammer was instantly hooked on oil. Shortly he was running the company with the iron hand (velvet fist) that he was famous for and continued to do so until he died in 1990. Incredible oil successes made on brilliantly acquired concessions aside from the L.A. basin were Nigeria, Libya and the North Sea, which gained Oxy major oil company status. All of this was accomplished within fifteen years of Hammer's takeover. The book contains many Hammer quotes and verbatim conversations concerned with important oil deals up to 1975. Excellent book on a unique businessman. British version retitled, LARGER THAN LIFE. The hardbound first edition is common.

**B51. (JOHN BUNN) Cornell, Doris, OIL FEVER: THE BIOGRAPHY OF JOHN R. BUNN, Oklahoma, Oklahoma Heritage Association, 1985, 181pp.**

Probably the most interesting biography of the Heritage Association's "Trackmaker Series", this one was edited by Kenny Franks. Bunn was born in Ardmore, Oklahoma in 1900 and spent sixty plus active years as a geologist, wildcatting in the U.S. and Mexico. His main contribution to the history and economics of the oil business was instigating the #1 Edwards, one of the discovery wells of the giant Oklahoma City Field. He was also essentially responsible for inventing the first hydraulic drilling rig (unusual accomplishment for a geologist). Several of Bunn's exciting oil plays are related, the good along with the bad. That is the way it is supposed to be done. Great geologist and very good reading. The hardbound first edition is common.

**B52. (ROCKEFELLER FAMILY) Corner, George W., A HISTORY OF THE ROCKEFELLER INSTITUTE: 1901-1953 ORIGINS AND GROWTH, NY, Rockefeller Institute Press, 1964, 635pp.**

Commemorating fifty years of Rockefeller Institute's part in scientific progress in the U.S. The Institute is essentially a medical research facility which is highly funded by the Rockefeller family and does basic and applied biological research. It is loosely associated with Johns Hopkins and has a low profile, except within the elite academic biologists ranks, who regard the Institute highly. Basically, the Institute was the idea of Reverend Frederic Gates, Rockefeller's constant advisor. It was organized by William Welch and originally funded by Rockefeller. It has a large graduate level facility in New York, including a hospital. The hardbound first edition is scarce.

**B53. (JIM HOGG) Cotner, Robert C., JAMES STEPHEN HOGG: A BIOGRAPHY, Austin, TX, University of Texas Press, 1959, xxvi, 617pp.**

Born near Rusk, Texas in 1851, he became a printers assistant, tried sharecropping and by the age of twenty printed and edited his own newspaper. He was admitted to the bar in 1875. Following political life as his grandfather and father before him, he went into law enforcement, first as a Justice of the Peace. He rapidly rose from District Attorney to Attorney General and then served as governor from 1891 to 1895. Chapter 19 discusses the Hogg- Swayne Syndicate. Hogg heard about the Spindletop discovery and immediately went there and began buying and selling oil and gas leases. In 1901 he went into partnership with Swayne an old political friend with considerable oil experience. The Hogg-Swayne Syndicate acquired a small lease tract from Guffey & Galey, right in the middle of the dome, and proceeded to drill a successful producer. Just why Guffey & Galey sold them the tract is not clear. Possibly a political motive. One hundred and twenty excellent producers were drilled on this tiny fifteen acre tract, causing the price of crude oil to drop to as low as three cents a barrel. Hogg-Swayne merged into Cullinan's Texaco in 1902 essentially ending Hogg's involvement in Spindletop. He died in 1906, but his oil career did not die with him. A speculative land purchase he had made years earlier came in as a significant oil discovery in 1917 as the West Columbia salt dome developed into an important oil field. Another oil-finding lawyer. This biography relates much political and economic history of Texas around one of the most important figures in Texas history. Cited in Swanson on page 174. The hardbound first edition is becoming scarce.

**B54. (JACK COX) Cox, L. W. "Jack", MY LIFE & HOW I LIVED IT: A TRUE STORY, Waco, TX, Texian Press, (1984), vii, 115pp.**

Born in El Paso, Texas in 1909, Cox grew up in the Eastland and Ranger boomtowns. He worked for various merchants on through the East Texas boom and relates stories about the people and the area. There is a gruesome discussion of the New London school disaster of 1937. He owned a store at nearby Overton at the time a natural gas explosion killed over two hundred children at the school. Cox and his wife were so shaken by the tragedy that they moved back to the Rio Grande valley and later to Waco, Texas. From the mid-1950's to the mid-1980's he suffered ill health in several forms, largely due to alcohol, which habit he had quit by the time he wrote this book. The book mainly contains firsthand glimpses of oil history as it unfolded in central and east Texas. The softbound first edition is becoming scarce.

**B55. (ROY CROSS) Cross, Roy;, FROM A CHEMIST'S DIARY., Kansas City, MO, Kansas City Testing Lab, (1943), 315pp.**
Cross was certainly a versatile chemist as his lab tested everything imaginable. His stories of idea-people who came to the lab with chemistry-related inventions in hopes of getting a marketable project or patent are often humorous, but also interesting. He personally filed 477 patent applications between 1917 and 1941, 86 of which were approved. Thirty of the Cross patents dealt directly with petroleum or drilling ideas. Quoting from Swanson: "Collection of items from diary of author, who devoted forty years to work as industrial and research chemist. Discusses results of cracking, with particular attention to Cross cracking process, and tells of testing anti-knock qualities of gasoline". Excellent presentation of stories by an enthusiastic and brilliant chemist. Cited in Swanson on page 103. The hardbound first edition is becoming scarce.

**B56. (VARIOUS OILMEN) Crowell, Evelyn Miller (editor), MEN OF ACHIEVEMENT: TEXAS EDITION, Dallas, TX, John Moranz Associates, 1948, 288pp.**
This large book includes a drawn portrait and biographical sketch of some two hundred prominent men of the 1940's in Texas. Of course, many oil personalities among them, being forty-five in number and including Everette DeGolyer, Frank Champion, Walter Lechner, Glenn McCarthy, Michael Halbouty, along with many other big time operators. For some reason a number of prominent characters were not included such as H.L. Hunt, Sid Richardson, W.A. Moncrief and Clint Murchison to name a few. Classy book. The hardbound first edition is becoming scarce.

**B57. (A. D. LLOYD) Cubage, Granville, OIL: A HANDBOOK FOR REFERENCE. A STUDY OF THE LLOYD OIL CORPORATION OF FT. WORTH, TEXAS, Fort Worth, TX, The author, 98pp.**
This book was not available for review by Neff, but it is regarded as important. Quoting from Swanson: "Career of Adelbert Durham Lloyd; his theory on origin of petroleum and geological examination of East Texas area for C. M. "Dad" Joiner; reproduction of map showing

recommended well location. Discusses Lloyd Oil Corporation". Cited in Swanson on page 169. The softbound first edition is very scarce.

**B58. (JIM CUMMINGS) Cummings, Jim, JIM CUMMINGS, San Marcos, TX, The Author, (1968), 99pp.**
Quoting from the cover: "He took the Pipe Line Industry from behind the mule's tail and put it on wheels". The first twenty pages of this biography concern Cummings childhood and growing up in rural Kansas. The next fourteen pages were written by author Wallace Davis and tells the story of how farmer Cummings got into the pipeline business and how he came to dominate the industry. The first 12" pipeline from Teapot Dome was being laid by Cummings' farm where he observed the inefficient methods being used. He decided to help the folks out. He created the world's first bull dozer and proceeded to back-fill their trench with incredible speed. From the first innovation in 1923, he designed, developed, manufactured and sold more pipeline equipment than anyone. Ever practical and efficient, Cummings later formed his company CRC to handle his pipeline interests. He invented several agricultural items as he continued farming in a big way. The softbound first edition is becoming scarce.

**B59. (RALPH DAVIS) Davis, Ralph E., STORIES OF NATURAL GAS, U.S, Privately Published, (1964), vi, 283pp.**
The introduction and editing were done by James A. Clark, Texas' most respected oil historian. Clark claims that Ralph Davis was one of the important Americans of the times and explains why. Davis was one of very few experts on natural gas reserves, pipeline performance, market and economics from the 1920's through the early 1960's. He was born in Virginia in 1884, interestingly the year that natural gas was first piped to Pittsburgh for use as industrial fuel. Davis graduated with honors from Wisconsin with a degree in Engineering Geology and in 1906 worked for mining companies. In 1917 he began consulting in oil and gas. Two of his classic papers are published in the book. One written in 1926 was a landmark presentation to the AIME which essentially caused the U.S. to begin turning from coal to natural gas for energy. The other paper written in 1964 at age seventy-three continued to tout natural gas economics and supply. Many good stories from his intimate knowledge of gas as it was initially supplied to major cities, and several foreign countries are included. The hardbound first edition is common.

**B60. (GLENN McCARTHY) Davis, Wallace, CORDUROY ROAD: THE STORY OF GLENN H. McCARTHY, Houston, TX, The Anson Jones Press, 1951, 282pp.**

Glenn McCarthy was born near the Spindletop field at Beaumont in 1907. He got his first job in the oil fields at age six as a waterboy. He was very industrious and started a few small businesses, held several interesting jobs and pursued a college education at Rice. He continued to start small businesses, one after the other, often selling out, sometimes at a profit and sometimes not. He made and lost a good deal of money as a drilling contractor. In 1935 he teamed up with Michel Halbouty (already a well known and successful geologist). McCarthy's trend of rich, broke, rich, broke, continued for several more years. He discovered several important Texas oil fields but made most of his millions in development drilling at the large Conroe, Anahuac and Hankamer fields. Quoting from Swanson: "Life of man hailed as King of the Wildcatters". Probably his biggest claim to fame was building the monumental Shamrock Hotel in Houston (a lifetime dream). The Shamrock was the finest luxury hotel in Texas throughout the 1950's and into the 1960's. McCarthy supposedly had no part in this biography, but it appears that someone looked after his interests as there is not one bad word about him. Very easy reading, but could have been much more interesting if the conflicts and controversies of this colorful character were examined in a little detail. Cited in Swanson on page 174. The hardbound first edition is scarce.

**B61. (GEORGE DEAN) Dean, George F., UNDYED YARNS, NY, Vantage Press, (1963), 186pp.**

The author was born in London in the late 1800's and served in the British Army in 1915. Barely surviving that, he spent thirty years in South America, mostly as an executive with oil companies in Venezuela, Ecuador and Colombia. His stories are of his adventures in the giant oil fields there from 1923, as he encountered one unusual character after another. Thirteen short stories are rendered with dry British wit, mostly humorous, but some frightening. Very entertaining reading. The hardbound first edition is becoming scarce.

**B62. (J. PAUL GETTY) DeChair, Somerset (editor), GETTY ON GETTY: A MAN IN A BILLION, Cassell, London, (1989), 170pp.**

At the publishing of this biography there had been at least a dozen "Getty" books on the market, over half of which were written by him. The author of this book states that it is something different in that his interview-conversations with Getty were taped and are presented word-for-word allowing us to see the real Getty. Getty authorized the book which was prepared from conversations in 1973 and 1974. Getty, the man, really does come to life in these pages. One excellent chapter concerns his roustabout and roughneck days in the Oklahoma oil fields of 1904. The hardbound first edition is common.

**B63. (NELSON ROCKEFELLER) Desmond, James, NELSON ROCKEFELLER: A POLITICAL BIOGRAPHY, NY, The MacMillan Company, (1964), xv, 330pp.**
Desmond's timing on the biography of Rockefeller was good in that Nelson's aspiration for the job of U.S president was becoming very well known as he sought the 1964 Republican nomination against the conservative Goldwater. An excellent story of Rockefeller, the first half describes his family life and upbringing in a world of money and power without public politics. The last half of the book describes how Rockefeller became a political animal, against the family wishes and tradition. Rockefeller almost made a run for the Republican nomination against Nixon in 1960 but for several reasons (not made clear) he dropped out of the race early. On reflection, after noting the narrowness of the subsequent Kennedy win, many observers still believe that Rockefeller gave up his best chance to become president at that time. The Nixon people wanted Rockefeller to go on the ticket as vice presidential nominee so as to balance the conservative Nixon with a perceived liberal Rockefeller. Rockefeller was indignant and rejected the offer. Desmond briefly mentions the unfortunate accidental death of Nelson's son, Michael as well as the divorce from Mary Clark and later marriage to Margaretta (Happy) Murphy as being his most difficult times. Generally, the book is up-beat and as the title says, "Political". No sources, notes or bibliography makes this work appear to be Rockefeller sponsored and edited. The hardbound first edition is common.

**B64. (ROGER SHERMAN) Destler, Chester McArthur, ROGER SHERMAN AND THE INDEPENDENT OILMEN., Ithaca, NY, Cornell University Press, (1967), ix, 305pp.**
Roger Sherman was born in Tennessee in 1839. His father was a medical doctor and Roger was raised on a good sized Mississippi River plantation in Arkansas. He studied law and passed the bar in 1860. He fought for a

year with the Tennessee cavalry but after being captured and released, became dismayed at the South's situation and moved to New York to continue his law practice. His property had been devastated in the south and he was rejected as a rebel in the North. Discouraged, but not out, he made his way to the fast action at Pithole, Pennsylvania, where the drilling boom was in full swing. He opened a branch law office for a firm out of Titusville and caught "oil fever". Three years at Pithole ended unhappily when one of the great oil fires destroyed his office, property and law library. He reopened for business at Pleasantville where he was instrumental in developing the "Mechanics lien" law which was designed to control unscrupulous oilfield contractors. He wrote many of the contracts for the fledgling Rockefeller and others creating railroad rebates, obviously in restraint of trade. Rockefeller formed South Improvement Company in an attempt to monopolize crude transportation. Freight rates doubled for all but the Rockefeller interests, causing near riots. In 1872, Roger organized the independent refiners and oil men to fight the growing Standard Oil, encouraging them to build their own railroads, storage facilities and pipelines. He was a key figure in the early opposition to the Standard monopoly. He died in 1897 after spending his last twenty-five years opposing every Standard move. Excellent history and story of a good attorney. The hardbound first edition is common.

**B65. (HENRI DETERDING) Deterding, Sir Henry, As told to Stanley Naylor, AN INTERNATIONAL OILMAN, NY, Harper and Brothers, 1934, 126pp.**

The book consists of selected memories and reflections of the author. For the most part they had been published as a series of articles in the Saturday Evening Post magazine. They had been solicited by the editor. In virtually all of Deterding's stories, he comes across as an even handed but strong personality with an acute business sense. The real Deterding does not seem to show through, as he did not become the most powerful man in the world simply by being a good tough businessman. He deplores Rockefeller's tactics of crushing competitors and gives his philosophy of life and business which was developed in the days of his youth. He speaks of the early fights with Standard and the giant merger with Shell while trying to relate his keys to success. He ranks simplicity as the number one attribute. Simplify your life on all fronts and you will be a success. In one story, Deterding tells of meeting with Mussolini in the early 1930's and finding that they were kindred souls, thinking and acting alike. This may be the only revelation of the real Deterding in the book, as Mussolini had not yet revealed himself to the rest of the world as a ruthless dictator. The book is interesting, but not too credible, somewhat like Getty's autobiographies. Cited in Swanson on page 174; Hardwicke page 45. The hardbound first edition is scarce.

**B66. (HENRY ROGERS) Dias, Earl J., HENRY HUTTLESTON ROGERS: PORTRAIT OF A CAPITALIST, Fairhaven, MA, The Millicent Library, (1974), 190pp.**
Rogers was born in 1840 and died in 1909 at Fairhaven Massachusetts. His father was a fairly well off independent thinker who had trouble sticking to a steady job. Rogers himself breezed through high school and found a job as store clerk in a local grocery. In 1861, at the age of twenty-one, he took his life's savings of $600 and headed for Titusville to establish a refinery with his partner Ellis, who also contributed $600. With good success in a short time, Rogers quickly gained the reputation of being a ruthless oil businessman. In 1867 he joined Charles Pratt in Boston to help run his Long Island refinery. Rogers patented a naptha recovery technique and developed an early fractionating tower for crude distillation, making the Pratt Refinery one of the most successful in the nation. When Rockefeller began to monopolize refineries, Rogers fought hard against him, however by 1874, he agreed to merge with Standard. Brilliantly, Rogers took stock in Standard, rather than cash. He joined Rockefeller and with his great oil knowledge became a spokesman for the company. When Rockefeller retired in 1882, Rogers had hoped to take over. However, Rockefeller evidently thought Rogers was too flamboyant and selected John Archbold instead. Historians treat Rogers roughly because his reputation for being ruthless followed him and was apparently justified. Many other Rogers stories are included in this unbiased biography. The hardbound first edition is scarce.

**B67. (S. C. T. DODD) Dodd, S.C.T., MEMOIRS OF S.C.T. DODD, WRITTEN FOR HIS CHILDREN AND FRIENDS, 1937-1907, NY, Robert Grier Cooke, 1907.**
This book was not available for review by Neff, but it is regarded as important. Quoting from Swanson: "Account of early contacts with oil business in Pennsylvania and with Standard Oil Company, including formation of Trust". Cited in Swanson on page 161. The first edition is very scarce.

**B68. (MICHEL HALBOUTY) Donahue, Jack, WILDCATTER: THE STORY OF MICHEL T. HALBOUTY AND THE SEARCH FOR OIL, NY, McGraw-Hill Book Co., (1979), 268pp.**
Halbouty has been the best known petroleum geologist in the world for the last forty years (1994). He has produced more political propaganda for the oil industry and for geology than anyone of this era. At eighty-five, he still makes national news when he speaks. Surprisingly Halbouty's biography was not written by his close personal friend James A. Clark. Clark was one of the most famous southwest oil historian-writers and had collaborated with Halbouty on several books. Halbouty's first experiences out of school were with the famous Yount-Lea Company in 1931. Within months Halbouty proved his own theory of oil accumulation beneath salt dome overhangs with a giant oil field discovery at High Island, Texas. After Yount-Lea sold out, he worked for Glenn McCarthy as chief geologist. He opened an independent consulting office in 1937. The story is wild and his career details are very interesting. He found oil and gas all over the gulf coast and drilled countless dry wildcats. He is best known for his promotion of the oil industry through his many speaking engagements. The man is charismatic, if not offensive. He predicted the 1973 Arab embargo fifteen years before it happened but had been accused of crying "Wolf!" by government and major oil company officials. This is an accurate picture of a modern legend. An enjoyable, colorful picture of the industry and Halbouty. Reprinted by Gulf Publishing, undated. The hardbound first edition is common.

**B69. (HOWARD HUGHES) Drosnin, Michael, CITIZEN HUGHES: IN HIS OWN WORDS - HOW HOWARD HUGHES TRIED TO BUY AMERICA, NY, Holt, Rinehart and Winston, (1985), x, 532pp.**
This book would not have been written except for some very unusual circumstances which occurred in 1974. A burglary at Hughes' Hollywood warehouse-office was designed and executed by perpetrators still unknown, specifically to steal his most private files. These files surfaced three years later and Drosnin was offered the opportunity to write a Hughes biography which would almost write itself. Thousands of documents and Hughes' own handwritten memos are synthesized and several reproduced for this work. By far, this is the most realistic picture of how powerful and how bizarre Hughes really was...and in his own handwriting. The hardbound first edition is common.

**B70. (HAYNES DUGAN) Dugan, Haynes W., SIGN HERE, PLEASE: THE STORY OF AN (ALMOST) BAREFOOT LANDMAN, Hicksville, NY, Exposition Press, (1974), 191pp.**
Dugan's grandfather was a petroleum landman and worked for Gulf Oil. His father and grandfather's brother both worked for Texaco around 1917 and as independents. Numerous family members over the generations worked as oil land leasing agents. Haynes Dugan was born in 1913, college educated and served in WW II. Returning, he jumped into the oil business wholeheartedly. After spending two years in Tampico, he concentrated his efforts in Alabama and Louisiana. Many oil country stories, including some tales, are designed to educate the reader as to the petroleum landman's essential service to the oil industry through their dealings with companies, farmers, bureaucrats, investors, etc. The hardbound first edition is common.

**B71. (LORECIA EAST) East, Lorecia, THE BOOMERS: THE AUTOBIOGRAPHY OF A ROUGHNECK'S WIFE, Baton Rouge, Legacy Publishing, 1976, 63pp.**
Born in Louisiana in the 1890's, Lorecia married John East in 1913. She does a good job of relating their family trials and travels through the oil fields. Shortly after their marriage, John began roughnecking on a wildcat well near the family farm at China, Texas and later proceeded to Sour Lake, Humble, Goose Creek and Mexia. He was handy with machines and materials and became an oil company field superintendent at Long Beach, California. There he patented and manufactured a pumping unit counterbalance which was very successful, however it was 1929 and he luckily got all of his money back before crashing like the rest of the country. He returned to the China area where he farmed and worked some for oil companies and as a shipyard machinist. There are many remembrances of tough times in the tent city boomtowns and of the boomers who followed the discoveries. Most interesting biography. The softbound first edition is common.

**B72. (BRIAN EBY) Eby, Brian J., MY TWO ROADS, Houston, TX, Pacesetter Press, (1974), xiii, 292pp.**
Born in Pennsylvania in 1896 to a farming family, Eby studied geology at Johns Hopkins and also worked as a journalist for the Baltimore

newspaper. Eby's "Two Roads" in the title refer to his internal conflict of whether to become a geologist or a journalist. He spends his life torn between the two. His first geological job was with the U.S.G.S. while working on his Ph.D. He worked in the Appalachians and Rockies, finally leaving for employment with Shell's U.S. subsidiary, Roxana. Eby did oil exploration geological work all over Oklahoma and Texas and was in charge of their geophysical department from 1925 to 1931, which was the beginning of the era of seismograph exploration. He left Shell and was a partner in The Fifteen Oil Company which through his efforts was responsible for discovering the large Charenton Oil Field on the Texas Gulf Coast. Later Eby did extensive oil research in Europe and Russia for "World Oil", becoming an international expert and writer. Thus his journalistic flair resurfaced and effectively continued with his preparation of this fine autobiography. The hardbound first edition is common.

**B73. (ERBY ELLIS) Ellis, W. Erby, THE YOUNG CRY AND THE OLD DIE, NY, Vantage Press, Inc., (1981), 96pp.**
Ellis wrote this autobiography while in his eighties. He was born at Chandler, Texas in 1898. By age two his father had died and his mother starved out, turning her four kids to an orphan home in Corsicana. Ellis quit school after the third grade and worked his way up from messenger boy to telegraph operator. He worked all over Texas during the oil boom years when the telegraph was the primary tool for rapid communication. Barely scraping by, he invested in a few oil leases with a cousin in the East Texas boom. Within days he had made and lost more money than he had made in his life. He worked as dispatcher for Magnolia for a few years and then started a successful oil company with partner Roy Allen. They found and developed good shallow oil production, mainly in central Texas. The hardbound first edition is common.

**B74. (H.V. FOSTER) Ellis, William Donohue, OUT OF THE OSAGE: THE FOSTER STORY, OK, Oklahoma Heritage Association, 1993, xiv, 309pp.**
This is the biography of H.V. Foster, who became an oilman almost by accident and quietly revolutionized America's oil industry. In 1904 at the age of twenty-seven, he took over operation of the giant Osage "Blanket Lease" which had been held by his father. He proceeded to discover three giant oil fields in Oklahoma: Osage, Seminole, and Oklahoma City. He made one brilliant exploratory move after another in the face of constant

criticism and ridicule by knowledgeable industry experts. His company was called Indian Territory Illuminating Oil Company (ITIO) which eventually became the core of Cities Service (Now Oxy). H.V. was comparable to Doheny, Marland, Bennedum, Barnsdall, Slick, and McCarthy in his intuitive oil finding ability. Excellent book. The hardbound first edition is common.

**B75. (ISABEL ELMER) Elmer, Isabel Lincoln, CINDERELLA ROCKEFELLER: AN AUTOBIOGRAPHY, NY, Freundlich Books, (1987), xiii, 217pp.**

Autobiographer Elmer describes her sheltered childhood in detail. She is the great granddaughter of William Rockefeller (John D.'s brother and partner). She had grandchildren of her own when she wrote this book. She gives a surprising look at how the money and power has decreased through generations but continues to have an overwhelming effect on the heirs and beneficiaries. Although she was subjected to the "stifling control" of the Rockefeller family, she seems unaffected by it today. Included is a current genealogical chart. The hardbound first edition is common.

**B76. (JOHN D. ROCKEFELLER) Englebrecht, Curt E., NEIGHBOR JOHN: INTIMATE GLIMPSES OF JOHN D. ROCKEFELLER IN ORMOND BEACH, NY, The Telegraph Press, (1936), 216pp.**

This unusual book is essentially a record of the last ten years of John D.'s life. The photographer, Englebrecht gained unique access to John D. largely through a flattering 1925 candid photo he took of Rockefeller walking down the Union Church steps at Ormond Beach, with a broad smile on his face. In a calculated move, Englebrecht developed and delivered the picture to the Rockefeller home, "The Casements", before John D. had even returned from church, accompanied by a note asking to publish the shot. It worked. They had a tidy business relationship wherein the camera-shy Rockefeller insisted that all photos had to be taken by appointment and were subject to his approval prior to release. Fifty-eight staged photos are included in the book and they present John D. at his very best. Excellent book. Cited in Swanson on page 216. Reprinted in softbound. The hardbound first edition and reprint are scarce.

**B77. (BILLIE SOL ESTES) Estes, Pam, BILLIE SOL: KING OF THE WHEELER DEALERS, Abilene, TX, Noble Craft Books, (1983), 205pp.**

Pam Estes is Billie Sol's daughter and she writes objectively about his childhood and character as she follows the hard- working, generous Billie Sol into a world of power and politics. As a child he had an innate ability to make money, dealing and improving his position with each maneuver or trade. By thirty he was a millionaire rancher, farmer and oilman living in Pecos, Texas. He was a great innovator and opportunist. After becoming involved, through donations, in high state and federal politics, he was allowed to take advantage of the numerous government farm crop programs. He made more millions. The fall of Billy Sol involved a scheme where he sold non-existent liquid fertilizer tanks to farmers, oilmen and others then rented them back to others, creating large tax benefits. The thing grew beyond imagination and finally came crashing down with Billie Sol going to prison at Leavenworth. Later paroled, he was constantly pursued by government tax people. Associating with the wrong crowd, he eventually got in trouble again and was sent back to prison. Not a pretty story from either side. Lots of photos and family correspondence along with good source notes. A con man of historic proportions. The hardbound first edition is common.

**B78. (JEFF EWING) Ewing, Jefferson S., OIL'S WELL: AN AUTOBIOGRAPHY, Guthrie, OK, Co-Operative Publishing Company, (1966), 115pp.**

Jeff Ewing was born in 1893 and became interested in railroading early in life. He was in the Navy during WW I and afterwards (1919) decided to go into another field. Through friends, he gained an introduction to Jake L. Hamon, a successful oil operator at Ragtown, Oklahoma. He got a job with Hamon and quickly worked his way up from swamper to surveyor to general field hand at Healdton, Hewitt, Dillard and Fox fields. Hamon died in 1920 and the estate was operating in the red in spite of owning significant oil properties. Ewing worked for the estate trustees as field clerk and farm boss. Jake L. Hamon, Jr. had been working for another drilling company for a few years when he realized that the trustees of his Dad's estate were allowing the production to decline rapidly by not paying for normal remedial maintenance. Hamon, Jr. took over trusteeship of the estate in 1923. It was heavily in debt and in dire straits. With Ewing as superintendent and the financial support of only one creditor, National Supply Company, the properties quickly regained profitability. In 1926 Hamon formed a fifty-fifty partnership with Edwin B. Cox, a financial genius. Cox and Hamon invested heavily in drilling and mineral rights acquisition in the known fields of southern Oklahoma and later in the East Texas Field. The partnership survived until the death of Cox in 1950. At that time, Hamon took over operation of the partnership properties. Ewing was Hamon's right hand man for drilling and production through most of the years until his retirement in 1958. Rumor has it that many copies of this book were destroyed by the Hamon family shortly after its release in 1966. Must have been a few family secrets let out. The hardbound first edition is very scarce.

**B79. (R.A.F. PENROSE) Fairbanks, Helen R. and Charles P. Berkey, LIFE AND LETTERS OF R.A.F. PENROSE, JR., NY, Geological Society of America, 1952, x, 765pp.**
Penrose died in 1931 and left his ten million dollar estate divided between the Geological Society of America and the American Philosophical Society. This book was prepared by the GSA as a memorial to the man. Born in Philadelphia in 1863, his father was a prominent surgeon and professor. Penrose Jr. (called Dick by his friends) was educated there and at Harvard where he studied under one of the early proponents of geology, Nathaniel Shaler. He worked for the U.S.G.S. after schooling and by 1888 was an early appointee to the Texas State Geology Survey, where he became an associate of Robert T. Hill. Later Penrose worked in several mining areas in the U.S. and then on a worldwide basis as his reputation grew. He published many enlightened articles in the geology and exploration field. Not directly associated with oil and gas, he was in continuous contact and directly influenced those who were such as Emmons, Arnold and Ashley. He was the first president of the Society of Economic Geologists, which promoted the practical money-making side of geology as opposed to the theoretical. Much of the book is his personal letters from childhood through the 1920's and shows his personality and character. A major work on an important geologist. The hardbound first edition is becoming scarce.

**B80. (ROBERT HEFNER JR.) Faulk, Odie B, and Laura E. Faulk, AN OKLAHOMA LEGACY: THE LIFE OF ROBERT A. HEFNER, JR., OK, Western Heritage Books, (1988), xiii, 274pp.**
Hefner, Jr., son of the pioneer oil man was born in 1909 at Beaumont, Texas and shortly moved with the family to Ardmore. He had a great time

growing up: Eagle Scout, Culver Military Academy, Cum Laude at Stanford and law school at Harvard and University of Oklahoma. He practiced law in Washington D.C., Illinois and Indiana, and in 1946 moved back to Oklahoma to take over the family business, The Hefner Company. He successfully ran the company and created Hefner Production Company to drill and further develop family properties. Interesting oil family and good oil history. The hardbound first edition is common.

**B81. (JOE LaFORTUNE) Faulk, Odie B, et al, THE GENTLEMAN: THE LIFE OF JOSEPH A. LaFORTUNE, Oklahoma, Oklahoma Heritage Association, (1979), viii, 224pp.**
Born in Indiana in 1894, LaFortune received a limited education and worked at odd jobs to help support the family. With only a seventh grade education, he was allowed to enter a special college program at Notre Dame. He got a job as accountant for Standard Oil of Indiana. He served in WW I and then moved to the booming Tulsa oilfields in 1920 where he worked for several different companies prior to joining Warren Petroleum in 1924. Warren was a broker-marketer of gasoline and hired LaFortune to edit and publish the company trade publication. LaFortune quickly took over the supply and marketing branch operations and soon joined Warren as a partner. After the purchase-merger of Hanlon-Buchanan in 1946, Warren became the largest independent natural gasoline producer and marketer in the U.S. LaFortune negotiated the Warren-Gulf Oil (now Chevron) merger in 1955 as the largest petroleum transaction to date, with Warren receiving forty-two million and LaFortune fifteen million dollars for their shares of the company. LaFortune continued to operate as an independent driller and producer until his death in 1975. The book discusses his philanthropies and government service in much detail. The hardbound first edition is common.

**B82. (HOWARD HUGHES) Fay, Stephen, et al, HOAX: THE INSIDE STORY OF THE HOWARD HUGHES - CLIFFORD IRVING AFFAIR, NY, The Viking Press, (1972), ix, 310pp.**
A well documented history and analysis of the titled hoax. From McGraw-Hill's 1971 pre-publication announcement of their so called authentic Hughes autobiography, to the conviction of author Clifford Irving on fraud charges the next year, this book pries into America's most

mysterious man's personal life and business. Co-conspirators and motives are explored in this very good piece of investigative journalism, although questions are still left unanswered. The hardbound first edition is common.

**B83. (VARIOUS GEOLOGISTS) Fenton, Carroll Lane and Mildred Adams Fenton, THE STORY OF THE GREAT GEOLOGISTS, NY, Doubleday, Doran & Co., 1945, xvi, 301pp.**

This is an exceptional book which gives accounts of the early geologists up until about 1900. Thirty-five of these brilliant scientists are covered in detail as the evolution of geological thinking progresses. Virtually all of these men were subjected to severe criticism, and often abuse for standing up for their various theses...mostly from their own colleagues. No oil content in the book, however these men such as John Wesley Powell and Benjamin Silliman greatly influenced the rising field of petroleum geologists. Excellent history and biography. Second revised and enlarged edition is retitled GIANTS OF GEOLOGY, 1952, xvi, 333pp. The hardbound first edition is common.

**B84. (HUEY LONG) Fineran, John Kingston, THE CAREER OF A TINPOT NAPOLEON: A POLITICAL BIOGRAPHY OF HUEY P. LONG, New Orleans, LA, The Author, ca 1932, 172pp.**

The first of several Long biographies. This is a most unfavorable book, as the author goes into much negative detail concerning Long's political career through his first term as governor. It is largely an attempt to discredit him to the voters in the pending election and borders on libel. Much on the impeachment attempt which Long blamed on Standard Oil and gambling interests. The hardbound first edition is becoming scarce.

**B85. (HOWARD HUGHES) Finstad, Suzanne, HEIR NOT APPARENT, Austin, TX, Texas Monthly Press, (1984), xi, 427pp.**

The author is an attorney and writer who acted as special counsel during the 1981 Hughes heirship trial in Houston. From the epilog describing the settlement: "...the Hughes fortune was divided three ways: to his family...to the U.S. government, and to lawyers". The author finds this interesting in that Hughes had no interest in his family, and hated the government as well as attorneys. This is a thorough look at the claimants and the Hughes estate. The author unravels Hughes genealogy and presents a chronology with biographical notes and detailed sources. The heirship drama was as unusual as the man who caused it simply by not leaving a will. The hardbound first edition is common.

**B86. (AMON CARTER) Flemmons, Jerry, AMON: THE LIFE OF AMON CARTER, SR., OF TEXAS, Austin, TX, Jenkins Publishing Company, 1978, 520pp.**

Carter arrived at Fort Worth, Texas in 1905 and the place has not been the same since. Along with two newspaper reporter friends, he started the Fort Worth Star, an afternoon newspaper. The Star later came out on top of a merger with the Fort Worth Telegram, making the Star-Telegram, which is the best known newspaper in West Texas. He had previously been a successful salesman for various kinds of advertising. Carter was born about 1880 and died in 1955. He was a zealous, uninhibited promoter of Fort Worth and West Texas. Some of his booster antics related here are hard to believe. He was respected for his power and position, but known all over the country as a Texas character. Oil is not mentioned until page 484, where it is revealed that Carter was a real life compulsive wildcatter, making a large part of his fortune through successful wildcat well drilling. Early success in Southeast New Mexico was followed in 1937 by his discovery of the giant Wasson Field in the Permian Basin and subsequent sale of that property to Shell Oil. Great reading. The hardbound first edition is becoming scarce.

**B87. (GRADY FLETCHER) Fletcher, Grady W., TALES FROM THE OIL PATCH: AND OTHER IMPORTANT TRIVIA CONCERNING THE OIL & GAS BUSINESS, Houston, TX, Alpha Publishing Group, 1991, 226pp.**

Fletcher was born in 1919 in Gainesville, Texas. He held a job typing legal instruments for a title and abstract company while in college in Cooke County. He soon became acquainted with oil people and lease trading...so he started buying and selling oil and gas leases for a profit. After serving in WW II he graduated from Texas Tech University and went to work for Texaco in Fort Worth as a lease rental clerk. After four years he went to work for Pan American in Houston as a division order clerk, being allowed to acquire mineral interests for his own account to fair reward. In 1951 he was employed by Sinclair as a landman leasing drilling blocks. In the late 1950's he went into partnership doing contract leasing for clients. Many good stories of land trades and other oil adventures. The last section of the book is entitled "Important

Information to be Considered When Negotiating or Before Signing an Oil and Gas Lease". It is exactly that. The softbound first edition is becoming scarce.

**B88. (GIOVANNI FLORES) Flores, Giovanni, ARC OF THE SUN: ADVENTURES OF A PETROLEUM GEOLOGIST, Tulsa, OK, Lion & Thorne Ltd., (1987), xi, 229pp.**

Edited by Sue Blakey, this autobiography of Flores has lots of historic photos which document his odyssey in petroleum exploration and in the oil fields. Born in 1917, Italian-English, he graduated in geology from the University of Naples in 1939. First working in Europe then after WW II in Cuba, then to Africa and back to Europe. He worked for the major oil companies and has been an independent consultant since 1972. His jobs and travels are related in first person with lots of dialogue and reminiscences and philosophy. Very interesting reading with a rare look at the life of a worldwide field petroleum geologist. The hardbound first edition is common.

**B89. (JOHN D. ROCKEFELLER) Flynn, John T., GOD'S GOLD: THE STORY OF ROCKEFELLER AND HIS TIMES, NY, Harcourt, Brace and Co., (1932), 520pp.**

This Rockefeller biography was written with a vengeance three years after Winkler's innocuous book. The title comes from Rockefeller's often used quote "God gave me my money" or "The Lord gave me my money." Published in the middle of the depression Flynn gained a wide readership with this sensational work which had been prepared without Rockefeller input. The book is over three times as long as Winkler's and contains twenty pages of sources. Flynn gives good detail to the early years leading up to the 1870 formation of the stock company, Standard Oil (Ohio) with John D., Henry Flagler, Samuel Andrews, brother William Rockefeller, S. V. Harkness and O. B. Jennings. The extremely shrewd John D. ran the show and had monopoly in mind with his first target being the Cleveland refiners. Frew and Lockhart's Standard Oil (Pennsylvania) organized the Philadelphia refiners. Warden and Waring organized those in Pittsburgh. Together they formed the South Improvement Company to tackle the refiners and producers at Oil Creek who had a distinct advantage over the other distant refiners. War was declared by the Oil Creek group against South Improvement which had gained railroad rebates causing Oil City and Titusville transportation fees to double. The state shortly revoked South Improvement's charter, but John D. lost only the battle, not the war. He controlled almost every facet of the business outside of Oil Creek, through on-going battles with refiners, railroads, pipelines, marketers as well as state and federal agencies. He kept this constant antagonism up for most of his life. Within a few years he controlled transportation and marketing and was becoming involved in producing. His stranglehold was broken by the gigantic Spindletop, Texas oil production which flooded the market in 1901. The government finished the job in 1911 after some fifteen years of trying to break his company up. In fact, they did not diminish him, they only spread him out among the several Standard Companies. Flynn relates good history of the times as he explores the fabulous character of Rockefeller. Rockefeller was ninety-three when this book was published and had outlived virtually all of his associates and enemies. Great reading about an incredible individual. Cited in Swanson on page 161; Hardwicke page 45. The hardbound first edition is common.

**B90. (JOHN D. ROCKEFELLER, JR.) Fosdick, Raymond B., JOHN D. ROCKEFELLER, JR.: A PORTRAIT, NJ, Lyle Stuart, Inc., (1975), xi, 477pp.**

Author Fosdick was a member of John D., Jr.'s inner circle of friends and advisors for many years. At the request of Jr.'s son Nelson, Fosdick agreed to do this biography. John D., Jr. was born in 1874 in Cleveland, being eighty while this work was written. His privileged childhood was molded in business education by the mandatory expense-income ledger his dad required him to keep to the penny. One basis of the Rockefeller fortune was knowing exactly where every penny went...every penny! He graduated from Brown in 1897, where he was elected to Phi Beta Kappa. He went to work in his dad's New York office, offering to serve in any capacity required. He was quickly directed into non- business family affairs such as church, charity and handling requests of the family from outsiders. He administered these requirements directly under the Reverend Frederick Gates so expeditiously that he was given more and more responsibility. Aside from these duties he began successfully speculating in the stock market. Within a short time he had lost a million dollars which he could not cover. Dad considered it a lesson well learned and covered the loss without saying a word. Jr. was so mortified he ceased speculation of any kind. He was frequently allowed to conduct major negotiations with other moguls and did that well. One of Jr.'s worst times was during a union strike at a Rockefeller owned coal company in Colorado where serious fighting and injuries were escalating. It was a serious public relations problem and Jr. was in charge of squelching it. He took a good deal of abuse from all directions, but finally managed to quiet the strikers. Jr. further proved his courage and toughness in a public confrontation with Colonel Stewart, chairman of the board at Standard of Indiana. Shortly after the Teapot Dome scandal, Stewart was found to have organized a kickback scheme on a Standard oil production purchase. Jr. was irate and eventually succeeded in ousting Stewart. He soon became responsible for a major phase of the Rockefeller domain with the mission of making the public aware of the good the family was doing, at the same time squelching the monopoly image. Jr. was responsible for the restoration of Williamsburg, a massive and very expensive job. He was the impetus to getting the Nevins biography written and published. He brilliantly directed the family foundations and charities work all over the world. Fosdick gives a lot of history and insight into the man, but is clearly biased towards him. The hardbound first edition is common.

**B91. (HENRY FLAGLER) Friends of Henry Flagler, HENRY M. FLAGLER MEMORIUM, Palm Beach, FL, His Friends, 1914, 53pp.**
Flagler was born February 2, 1830 and died May 20, 1913. This is a memorial to the remarkable man. He was born in New York, the son of a traveling preacher. As a young man he was very unsuccessful in business, trying several ventures. He joined up with John D. and William Rockefeller and Samuel Andrews in a refining partnership that eventually merged into Standard Oil. Flagler organized the new company and developed its method of analyzing the worth of competitors as they schemed to acquire or absorb all of them. He left Standard in 1883, after having been John D.'s right hand man for several years and essentially directing many of the company's activities. He moved to Florida where he spent the next thirty years and most of his effort and much money developing the east coast of that state. Comments of the press from around the U.S. are included. The hardbound leather edition is scarce.

**B92. (BENJAMIN SILLIMAN) Fulton, John F., BENJAMIN SILLIMAN 1779-1864: PATHFINDER IN AMERICAN SCIENCE, NY, Henry Schuman, (1947), ix, 294pp.**
Silliman was born in Connecticut in 1779 where his father was a Revolutionary War general and wealthy landowner at Fairfield. He

entered Yale College at thirteen and graduated in 1796. He studied law and passed the bar, however, instead of practicing, he was appointed chairman of the newly established department of chemistry at Yale. He also came to be the first professor of geology in America. He was no great research scientist, but contributed much to the literature in chemistry and geology. He started "The American Journal of Science", for free exchange of ideas with the world. His main contribution to the oil industry was through his son, Benjamin, Jr. who succeeded him at Yale and carried on the tradition. It was Junior who evaluated the potential of petroleum for promoters Brewer and Watson and published the famous "A Report on the Rock Oil..." which was a brilliant work describing the makeup of crude and its fractional separation. The Drake well was successfully promoted, based on Junior's work. Benjamin Senior died in 1864 and is much revered in chemistry and geology. The hardbound first edition is common.

**B93. (VALENTIN GARFIAS) Garfias, Valentin R., GARF FROM MEXICO, Mexico, D.F., Editorial Jus, 1950, 261pp.**
This book was not available for review by Neff, but is regarded as important. Quoting from Swanson: "Author worked in California oil fields; was first professor of petroleum engineering at Stanford; worked as geologist in many foreign countries; trained many to look for oil". Cited in Swanson on page 174. The hardbound first edition is very scarce.

**B94. (HOWARD HUGHES) Garrison, Omar V., HOWARD HUGHES IN LAS VEGAS, NY, Lyle Stuart, Inc., (1970), 293pp.**
This is a typically sensational Lyle Stuart published book. Author Garrison tries to answer the pressing question of why Howard Hughes, the richest industrial genius in America went on a spending spree, buying up Las Vegas. Not too obvious. Much on Hughes' running battle with the Atomic Energy Commission in an effort to halt the underground nuclear testing in Nevada. A few new stories are revealed here and the book is informative although written more to entertain. The hardbound first edition is common.

**B95. (BENJAMIN KAVANAUGH) Geiser, S.W., BENJAMIN TAYLOR KAVANAUGH AND THE DISCOVERY OF EAST TEXAS OIL, Field & Laboratory, June 1944, Vol. XII, No. 2, p46-55.**
The author wrote this biographical sketch to tell the world that one of the first East Texas oil explorers was Kavanaugh. He rehashes much early

Texas oil history, taken largely from Warner's "Texas Oil and Gas Since 1543". Dr. Kavanaugh was born in 1805 in Kentucky, he was apprenticed as a book-binder but became a Methodist minister, later gaining an M.D. degree from DePauw. He practiced medicine in St. Louis until 1857 and then began preaching. In 1865 he transferred to Chappell Hill, Texas, and being scientifically oriented, spent much of two years time exploring Eastern Texas for petroleum and other mineral deposits. He evidently explored a large part of East Texas with his divining rod which he called a mineraloger. He was one of the incorporators of San Augustine Petroleum Company in 1866 and the author believes that Kavanaugh's efforts and publicity gained thereby were largely responsible for the oil interest beginning in East Texas. He worked for the church in Texas until 1881 and died in Kentucky in 1888. No success in the oil field, however he was definitely an early explorer although remaining an obscure character. The softbound first edition is becoming scarce.

**B96. (DUX GENTRY) Gentry, Winalee, ONE MORE RIVER TO CROSS, Philadelphia, Westminster Press, 1955, 237pp.**

This book was not available for review by Neff, but it is regarded as important. Quoting from Swanson: "Experiences in the field by a wife who helped her oilman husband, Dux Gentry, Jackson, Michigan". Cited in Swanson on page 174. The hardbound first edition is very scarce.

**B97. (HOWARD HUGHES) Gerber, Albert B., BASHFUL BILLIONAIRE: AN UNAUTHORIZED BIOGRAPHY OF HOWARD HUGHES, NY, Lyle Stuart, Inc., (1967), 384pp.**

Gerber was the second to break through the biography defense set up by Howard Hughes. His book takes phases of Hughes' life by chapters and details each from start to finish. Hughes was born in 1905. His life as a multimillionaire entrepreneur began at age nineteen after his fathers death. He went from being a hard working businessman to reckless risk taker (both machines and women) and finally to a bizarre, paranoid, reclusive existence. An excellent chapter on Hughes' father's development and patenting of the rock bit that revolutionized the drilling industry. The brilliant exploitation of his patent and zealous protection of it is explained in detail. There is a chapter on Hughes' would-be biographers and how he kept his privacy through the years by effectively squelching each attempt to publish a biography on his life. The author's format is to give important news headlines of the time, then Hughes age, then explain how the two fit together in history. This is a good biography, particularly for the time and considering the total lack of Hughes support. The hardbound first edition is common.

**B98. (NELSON ROCKEFELLER) Gervasi, Frank, THE REAL ROCKEFELLER: THE STORY OF THE RISE, DECLINE AND RESURGENCE OF THE PRESIDENTIAL ASPIRATIONS OF NELSON ROCKEFELLER, NY, Atheneum, 1964, xii, 271pp.**
The author looks at Nelson's background and political history and tries to assess whether the public will accept his great wealth, his divorce and his record. Nelson was running a hard campaign for the Republican presidential nomination against his antithesis, Goldwater. This Rockefeller sponsored book gives details of his business and public service career, and in a positive light. The hardbound first edition is common.

**B99. (J. PAUL GETTY) Getty, J. Paul, AS I SEE IT: THE AUTOBIOGRAPHY OF J. PAUL GETTY, Englewood Cliffs, NJ, Prentice-Hall, Inc., (1976), 360pp.**
Getty died in 1976 at the age of eighty-four as this book was going to press. He had previously written numerous autobiographical books and articles defining his interests and philosophies. As in most of his works, he states exactly how he views himself, his life, his work and the world. He presents many valid slogans and credos which he followed in business, allowing him to accumulate and keep his great wealth. It helped that his father was a successful attorney and oil man. Getty largely discusses some of his incredible deals such as acquisition of Pacific Western, Tide Water and Skelly. His master negotiation of the neutral zone Kuwait concession became a pattern to be followed by other independent companies (such as Occidental and Hunt) to break into major oil company areas of dominance. Getty admits to having at least five wives and chasing women. His reputation was that of a world-class womanizer. Getty admits to frugality but he was better known as miserly. He admits to being shrewd, but was better known as brilliantly so. This is a long book with much the same material as his others. The standard Getty propaganda. The hardbound first edition is common.

**B100. (J. PAUL GETTY) Getty, J. Paul, EUROPE IN THE EIGHTEENTH CENTURY, Noplace, Privately Published, 1949, 117pp.**

Getty had a brilliant mind and had studied European history for years. He was considered an authority on the subject. He prepared this short history before WW II, but did not publish it until 1949, when the smoke had cleared. Many have accused him of pro- German sympathy, even though he lived in a large castle in England which may have had something to do with the publishing delay. The book is organized in sections: Geography and History, Science, Literature, Music, Art, A Contemporary Letter and Daily Life. Photos include many paintings of historic scenes and furniture of the period which belongs to the Getty Collection and other museums. The hardbound first edition is becoming scarce.

**B101. (J. PAUL GETTY) Getty, J. Paul, HOW TO BE RICH, Chicago, Playboy Press, (1965), xii, 264pp.**

Basically the book is a compilation of Getty's articles which were originally printed in Playboy from 1961 to 1965. The fundamental theme was supposed to be "Men, Money and Values in Todays Society" as per Hefner's request. Instead of following the theme on how to conduct yourself after acquiring wealth, Getty goes to his standard format of detailing his various successes and how he got to be "The World's Richest Man". If you want to know how shrewd and smart Getty was, this book explains it first hand. Boring after a while. The hardbound first edition is common.

**B102. (J. PAUL GETTY) Getty, J. Paul, MY LIFE & FORTUNES, NY, Duell, Sloan and Pearce, (1963), 300pp.**

Good solid history of Getty and the oil industry. The saga starts with George F. Getty, J. Paul's successful attorney father inadvertently getting involved in the oil business in 1903. The family moved to California and began expanding their oil business. As a teenager, J. Paul spent his summers roughnecking in the Oklahoma oil fields. He was sent to be educated at Oxford and during that time became a world traveler. Barely missing the fateful voyage of the Lusitania, he went to work in Tulsa for his Dad's company Minnehoma (1914). Most of their production was in California, but Getty Oil, a 70/30 father and son partnership, was formed in 1916 with both men reserving the freedom to prospect and drill on their own. Getty gives his first-hand impression of many of the major financial and political events up to the early 1960's. The acquisition of Pacific Western, Tide Water, Skelly and the incredible Kuwait Neutral Zone Concession are all explained in the standard Getty style. The hardbound first edition is common.

**B103. (J. PAUL GETTY) Getty, J. Paul, THE GOLDEN AGE, NY, Robert Collier Book Corp., (1968), 220pp.**
By 1968, at the age of seventy-six, Getty had so much money, he did not known what to do with it, aside from buying fine art. From Sutton Place, his castle in England, he takes time to reflect on the golden age of life and tell all who will listen. Although he speaks of the successful businessmen and professionals as being the affluent and thus able to enjoy this "Golden Age", he appears to be speaking to a select, less than one tenth of one percent of us. He gives the Getty philosophy on how to get rich and keep it, then how to retire and enjoy it. Typical Getty, he is in his own world, but such a unique character that he grows on you. The hardbound first edition is common.

**B104. (J. PAUL GETTY) Getty, J. Paul, THE JOYS OF COLLECTING, NY, Hawthorn Books, Inc., (1965), 167pp.**
Getty began collecting art in 1912 and became more enthusiastic about it each year. The first forty-two pages of the book give his philosophy of collecting art which is to buy great art at exceptionally low prices and never sell it. Most of the Getty discussion concerns his incredible art purchases and how much smarter he was than the seller. He conveys "the excitement, thrills and triumphs that make art collecting one of the most exhilarating and satisfying of all human endeavors". The remainder of the book is photos and narration of the sculpture, paintings, etc. and they are wonderful. The hardbound first edition is becoming scarce.

**B105. (JOHN D. ROCKEFELLER) Gilbert, Miriam, MONEY AND MUD: THE LIFE OF JOHN D. ROCKEFELLER, Montgomery, AL, American Southern, (1964), vii, 95pp.**
This pleasant book probably belongs in the fiction category but it is done with such regard for history and with such sincerity that it deserves reading, particularly after seeing Rockefeller raked over the coals time and time again. In what is probably considered a young adult novel, Miss. Gilbert portrays John D.'s life from high school through the work of his philanthropic foundations in 1913. She uses dialogue to bring Rockefeller

to life in various situations and meetings. She presents him as a man trying to unify the oil industry. She continually describes his humanity, religiousness, sensitivity and charity. The hardbound first edition is becoming scarce.

**B106. (GUS GLASSCOCK) Glasscock, Lucille, A TEXAS WILDCATTER: A FASCINATING SAGA OF OIL!, San Antonio, TX, The Naylor Company, (1952), ix, 126pp.**
As related by his wife, Lucille, Gus was born to an East Texas farm family in 1896 and left home at age fourteen to join his brother Alex's traveling circus. He had a good six years as an acrobat, traveling the small towns of Louisiana, Arkansas and Texas. As WW I broke out in 1917, the act broke up and Gus was out of a job. Returning to Texas, he married the author and began work in the oil fields. They first went to Ranger for a couple of years, living in expensive tents and shacks. There, the lease speculators made more money than the drillers and producers. They then moved to Wichita Falls. Gus got his brothers to invest with him in drilling a lease near San Antonio which already had some poor production. They had moderate success which kept them inspired to look for the big one. As a drilling contractor, he had wildcat success at Lytton Springs. Gus and his brother Lonnie had success at East Texas and on the gulf coast. They bought more drilling rigs and drilled more wells and found more oil. They contributed to increasing U.S. oil production for the WW II effort and went on to pioneer barge drilling in the bays and then offshore Gulf Coast. Lucille describes the living and social conditions up to the early 1950's as they became more bearable. Maybe not fascinating like the title indicates, but a very good history and biography. Cited in Swanson on page 174. The hardbound first edition is scarce.

**B107. (CHARLES GOULD) Gould, Charles N., COVERED WAGON GEOLOGIST, Norman, OK, University of Oklahoma Press, (1959). xiii, 282pp.**
Born in a log cabin near Duck Creek, Ohio in 1868 and reared in a dugout on the plains of Kansas, Gould became interested in geology as a teenager. He managed to work his way through to a degree in geology from the University of Kansas. He went almost directly (1900) to the University of Oklahoma, where he became the first chairman of the school of geology. He proceeded to build the geology school from nothing and train the best petroleum geologists of the day. He spent time

as a consultant from 1911 to 1924, later returning to O.U. to organize the Oklahoma Geological Survey. He received a Ph.D. from Nebraska. He was an enthusiastic geologist spending as much time mapping geology in the field as possible. Very early he came to recognize potential oil bearing structures from their anticlinal attitude. He admitted to hating "oil geology", but would eagerly tell anyone who would listen where the most likely drillsites were. The beauty of Gould's autobiography is not in his incredible oil finding ability (described only in Beebe's introduction), but in his color description of early geological field mapping which really took place in covered wagon days. It is difficult for modern geologists to comprehend the magnitude and historic effects of the efforts of this man. He died in 1949. He was a very important geologist, although it is difficult to tell that by reading his self-effacing autobiography. Not in Swanson. The hardbound first edition is becoming scarce.

**B108. (CHARLES GOULD) Gould, Charles N., TRAVELS THROUGH OKLAHOMA, Oklahoma City, Harlow Publishing Company, 1928, 174pp.**
Gould takes the reader for a tour of Oklahoma in first person dialogue. He covers much of the state and enlightens the reader with his first hand knowledge of geology, water, flora, fauna, crops, minerals, roads, cities and people. Also some oil and gas. Good photos and some maps. Gould knew his adopted state well. Great history. The hardbound first edition is scarce.

**B109. (DON GRABLE) Grable, Donovan B., ANYONE FOR DIAMONDS? MEMOIRS OF AN OILMAN IN THE AMAZON BASIN, Madera, CA, Reflections Press, (1988), xi, 81pp.**
Grable's story has little to do with diamonds and a lot to do with an important, remote wildcat oil discovery in the Amazon Basin in 1939. The story covers his two years of work for Drilexco, from salvaging a sunken drilling equipment barge to drilling the wildcat which was brought in for 2,000 barrels per day at 1,175 feet. The jungle discovery was drilled on the Agua Caliente anticline in Peru, an eight by twenty mile surface feature with 3,000 feet of structural closure. In 1938, Grable also gained fame and income by inventing and patenting power tongs, which are now in use on many drilling and workover rigs. Told in the first person, this is a good travelog, adventure and oil history. The softbound first edition is common.

**B110. (HUEY LONG) Graham, Hugh Davis (editor);, HUEY LONG., NY, Prentice-Hall, Inc., (1970), vii, 184pp.**

This is a compilation of Long's speeches and other published material which is brought into an autobiographical type format. The editor presents Long in three parts: How Long viewed the World; How Long's contemporaries viewed him; and How Long will be viewed by history. A fabulous, turbulent political career fashioned by a brilliant mind and character make for great reading. History from the 1920's and 1930's, much of it on Long's continual bouts with Standard Oil. Standard was actively involved in Long's unsuccessful impeachment proceedings in 1929 and probably unwittingly in his assassination in 1935. The hardbound first edition is common.

**B111. (JAKE SIMMONS, JR.) Greenberg, Jonathan, STAKING A CLAIM: JAKE SIMMONS, JR. AND THE MAKING OF AN AFRICAN-AMERICAN OIL DYNASTY, NY, Atheneum, 1990, 311pp.**

One of the most surprising oil biographies of recent vintage, Greenberg did considerable research over eight years to prepare this history. Much Indian and Black Oklahoma history is related to set up Jake Simmons, Jr.'s remarkable, but unheralded career in U.S. and international oil wheeling and dealing. Jake's parents were African-American, but at least one-eighth Creek Indian. His great grandfather was a Black Creek chief of great renown. Jake Jr. was born in 1901 about twenty miles west of Muskogee, Oklahoma (then Indian Territory). Although poor, he was ambitious and had a chance to study with Booker T. Washington at Tuskegee Institute. He spent five years there and emerged with high self esteem and set a course for himself to become an oilman. He was smart and a good salesman. He began by helping to negotiate oil leases for Black families in Oklahoma and Texas. Many Blacks found themselves unable to deal with the white oil operators. He later went to work the East Texas field boom and again played the middle man, or broker, risking none of his own funds, only his time and expertise. By the mid-1960's, his reputation as a deal maker had grown to such an extent that after Ghana and Nigeria emerged from British control, he was able to work his way in to broker the major oil concessions from those countries to such as Amoco, Oxy, Signal and Phillips. His low key, low risk dealings for The Simmons Royalty Company generated incredible success. He became involved in Civil Rights and was elected president of the Oklahoma

NAACP. Simmons died in 1981 and the company continues in Muskogee under the leadership of son, Donald. Great success story and well worth reading. The hardbound first edition is common.

**B112. (NUBAR GULBENKIAN) Gulbenkian, Nubar, PORTRAIT IN OIL: THE AUTOBIOGRAPHY OF NUBAR GULBENKIAN, NY, Simon and Schuster, (1965), 383pp.**
Born in 1896 and still going strong as he wrote this autobiography, Nubar was the first son of Calouste Gulbenkian, Mr. Five Percent. Nubar gives an entertaining rendition of his childhood and subsequent education at London, Bonn, and graduation at Cambridge in 1917. Working for his Dad, Nubar helped negotiate and draft the merger agreement between Mexican Eagle and Shell Oil. A big one. He admits to being terrified of his powerful father all of his life, but was also his greatest admirer. He worked with Henri Deterding at Shell in 1922 and was exposed to many of the powerful oil people in the world. A falling-out between both of the Gulbenkians and Deterding caused Nubar to leave Shell. He did diplomatic consulting in the Middle East and was an advisor to his Dad. He later spent much of his time directing the activities of the Gulbenkian Foundation which had been formed to manage the family fortune upon the death of Calouste. Major wealth. Entertaining reading for Middle East and world oil fans. The hardbound first edition is common.

**B113. (LUCY GWIN) Gwin, Lucy, GOING OVERBOARD: THE ONLIEST LITTLE WOMAN IN THE OFFSHORE OILFIELDS, NY, The Viking Press, (1982), 288pp.**
Gwin writes her story of one year in the man's world of offshore oil, with humor and sensitivity. Vernacular dialogue and often Cajun accents are effectively used throughout. In 1979, at the age thirty-five, Gwin was divorced and her business had failed...so she ran away to sea. An adventurous soul, she took a job as cook on an offshore oil field supply boat out of Morgan City, Louisiana. She made her way in the dirty, dangerous offshore oil world, overcoming one obstacle after another, largely the men....not the job itself. Thanks to toughness and determination she achieved her goal of becoming a deck hand. Her gutsy personality helped create several critical friendships along the way, without which the ending of this story would not have been happy. The hardbound first edition is common.

**B114. (ADAH HADLOCK) Hadlock, Adah, MY LIFE IN THE SOUTHWEST: EARLY DAY EL PASOAN, AMATEUR ARTIST, CHAMPION GOLFER, AVID WILDCATTER AND GOLD SEEKER, El Paso, TX, Texas Western Press, 1969, xvii, 113pp.**

Born in Missouri in 1883, Adah Robertson was orphaned at age seven and raised by other family members there. She moved to El Paso in 1902 because of her grandmothers health. She quickly adjusted to the new environment and married Frank Hadlock, an artist in 1903. They had five children over the next twelve years, all of whom died in infancy. Adah was talented in many fields and Frank became a successful real estate investor, having a touch of compulsive gambler in his personality. In the mid 1920's, he invested in his first oil deal at Shreveport. The well returned several times its investment within weeks. He was hooked on oil. Adah instantly latched onto anything new and exciting and oil fell in that category. She relates several stories over the next forty years of exploring for oil and gold and other forms of exciting adventure. Most of their exploratory efforts went into the Rockies and Texas, fairly successfully. She was an accomplished artist, golfer, and writer. She also provides much on the early history of El Paso. The hardbound first edition is becoming scarce.

**B115. (HUEY LONG) Hair, William Ivy, THE KINGFISH AND HIS REALM: THE LIFE AND TIMES OF HUEY P. LONG, Baton Rouge, Louisiana State University Press, (1991), xvi, 406pp.**

Scholarly, well researched work. From the DJ: "..dispels many of the myths that have surrounded him and also challenges T. Harry Williams' influential interpretation. Hair emphasizes Huey's darker side, portraying him as a gifted, charismatic leader who craved power above all else". Much on Long's constant involvement with the state oil industry and his battles with Standard Oil. The hardbound first edition is common.

**B116. (ALFRED NOBEL) Halasz, Nicholas, NOBEL: A BIOGRAPHY OF ALFRED NOBEL, NY, The Orion Press, (1959), 281pp.**

Alfred was born in 1833. His father was a prominent architect, engineer, inventor, etc. Alfred was a major part of Nobel Brothers Petroleum, which dominated Russian oil until the 1917 revolution, However, this book mainly concerns other aspects of his life. Halasz follows Alfred from his childhood in Sweden and Russia, as a lonely and solitary child, to a complicated career where he had gained the title of Explosives King of the World. As an adult, Alfred became a pacifist, primarily due to an association with a German woman. The author uses Alfred's poetry and letters to punctuate his interpretation of the man's complex personality and philosophy of business and life. The shock of reading a premature obituary of himself (one of his brothers had died elsewhere) traumatized him as it revealed the public picture of him as the author of mass destruction, "The Dynamite King". Evidently he had been looking upon himself as some kind of a world peace promoter. From that day, he vowed to give his entire wealth to the promotion of peace and proceeded to establish the foundation which bears his name. The hardbound first edition is common.

**B117. (MICHEL HALBOUTY) Halbouty, Michel T., AHEAD OF HIS TIME: MICHEL HALBOUTY SPEAKS TO THE PEOPLE, Houston, TX, Gulf Publishing, (1971), xxii, 354pp.**
James A. Clark is a good friend and associate of Halbouty. He compiled and edited this work. In the opening biographical summary, he states that Halbouty "...is internationally recognized as one of the most outstanding geologists and petroleum engineers in the world". Clark briefly recounts Halbouty's education, associations, accomplishments and honors. No question that he is a unique individual. The book is a collection of some of the hundreds of speeches on oil, all written and presented by him. He is a charismatic and dynamic speaker, always with a message. Any time he felt something was wrong with the oil industry and needed fixing, he did not hesitate to attack whoever he thought was responsible or could do something about it, the independent, the major companies, the state or the feds. His major topics have always been foreign oil and domestic complacency. If not the first to recognize these problems, he was the most vocal. Fine collection of papers by a legend in his own time. The hardbound first edition is common.

**B118. (ARMAND HAMMER) Hammer, Armand, with Neil Lyndon, HAMMER, NY, G. P. Putnam's Sons, (1987), 544pp.**
Done in the first person, throughout, this book expands Considine's 1975 biography and then brings it up to 1986. Hammer's part in humanitarian aid and relief effort for the Russian Chernoble nuclear reactor disaster is the lead-in. Many details are related of his historic dealings with the Russians, beginning while he was in his early twenties and extending to

his efforts towards world peace in personal negotiations with Gorbachev. His brilliant acquisition of a North Sea concession is only skimmed here, but was critical in saving Oxy's world position as an oil producer after Libya nationalized their oil properties. Hammer was an excellent writer and although he appears to be blowing his own horn a lot, in many cases he down-plays his role in important world happenings. Excellent book on an important oil man. The hardbound first edition is common.

**B119. (JACK DRUMMOND) Hammons, Terry, RANCHING FROM THE FRONT SEAT OF A BUICK: THE LIFE OF OKLAHOMA'S A. A. "JACK" DRUMMOND, Oklahoma Historical Society, (1982), xii, 258pp.**
Jack was born in 1896 on the Osage Indian reservation, Indian Territory. His father was a well respected Indian trader and the family lived in the small town of Hominy. As Jack grew up, he developed an obsession to own the largest ranch in the county and by 1928 he owned and leased some 40,000 acres there. By 1931, he had 70,000 acres and was being called a land hog. By the 1940's he started another spread in Marshall County, near the Red River and this led to much oil and gas leasing and speculation. He eventually acquired oil interests in several states, although mostly Oklahoma and Texas. Through the 1950's and into the 1960's, his oil interests became a stabilizing financial influence, balancing the ups and downs of his real estate holdings. Interesting Oklahoma character. The hardbound first edition is common.

**B120. (GODFREY CABOT) Harris, Leon, ONLY TO GOD: THE EXTRAORDINARY LIFE OF GODFREY LOWELL CABOT, NY, Atheneum, 1967, xiv, 361pp.**
Harris received full cooperation of the Cabot family when writing this biography and they held no editorial reservation. Godfrey Cabot was born in Boston in 1861 and he died there 101 years later. He played a significant role in America's growth during that period. His ancestors settled in Massachusetts in 1700. Godfrey's Dad was a Harvard graduate surveyor of some fame. Godfrey was the smallest of seven children and quickly became self assertive and ambitious, to the extreme. After Harvard graduation, he went to work for his brother doing chemical experiments. His first work was in the field of burning Kerosene to form carbon black (soot) which was used in ink and various polishes. He and brother Sam, recognized the potential of producing carbon black from

excess natural gas of the Pennsylvania oil fields which was being burned off or vented, as a useless byproduct. During 1883, he spent much time experimenting and enhancing the production technique and quality of his product. He proceeded to locate and acquire rights to cheap acceptable gas. Many interesting and humorous stories of their somewhat arrogant and hard-nosed business dealings as he and the family expanded to form Cabot Corporation which would aggressively pursue oil and gas exploration and development. He was knowledgeable in geology as he had studied the subject at Harvard and in Zurich. After finding out his sons wanted to use trained oil and gas geologists to help find oil, he quickly squelched the idea. Beginning about 1890 he began spotting his own wells utilizing his own methods. The hardbound first edition is common.

**B121. (ROBERT O. ANDERSON) Harris, Kenneth, THE WILDCATTER: A PORTRAIT OF ROBERT O. ANDERSON, NY, Weidenfeld & Nicholson, (1987), x, 180pp.**
Author Harris works for The Observer, a London newspaper which Anderson owns. He is also a good friend and this biography reflects that. Robert O. was born in 1917 to wealthy, hard-working parents in Chicago. He had a normal upbringing and earned a scholarship to the University of Chicago. He was torn between a career in teaching or going into business. His father's influence prevailed as Dad was a banker who specialized in lending to the independent oil men, a high risk but rewarding field. Robert graduated in 1939 after spending summers in the Texas oil fields learning drilling, producing and refining methods. After working two years for a refining-chemical company, he and two partners bought a 1,500 barrel per day refinery called MALCO at Artesia, New Mexico. He took half interest and moved wife and child to the small town of 4,000 people. Robert quickly took over operations, tripled the output and began producing high octane aviation gasoline. Five years later he bought out the partners. He drilled twenty shallow oil producers in the Caprock Queen field east of Roswell in the late 1940's and continued to expand and diversify, while wheeling and dealing. As a partner of Gulf Oil, he bought the largest independent refinery in California. In 1957, Stanolind Oil drilled a giant oil discovery well near Artesia under which Robert's Hondo Oil & Gas Company held considerable leasehold. The new field was named Empire Abo, and Robert eventually drilled fifty strong oil producers there. Many of the wells exceeded one million barrels per forty

acre drillsite. Robert later came out on top of a merger with Atlantic Refining, a crude-poor refiner who needed his reserves and exploration expertise. A brilliant merger with Richfield Oil for their North Slope of Alaska oil leases in the Prudhoe Bay field was then negotiated. He later had a big part in the Trans-Alaska Pipeline as head of ARCO. As the book was written, Robert was seventy years old and had left the running of ARCO to others. He was preparing to reestablish his small Hondo Oil & Gas Company which would acquire producing properties and again become a factor in the southeast New Mexico oil industry. This is a simple biography of a complex man along with his corporate history. The hardbound first edition is becoming scarce.

**B122. (JOHN D. ROCKEFELLER) Hawke, David Freeman, JOHN D. THE FOUNDING FATHER OF THE ROCKEFELLERS, NY, Harper & Row, Publishers, (1980), vi, 260pp.**
Hawke claims to be the first to use John D.'s personal scrapbooks in preparing a biography. He tries to answer the often asked questions: Why did Rockefeller, being a religious man, break the law so often and in so many ways? Also, how did he manage to become so much richer than anyone else? Both answers are fairly simple for Rockefeller students. Easy reading story with good notes and index. The hardbound first edition is common.

**B123. (JOHAN UDDEN) Heiman, Monica, A PIONEER GEOLOGIST: BIOGRAPHY OF JOHAN AUGUST UDDEN, Kerrville, TX, Privately Published, (1963), xiii, 210pp.**
This biography was originally prepared as a Masters thesis by Mrs. Heiman. The Udden family later published the polished work in a limited edition of 200. The biography is brief but concise. It is half notes, bibliography, appendices and chronological descriptions of Dr. Udden's 130 publications from 1891 to 1941. He was a pioneer geologist of great stature, born in Sweden in 1859 and raised in Minnesota. He studied science at Augustana College. He later taught geology there and became expert in all branches of that field. As a consultant in the early 1900's, he began working in Texas for the University. He started in the Big Bend area and later performed studies throughout the state. As oil was found, he became a proponent of geological record keeping of oil well information, then concentrating his efforts on subsurface geology and its interpretation. He helped to evaluate the potential of the vast University

of Texas lands in West Texas prior to the Big Lake discovery there in 1923. He kept company with the great geologists of the day until he passed on in 1932. Excerpts from his publications are given throughout this book. The hardbound first edition is scarce.

**B124. (MARCUS SAMUEL) Henriques, Robert, MARCUS SAMUEL: FIRST VISCOUNT BEARSTED AND FOUNDER OF THE "SHELL" TRANSPORT AND TRADING COMPANY 1853-1927, London, Barrie and Rockliff, (1960), xi, 676pp.**
Marcus Samuel was born in the "East End" of London in 1853. He was raised in a hard-working Jewish family and received an excellent education in England and Belgium. His family had moved from poverty to moderate wealth by trading in semi-precious stones and rare sea shells from the Orient. They also manufactured small "shell boxes" which were of unusual style and quality, which became the basis of the family fortune. Upon their father's death, Samuel and his younger brother formed a partnership and began trading in the East. By 1883, they bought out older brother who had inherited the bulk of the family business through primogeniture, thus they became owners of Marcus Samuel Company (founded 1834). By 1888 they were transporting small amounts of oil in metal containers. At the time Standard controlled almost all oil transportation and marketing in the world. With the uses of oil and its products increasing daily, Samuel recognized the potential and in 1892 contracted for oil from the Rothchild's Russian oil fields, built shallow-draft tankers and moved the oil cheaply through the Suez Canal, breaking Standard's hold on the Eastern market. In 1897 he formed Shell Trading and Transport. Shell and Royal Dutch had a symbiotic relationship in their constant fight against Standard, but Shell's discovery of oil in the Dutch East Indies was a prelude to their eventual merger. Samuel was slighted in the merger, but had an incredibly productive career. He became mayor of London in 1902. He was knighted for his WW I service and in 1925 was given the title of Viscount for his exceptional British efforts over the years. He died in 1927. Great book about a great oilman. The U.S. edition by Viking is titled BEARSTED: A BIOGRAPHY OF MARCUS SAMUEL. The hardbound first edition is becoming scarce.

**B125. (MELLON FAMILY) Hersh, Burton, THE MELLON FAMILY: A FORTUNE IN HISTORY. NY, William Morrow and Company, Inc., 1978, 640pp.**

One of the two major family biographies, Hersh published this book at about the same time as Koskoff's THE MELLONS. He essentially provides the family history, while shedding new light here and there gained from the release of private information from the younger Mellons. A little shorter than Koskoff's book, this one reads easier and seems very fair to history and the family as well. The last chapter describes how the Mellon family has been divided again and again by deaths and inheritances and the author asserts that by 1978 there was virtually no family cohesion except for the primary foundation. Excellent history and analysis of the family. Probably the best of the five Mellon family books reviewed here. The hardbound first edition is common.

**B126. (J. PAUL GETTY) Hewins, Ralph, THE RICHEST AMERICAN: J. PAUL GETTY, NY, E. P. Dutton & Co., 1960, 404pp.**

This is one of the early biographies of Getty, done when he was sixty-seven years old. Hewins pursued Getty for several years before finally getting the go-ahead and a multiple interview agreement. The book is pretty much a whitewash of the family and digs no dirt beyond general public knowledge at the time. Getty was known as a mysterious oilman even though it was established that he had been worth over a billion dollars by the mid-1950's (largely from the Kuwait oil reserves). The author takes us from Getty's childhood to his early retirement at age twenty-three after acquiring his first million, through his playboy years and on to the incredible Kuwait oil discoveries and development. The author compares Getty to Rockefeller and finds several character similarities such as the penny pinching and hands-on management style. However, the differences are more obvious, such as the gregarious Getty versus the semi-reclusive Rockefeller. Also, Getty's family was wealthy and Rockefeller's poor. Hewins does not point out a primary difference in the men, being the way they made their fortunes. Getty was a lucky and very sharp trader in oil production and exploration, breaking into the major oil company territory, while Rockefeller went to extremes bending laws and brutalizing his small independent competition to corner the new market in oil refining and transportation. British edition titled J. PAUL

GETTY: THE RICHEST AMERICAN. The hardbound first edition is common.

**B127. (CALOUSTE GULBENKIAN) Hewins, Ralph, MR. FIVE PERCENT, NY, Rinehart & Co., (1958), x, 261pp.**

Born near Constantinople in 1869, Calouste was a precocious child and a risk-taker from the start. His Dad became wealthy trading in Baku oil products. Gulbenkian graduated with honors in engineering and science from Kings College in London in 1887. His brilliant one-track mind took control when he returned to Constantinople. He decided he would be a worldwide oil industry exploiter. He knew geology and engineering and was one of the first to recognize the tremendous oil potential of Iraq, writing articles about it while trying to interest investors in a drilling venture there. Gulbenkian doggedly pursued the giant Iraq concession and eventually was instrumental in getting it, receiving five percent royalty on all of its production for his share. Much high financial intrigue by and against him took place primarily due to his early persistent belief in the strategic importance of oil as a political and economic weapon. Like Gutty's later attempt, he broke into the major oil company territory in a way which had not been done before. He was also a voracious art collector and womanizer. Complex international oil dealings with Gulbnenkian in the middle are fairly well sorted out and make very interesting reading. He died, unbelievably wealthy in 1955, at age eighty-six with his estate in a legal mess. This man was a big part of oil history. Cited in Swanson on page 175. The hardbound first edition is becoming scarce.

**B128. (GEORGE HILL) Hill, George P., 70 YEARS, Fort Worth, TX, Branch-Smith, Inc. (1968), 95pp.**

Hill was born at Claude, Texas in 1898. His remembrances from childhood to 1968 are recounted here in well written first person style. Much of his book is on his growing up, friends, family and travels, but the downplayed oil sections are revealing as to his successful business abilities. He graduated from the University of Texas in 1922 and went to work as manager of the University Club at Fort Worth. He had been exposed to the oil business since observing a discovery well blowout in Stevens County, Texas. He invested in the oil business with his Dad from the early 1920's and relates the story of their discovery of the first oil produced in Runnels County, Texas. He describes Fort Worth as being the mail order, oil scam capitol of the U.S. He was involved in a wildcat well which just missed discovering the giant Hendricks field of West Texas. After he and his father sold a substantial lease to the Pure Oil Company in Van Zandt County, Texas, they proceeded to buy and successfully drill leases wherever the action took them in Texas. As of the writing of this autobiography he is still searching for oil. Family photos. The hardbound first edition is scarce.

**B129. (H.L. HUNT) Hill, Margaret Hunt w/Burt and Jane Boyar, H. L. AND LYDA: GROWING UP IN THE H.L. HUNT AND LYDA BUNKER HUNT FAMILY AS TOLD BY THEIR ELDEST DAUGHTER, Little Rock, AR, August House, (1994), 276pp.**
Of the numerous biographies of H.L. Hunt, this one is unique in that it is written by his oldest daughter, Margaret. She was about fifteen when the East Texas field came in and was well aware of oil business happenings from that time forward. H.L. and Lyda always discussed details of oil dealings with their children during family meals. Margaret first worked on the company switchboard and later as H.L.'s executive assistant, thus getting to know many of the big time oil players from the 1930's on. As she tells her story, she makes it sound like she grew up in a normal family....really! The hardbound first edition is common.

**B130. (ALFALFA BILL MURRAY) Hines, Gordon, ALFALFA BILL: AN INTIMATE BIOGRAPHY, Oklahoma City, Oklahoma Press, 1932, ix, 308pp.**
From the dust jacket: "It is perhaps safe to say that no other American ever sky-rocketed so suddenly into popular prominence as Alfalfa Bill Murray has done". This biography was written as Murray was still governor and just as the price of crude oil was recovering from a low of fifteen cents a barrel, which had been due to flush production from the East Texas and Oklahoma City fields. Murray declared military law in August 1931 and closed-in the massive Oklahoma production in defiance of the producers. Only stripper wells were allowed to produce until the price rose to Murray's accepted level of $1.00 per barrel (theoretically making them profitable). This and other measures he brought about have been crucial to the proration and conservation of petroleum, saving the industry irreparable damage. Born in 1869 in Grayson County, Texas, Murray became a lawyer and in 1906, president of the Oklahoma Constitutional Convention, prior to statehood. He was later a

congressman. Murray became the most unusual governor Oklahoma ever had and was held in great esteem by the masses. He was crude, smart and tough, always with his efforts towards the common people's causes. He was one of the most important characters in the history of Oklahoma and he had a big impact on the nationwide oil business. The hardbound first edition is becoming scarce.

**B131. (DAVID ROCKEFELLER) Hoffman, William, DAVID: REPORT ON A ROCKEFELLER, NY, Lyle Stuart, Inc., (1971), 192pp.**

The combination of Hoffman as author and Lyle Stuart as publisher makes certain you will get a great deal of information as well as insight into the subject. From the dust jacket: "David is a report on a man, his money and the ways he uses it. It is a biography of the Chase Manhattan Bank, which David Rockefeller heads and controls, which in turn has a controlling interest in fifty-six of the largest corporations in the U.S." David is portrayed as "savagely disliked" and "one of the three or four most powerful men in the world". David is the last of the five sons of John D. Rockefeller, Jr. Hoffman received no help or anything else from David or any of the Rockefellers in completing this biography, so it was done largely from public records and interviews. With humor and satire, Hoffman paints a frightening picture of David as a wielder of incalculable power. The hardbound first edition is common.

**B132. (PAUL MELLON) Hoffman, William S., PAUL MELLON: PORTRAIT OF AN OIL BARON, Chicago, IL, Follett Publishing Company, (1974), 204pp.**

Paul Mellon was Andrew Mellon's son and heir to the power of the family fortune. To get the author's tone of this 1974 book, a quote from the dust jacket reads: "In an extremely skillful way, the author juxtaposes the story of an eccentric dilettante philanthropist with the saga of Gulf Oil's systematic exploitation of oil rich, but underprivileged nations". The 1973 Arab oil embargo and major oil company price gouging probably motivated Hoffman to chastise the industry. Paul Mellon was an easy target. He was born in 1906 and was alive and well as recently as 1993, and in attendance when one of his horses won the Kentucky Derby. Hoffman rehashes the family history fairly, but goes on to berate Gulf Oil as exploiter of people and nations. Andrew Mellon's donation of fifty million dollars to the National Gallery of Art in 1935 relieved him of the federal tax litigation which he had been fighting and got Paul appointed director of the gallery. Hoffman describes Paul's ulterior motives in many of his philanthropic or charitable contributions. As the book progresses, more and more paragraphs drip with snide innuendo starting with established historical facts and then degenerating to speculation. Hoffman obviously sees no redeeming value in Paul. More good reading for mogul haters. The hardbound first edition is common.

**B133. (JIM HOGG) Hogg, James Stephen, SPEECHES AND STATE PAPERS OF JAMES STEPHEN HOGG EX-GOVERNOR OF TEXAS WITH A SKETCH OF HIS LIFE, Austin, TX, The State Printing Co., 1905, 453pp.**

The book opens with a biographical sketch. Born in Texas in 1851, he was orphaned at age twelve. The considerable family property was lost in the Civil War and he went to work early. Typesetter, publisher, journalist, lawyer, Wood County Attorney, State Attorney General, and in 1890 he became the first native born Texan to be elected governor. A brilliant and powerful speaker, his many speeches included here are worth reading today, as much of the content is valid still. The speeches contain the right amount of humor and sarcasm to get his points across. Much about his work in establishing the Texas Railroad Commission and legislation to protect the state from various scams and corporate abuse. Only oil mention is his move from Austin to Houston to keep up with his oil interests (The Hogg-Swayne Syndicate had a major development at Spindletop). Not in Swanson. Excellent book. The hardbound first edition is scarce.

**B134. (BOYCE HOUSE) House, Boyce, HOW I TOOK HOLLYWOOD BY STORM, Dallas, TX, Banks Upshaw and Co., (1942), 26pp.**

The literary expertise of newspaperman House was nationally recognized from his many articles and books on the Texas oil business and Texas in general. Sam Zimbalist called him on the phone one day and insisted that he help them out in the filming of the now classic movie, BOOMTOWN. Humorous trials and tribulations of his four month stay in Hollywood while working on the film are told in this booklet. Stars Clark Gable, Spencer Tracy, Claudette Colbert, Hedy Lamar, Frank Morgan and Chill Wills made this the blockbuster hit of 1940. House says that almost all of his input and suggestions were ignored, however, he does claim credit for writing Spencer Tracy's speech in the final courtroom scene in which the oilman and his contribution to civilization are eulogized. Typical entertaining work by House. The hardbound first edition is scarce.

**B135. (JOHN D. ROCKEFELLER) Hubbard, Silas, JOHN D. ROCKEFELLER AND HIS CAREER, NY, The Anchor, 1904, 192pp.**

This early diatribe against Rockefeller and the Standard Oil Trust is rarely cited in bibliographies, however it is well worth reading, just as Josephson's and Klien's attacks. Wild assertions and interpretation of history all are given here in a serious work that accuses Rockefeller of being a money-mad monopolist who not only controls the oil industry, but is attempting to monopolize the banking and railroad industries as well. Quoting from Swanson: "Rockefeller's career as an example of the origin and growth of monopoly". Interesting book. Cited in Swanson on page 162; Hardwicke page 45. The hardbound first edition is scarce.

**B136. (HOWARD HUGHES) Hughes, Howard, MY LIFE AND OPINIONS: BY HOWARD HUGHES, USA, Best Books Press, (1972), 244pp.**

Eaton, the editor, is a well-known author, producer, promoter and jet-set personality. He writes an exciting story, first of meeting Howard Hughes on several occasions, then of the unusual circumstances that led to Hughes' request that he write a book on his life and opinions on important events in which he played a key role. According to Eaton, Hughes gave him the manuscripts, written in long hand, and asked him to edit them. He wanted the writings formed into a viable format for publication. Doesn't really sound like Hughes. The book shows that Hughes felt that his reasoning on many controversial events had been misinterpreted. Not too surprisingly, his motives, reasoning, intuition etc. are not in line with public and media thinking on the various subjects. The book is written in the first person and contains much on Hughes Aircraft and flying, as well as Las Vegas days. There is also some information on the cryogenic preservation technique used to preserve his body. No new revelations, but very interesting. The hardbound first edition is common.

**B137. (H. L. HUNT) Hunt, H. L., ALPACA, Dallas, H.L. Hunt Press, (1960), 183pp.**

Hunt wrote this right-wing utopian novel (with a very thin story line) in order to put forth his idea of how a perfect constitution should be written.

The lead character in his story, Juan Alchala seeks input from the wisest minds in the world in hopes of gaining their knowledge as to what form of government works best. He needs this wisdom to prepare a constitution for his own country of Alpaca, which has just thrown out its dictator. Much on voting rights, taxes and governmental term limits. Hunt actually played a big part in the no-third-term amendment for the office of president. Right wingers unite! Revised and retitled ALPACA REVISITED edition in 1967 w/199pp. The softbound first edition is scarce.

**B138. (H. L. HUNT) Hunt, H. L., HUNT HERITAGE: THE REPUBLIC AND OUR FAMILIES, Dallas, Parade Press, (1973), 153pp.**

Hunt was born in Illinois in 1889. Writing this short book at age eighty-four, he gives a brief biographical sketch of his upbringing, young adult wanderings, start in the oil business and finally his participation in developing the great East Texas Field. Hunt was a precocious child and relates the now famous story of how he was reading commodity market quotes from the local newspaper at age two, and understanding them (Hunt does not dispute this family legend). The best part of the book is the appendix where six former Hunt employees recall the old times. The book is just a quick look at Hunt and excludes anything negative or controversial. Probably just the way he remembered it all. The hardbound first edition is becoming scarce.

**B139. (H. L. HUNT) Hunt, H. L., FABIANS FIGHT FREEDOM, Dallas, H.L. Hunt Press, (1964), 188pp.**

Fabians are an intellectual, left-wing international organization which offended Mr. Hunt. From the cover: "... shows graphically how we are being subtly infiltrated by alien ideas and how our freedom is fast being sold down the river!". This is Mr. Hunt at his best. The paperback original first edition is scarce.

**B140. (H. L. HUNT) Hunt, H. L., WHY NOT SPEAK?, Dallas, H.L. Hunt Press (1964), 192pp.**

Useful methods in preparing and delivering speeches, which Hunt did continuously. How to speak for freedom. Included are several classic speeches such as MacArthur's Old Soldiers Never Die and many others.

Also includes scripts from Hunt's "Life Line" right wing radio show. The paperback original first edition is scarce.

**B141. (H. L. HUNT) Hunt, H. L., HUNT FOR TRUTH, Dallas, H. L. Hunt Press, (1964), 254pp.**

This is the first of five Hunt paperbacks which consist of his letters to the editor. All are pure Hunt and right wing conservatism. This one was followed by: H.L.H. COLUMNS, (1966), 191pp.; RIGHT OF AVERAGE, (1967), 192pp.; WEEKLY STRENGTH, (1968), 192pp.; and CONSTRUCTIVELY, H.L. HUNT, (1969), 191pp. All paperback originals are scarce.

**B142. (H. L. HUNT) Hurt, Harry III, TEXAS RICH: THE HUNT DYNASTY FROM THE EARLY OIL DAYS THROUGH THE SILVER CRASH, NY, W.W. Norton & Co., (1981), 446pp.**

Published two years after the death of H.L. Hunt, this unauthorized Hunt biography is done in very objective style. He was able to interview Bunker, Herbert and Ray Hunt, the major family players, as well as two hundred and fifty friends and enemies in preparation of this work. Hunt's precocious childhood, his gambling, and drift into oil drilling are all rehashed in very good detail with copious reference notes at the end of the book. Much concentration is on the family's efforts to corner the soybean and silver markets. Most of this action takes place after H.L. Hunt's death. Interesting accounts are given of the three family's legal fighting along with the intricate settlements that were worked out. Probably the best and most comprehensive of the Hunt biographies. The hardbound first edition is common.

**B143. (THOMAS BARD) Hutchinson, W.H., OIL, LAND AND POLITICS: THE CALIFORNIA CAREER OF THOMAS ROBERT BARD, Norman, OK: University of Oklahoma Press, (1965), Vol 1 xv, 349pp, Vol 2 viii, 394pp.**

Bard was born in Pennsylvania in 1841 to wealthy and religious parents. His father was a successful attorney and Bard was trained in law, business and surveying. During the Civil War he was a railroader. Along with a cousin and family friend he decided to go to California to make his fortune in 1865. He was convinced that land and oil were the future of the country. He drilled and tunneled for oil in the Ojai area for years, meeting moderate but erratic success. Finally in 1890 he got together with a successful oil man from Pennsylvania, Lyman Stewart, and through very complex maneuverings merged his several companies into a new entity called Union Oil of California. Much early oil history is related as Union Cal took off to become a major oil company. Bard brought in California's first gusher and he became a critical factor in helping develop its oil industry. He had a bitter feud with Stewart and left Union Cal with a considerable fortune in 1900. He later participated actively in state and national politics where he consistently broke new ground. The slipcased hardbound first edition is common.

**B144. (C. C. JULIAN) Hutson, William Gardiner, MY FRIENDS CALL ME CC: THE STORY OF COURTNEY CHAUNCY JULIAN, Santa Fe, NM, Sunstone Press, (1990), 126pp.**

Any account of oil con men will mention something about C.C. Julian. He ranks right up there with others such as S.E.J. Cox, the mail order flim-flammer and Robert Trippet of the Homestake Swindle. The odyssey of Julian's remarkable oil and mining promotions takes place largely in California but continues in Oklahoma. The two main characters are: his friend, Bill Feeney, a journalist and supporter to the end; his antagonist, postal inspector Manny Madeira, who pursues Julian as far as he can. Much of the book is concentrated in a ten year period (1923- 1933), where Julian gets into the oil business and promotion, unscrupulously bilking investors and making millions of dollars. The feds finally managed to shut him down and break his spirit by pressing mail fraud charges. Julian was largely done-for by the time the Depression hit and his situation became unbearable. He had made several million dollars and lost it all, finally coming to a sordid end in Shanghai, China. Lots of authentic dialogue is included in the text which was well researched by the author, a criminologist. This is a first class oilman biography, unfortunately the oilman was not first class. The softbound first edition is common.

**B145. (HAROLD ICKES) Ickes, Harold L., THE SECRET DIARY OF HAROLD L. ICKES, NY, Simon and Schuster, (1954), xi, 2192pp.**

Vol I titled THE FIRST THOUSAND DAYS 1933-1936, xi, 738pp; Vol II titled THE INSIDE STRUGGLE, 1936-1939, 759pp; Vol III titled THE LOWERING CLOUDS 1939-1941, 695pp. The original diary contains 6,000,000 words. It has been selectively reproduced here. Ickes was appointed Secretary of the Interior under Roosevelt and spent most of his time in office dealing with oil problems, either created by the industry or involving the public lands. Later the logistics of oil in wartime, which was just beginning as these volumes end, would turn out to be his most impressive effort. Ickes had been unsuccessful in politics, most often supporting candidates who were out of synch with the public favor. He even switched parties several times. Roosevelt picked him for this important cabinet job because he was known as being honest, smart, loyal, dedicated and hard working. In his late 50's when he took the job, Ickes gained many friends and enemies while creating an exceptional public service record. Not cited in Swanson. Reprinted in 1974 by DaCapo. The hardbound first edition is scarce.

**B146. (NORMAN JACOBSON) Jacobson, Norman H. OIL FIELD DIVING: THE DIVER'S STORY., Westwego, LA, Privately Published, (1983), xii, 113pp.**
Jacobson was born in 1947 in Massachusetts. After high school he joined the army and emerged as a captain, serving four years in Vietnam. After the service he gained certification as a commercial diver and moved to Louisiana as an oil field diver. In October 1982, he undertook a twelve day job which is the substance of this book. The story is real with a lot of dialogue, photos and explanations including appendices describing this risky but lucrative business. The assignment required a diving bell and entailed the job of opening and closing undersea valves for a pipeline cleaning procedure at a depth of 400 feet. He learned a lot while resolving one crisis after another and you are there. Much action and excitement. The softbound first edition is common.

**B147. (ERNEST THOMPSON) Johnson, T.E., FRIENDS HONOR COLONEL THOMPSON, Amarillo, TX, Amarillo Daily News, 1935, 44pp.**
This testimonial dinner took place at the Baker hotel in Dallas, Texas after Thompson had done his monumental work on the Texas Railroad Commission and had been elected chairman. Numerous speakers rehash the great job Thompson did in keeping the feds out of Texas by establishing conservation practices and effective proration of oil production. Thompson still had many enemies in the ranks of independent hot oil runners, but within a few years almost everyone realized that his

work probably saved the oil industry from a form of nationalization. Good reading about a good man. The softbound first edition is scarce.

**B148. (L. E. PHILLIPS) Jones, Billy M., L. E. PHILLIPS: BANKER, OIL MAN, CIVIC LEADER, Oklahoma City, OK, Oklahoma Heritage Association, (1981), xi, 206pp.**
From the dust jacket: "Never one to forget his heritage, he used the lessons he learned as the son of a dirt farmer - hard work, honesty, thrift, and self reliance - and, along with his brother Frank, built the largest independent petroleum company in America..." L. E. was never as famous as his older brother Frank, but he worked at the successful oil and banking empire they constructed just as hard. He held several diverse jobs before Frank convinced him to go to Bartlesville and drill some wells with him. Their persistence, plus a certain amount of luck eventually allowed them to grow into a large independent oil company and banking business. L. E. died in 1944 after retiring in 1933. Much coverage is largely of L. E.'s early struggles and his family. Also, much is included on his rise in the Oklahoma banking industry. The hardbound first edition is common.

**B149. (JOHN D. ROCKEFELLER) Josephson, Emanuel M., ROCKEFELLER "INTERNATIONALIST": THE MAN WHO MISRULES THE WORLD, NY, Chedney Press, (1952), 448pp.**
Josephson was a medical doctor, born in 1895. He practiced medicine and conducted medical research. He also wrote extensively on medicine, economics and politics. The dust jacket describes him as: "fearless in his exposure of sham and corruption in those fields." He is a right-wing witch-hunter and makes his case that Rockefeller, Standard and the Communists either now control, or are trying to establish control over everything from education to the United Nations. This weak, rambling tirade probably was well received by the hard core right wingers and a few gullible innocents. Boring. The primary value of this book is that it provides a historical look at the anti-Communist hysteria of the 1950's. Cited in Swanson on page 162. The hardbound first edition is becoming scarce.

**B150. (ALINE BARNSDALL) Karasick, Norman M. & Dorothy K. Karasick, THE OILMAN'S DAUGHTER: A BIOGRAPHY OF ALINE BARNSDALL, Encino, CA, Carleston Publishing, (1993), xiv, 268pp.**

William Barnsdall's family came to America and settled on a farm in Pennsylvania. He had been a shoe maker in England, and the 250 residents of Titusville welcomed him by building him a boot and shoe shop in 1832. By the time Drake drilled his famous well in 1859, the Barnsdall clan was very prominent and owned a good deal of the north part of town. William and partners Meade & Rouse drilled the second well on Oil Creek over 100 feet deep and flowing 50 barrels per day. Barnsdall immediately built the world's first refinery. Area landowners soon became very wealthy because of the oil activity and population grew to 10,000 by 1865. Second son, Theodore was born in 1851 and joined his Dad in the business. Theodore had two daughters Aline & Frances, of a very unhappy marriage. Aline was twenty when her parents were finally divorced. Theodore went on to develop significant oil and gas reserves across the U.S., particularly on the Osage Indian reservation of Oklahoma. He died in 1917, leaving a fortune to the two daughters. Aline had become a wealthy free spirit, going where she wanted and doing what she pleased. She hit Los Angeles like a storm. She began running with the left leaning actors and politicians, and developed much clout there through her funding of civic and art projects. Her life from 1916 to 1938 was consumed with the plight of a jailed labor leader, Tom Mooney, who had been convicted of bombing an anti-labor rally and killing ten people in San Francisco. Constantly in conflict with the city, she became known as a leftist propagandist. The strange lady came to a strange end in 1946. The hardbound first edition is common.

**B151. (HOWARD HUGHES) Keats, John, HOWARD HUGHES, NY, Random House, (1966), 304pp.**

This is a very well written and researched biography of Hughes. The dust jacket and forward have no mention of the three years of court battles, between publisher Random House and the Hughes organization, which preceded the release of this book. Until 1966, no biography of one of the richest, most interesting and most bizarre subjects had been allowed to reach publication. Several had been squelched by threat or payoff. This work is a mild and restrained story of a strange and unique personality. Born in 1905, Hughes grew up fast and inherited his father's multi-million dollar oil tool business at nineteen. Hughes was an acknowledged genius, but almost immediately he began acting like a spoiled brat, doing whatever he felt like. Everything was done to extremes. Movie making, chasing movie starlets, aircraft research and development, and a relentless pursuit of air speed records consumed his life for several years. Noah Detrich ran much of the very profitable oil tool company and Hughes took the profits which he invested in diverse places. With incredible persistence he made success of almost everything he tried, although his WW II effort with the Spruce Goose (a massive amphibious wooden plane) as well as several other air contracts for the government were called off at the end of the war and never completed by Hughes, costing the government sixty million dollars. He suffered nervous exhaustion at the end of this episode at age forty. At the writing of this biography, Hughes was sixty years old and involved in serious, litigation over his TWA Airlines (77% Hughes owned) which threatened to cost him 135 million dollars, roughly ten percent of his net worth. Keats was denied access to Hughes or his close associates while he was preparing this unauthorized book, so he was probably unaware that his subject had already gone beyond merely eccentric, and that he was already involved in a most bizarre lifestyle. A classic. The hardbound first edition is common.

**B152. (JOHN KELLY) Kelly, John S., OIL HUNTERS, Detroit, Harlow, (1983), 184pp.**

Kelly was born in Pennsylvania around 1900 and had some early college work at Lehigh and Penn State. Family finances dictated that he leave school, so in 1925 he went to Texas, then Oklahoma to survey the oil industry and see if there was an opportunity for him. He gives much history of oil men and oil events from the 1920's to the 1970's and interweaves his own story. His first job was with Gypsy Oil as a roustabout in the Tulsa area. He took a job with Ohio Oil during the 1930's at Houston and through much effort was able to acquire Bachelor's and Master's degrees in geology. Several other jobs and exciting geological experiences later (1951), he went to work for the famous explorationist and entrepreneur Josh Cosden at Big Springs Texas. He became chief geologist and was responsible for several successful and otherwise exploratory and development projects for the company. Thereafter Kelly became a consulting and independent geologist, spending most of his time well sitting and mud logging around the U.S., mainly in the Permian Basin. He also put drilling deals together and sold them. This is a fairly typical picture of an oil geologist's career. Kelly says he learned something new about his field every day and was exposed to important oil figures and events routinely. The hardbound first edition is becoming scarce.

**B153. (R.E. KELSAY) Kelsay, R.E., MEOW AND THE MULLETS, Jacksboro, TX, The Author, (1994), viii, 134pp.**
Subtitled: An Oil man's observations on the late, and by some lamented, energy crisis, and the fleeting bonanza it created for a few of us. Kelsay was a promoter during the OPEC induced oil boom of the late 1970's and early 1980's. He is intimately familiar with the frenzy of private investing that took place in drilling ventures during that time. Unwary investors are the "Mullets" of the title. MEOW of the title is an abbreviation for President Carter's "Moral Equivalent of War" slogan for energizing the oil industry to find and develop new reserves. Everyone was upset and somewhat frightened by the Arab embargoes. The book appears to be written with good intentions as Kelsay relates the details of his personal experiences in dealing with the frantic, uninformed investors. Those readers who were involved in the boom will not be surprised at the waste and incompetence of the boom-ethic promoters. Those readers who were not involved will probably be shocked. Much useful information is provided for investors who might be tempted to participate in oil drilling ventures during the next boom. And there may be another one. The softbound first edition is common.

**B154. (CHARLES KEPLINGER) Keplinger, H.F., WITHOUT FEAR OR FAVOR: THE STORY OF CHARLES HENRY KEPLINGER, Houston, Gulf Publishing Co., (1982), 164pp.**
Author, Keplinger (nicknamed Kep) is the son of Charles and currently president of Keplinger Companies. Charles was born at Lyons, Kansas in 1910 and was a precocious child. The family moved to El Dorado in 1915 so his father could keep his railroad job at the oil boomtown. One of the giant U.S. oil booms had occurred there, following Wichita Natural Gas Company's #1 Stapleton discovery well. In high school, Charles got a summer job with Roxana (Shell) where he learned a lot about drilling and exploration. He attended college at the University of Tulsa and worked during the summers for Amerada and National Tank Company. He graduated Cum Laude in engineering and went to work for National Tank

at Kilgore, right after the East Texas field had been confirmed. He did innovative work on oil separators and sold them along with storage tanks which were in big demand. He received a Masters Degree from George Washington and in 1933 went to work for Shell. In 1944 he began his now famous consulting business at Tulsa where he established a first-class reputation. He and son Kep ran one of the top energy consulting firms for years as Charles concentrated on his favorite pursuit, expert witness testimony. Charles died in 1981, but the work goes on largely out of Houston. Good biography by a respectful son. The hardbound first edition is common.

**B155. (ROBERT S. KERR) Kerr, Robert S., LAND, WOOD & WATER, NY, Fleet Publishing, (1960), 380pp.**
Senator Kerr was one of Oklahoma's top oilmen and an expert in the field of natural gas conservation. He wrote this book with considerable help from other specialized experts. Engineers, geologists, environmentalists, agriculturalists, hydrologists all contributed to what Kerr established as a comprehensive conservation plan for a watershed area called A-W-R (Arkansas, White & Red River). The area was centered in Oklahoma but covered parts of adjoining states. He received national recognition as a freshman congressman by gaining approval for study of his plan. As of this writing it was still being studied, although many of his ideas had already been implemented. This is something of a political book, but there is much of historical interest as well. The hardbound first edition is common.

**B156. (HUGH CULLEN) Kilman, Ed & Charles E. Gilbert, Jr., THE CULLENS: TWO GREAT TEXANS, Houston, University of Houston, (1949), 82pp.**
Quoting from the cover: "One was a power in Texas in the days of The Republic. The other became the South's foremost philanthropist". Kilman was editor of the Houston Post and Charles Gilbert was editor of Houston, the magazine. They prepared this booklet at the request of the University of Houston. The Cullen family had donated an administration building and a copy of this book was to be deposited in the cornerstone. Much on Ezekiel Cullen's historic report to the Third Congress (1839) of the Republic as Chairman of the House Committee on Education. The report is reproduced here and was important to establishing the Texas school system. Nothing on the exceptional career of Hugh Roy Cullen

and his Quintana Oil Company, however, the philanthropic work he did is well documented. Photos and awards are also presented. The hardbound first edition is very scarce.

**B157. (HUGH CULLEN) Kilman, Ed & Theon Wright, HUGH ROY CULLEN: A STORY OF AMERICAN OPPORTUNITY, NY, Prentice-Hall, Inc., (1954), viii, 376pp.**
Cullen was born in 1881 near Dallas. He had little education but became incredibly self-confident. He was first a cotton broker, then real estate man. From the real estate, he became interested in oil and was a moderately successful operator during the 1920's in West Texas and the Texas Gulf Coast. Five short chapters in the middle of the book outline his remarkably successful wildcatting and brilliant oil exploration work. He had a rank-wildcat discovery in 1930 which really stimulated his oil career and soon formed Quintana Oil Company to manage the family fortune. The last half of the book relates Cullen's politics and many philanthropies. An exceptional oil finder. Cited in Swanson on page 175. The hardbound first edition is common.

**B158. (JOSEPH CULLINAN) King, John O., JOSEPH STEPHEN CULLINAN: A STUDY OF LEADERSHIP IN THE TEXAS PETROLEUM INDUSTRY 1897- 1937, Nashville, Vanderbilt University Press, 1970, ix, 229pp.**
Born in Pennsylvania in 1860 to a farm family, he went to work in the local oil fields as a fourteen year old messenger boy. He worked at several different jobs and by 1882 was employed by Standard Oil where he eventually became a division superintendent. In 1895 he started his own successful oil equipment manufacturing firm called Petroleum Iron Works. Disorganized operators and logistical problems at the developing Corsicana oil field in Texas (700 wells at the time) required management expertise. With the help of Standard, Cullinan was hired by a consortium to save the day and he did. Cullinan left his Standard support (apparently) and was at Spindletop in 1901 the day the Lucas gusher came in. He founded Texas Fuel Company (later Texaco) to buy crude and joined efforts with the Hogg-Swayne Producing Group. He also started several other exploration and development companies. Much about Cullinan's many companies which were designed to allow him to perform like an integrated oil company. Much on his big fights with the Texas anti-trust laws, and his winning. He resigned from Texaco in 1913 complaining that

eastern investors had taken over management of his company. The hardbound first edition is common.

**B159. (HOWARD HUGHES) Kistler, Ron, I CAUGHT FLIES FOR HOWARD HUGHES, Chicago, Playboy Press, 1976, 217pp.**
The author was hired by Hughes Productions, an inactive movie company in 1957. Up to that time he had been an ordinary citizen, born in 1926, served in WW II, and holding several different jobs. This first job with Hughes paid $7.00 per hour for eight hours a day, seven days a week, and consisted of protecting an old Convair 440 aircraft at a remote airport. He never had a clue why the plane had to be protected for Hughes. He was actually being groomed to become one of three personal bodyguards to Howard Hughes at his Beverly Hills bungalow. Numerous stories of his relationship with the bizarre, eccentric Hughes and the unusual job are related here. His main duty was to swat flies and run the movie projector. From this strange situation, Kistler gained his considerable knowledge of Hughes and the Hughes empire. Funny and strange. The hardbound first edition is common.

**B160. (RAY SIEGFRIED) Klappenbach, E.S., THE EXTRAORDINARY LIVES OF RAY H. SIEGFRIED, Tulsa, OK, Privately Published, (1978), vi, 115pp.**
Born near Pittsburgh in 1894, Ray suffered a tough childhood, but was industrious and survived with a desire to succeed. He left home at sixteen and decided to head for the oil boom. Of the several choices, Tulsa was the cheapest rail fare so he went there in 1911. He worked for Texaco for eight years in pipeline construction and supervision, later, for Pierce Oil Company (Sinclair). In 1920 he began his entrepreneurial career as an oil related insurance agent. He rapidly expanded into real estate development and oil and gas production then many other diverse ventures. Most of his successful ideas were oil related. Ray's son and grandson carry on the family business in Tulsa. The hardbound first edition is becoming scarce.

**B161. (MELLON FAMILY) Koskoff, David E., THE MELLONS: THE CHRONICLE OF AMERICA'S RICHEST FAMILY, NY, Thomas Y. Crowell, Publishers, (1978), v, 602pp.**
This is a very fair and comprehensive biography of the Mellon family. The author stresses the role of much ignored W.L. Mellon, who was the key man in structuring the future of Gulf Oil and acquiring the Guffey &

Galey interests in the Spindletop field. He ran it for years. The last half of the book explores the lives of the younger generations of the family. An incredibly rich and diverse group of people, their stories provide more material than one can easily absorb. It is a classic American saga. This is one of the most popular oil biographies, with good reason. The hardbound first edition is becoming scarce.

**B162. (NELSON ROCKEFELLER) Kramer, Michael & Sam Roberts, "I NEVER WANTED TO BE VICE-PRESIDENT OF ANYTHING": AN INVESTIGATIVE BIOGRAPHY OF NELSON ROCKEFELLER, NY, Basic Books, Inc., (1976), xii, 419pp.**
This biography is definitely unauthorized as the two authors (investigative reporters) put their research and interpretation of events on paper. Largely anecdotal, the book fairly covers Nelson's drive to governorship of New York, the work he did there and in the vice presidency. Power and money got him almost everything he wanted, except the Presidency. Timing was probably the only thing that precluded that. From the dust jacket: "...Rockefeller's complex career, surely the most fascinating example in our political history of unparalleled wealth put at the disposal of unlimited ambition". The book was published after a congressional committee had raked him over the coals in public. Nelson was rich, ambitious and very political. Not a very favorable biography. The hardbound first edition is common.

**B163. (ROCKEFELLER FAMILY) Kutz, Meyer, ROCKEFELLER POWER: AMERICA'S CHOSEN FAMILY, NY, Simon and Schuster, (1974), 288pp.**
Kutz presents no credentials other than degrees from MIT and Rensselaer. He pursues his thesis that the Rockefeller philanthropies and other civic projects are designed not for the public good, but only to create more Rockefeller power. He did considerable research on the philanthropies and describes them in an appendix. His closing statement expresses great fear that if Nelson is elected president in 1974, he will gain control of the armed forces and the federal administrative apparatus, giving him and his family more power than any American family....ever. The hardbound first edition is common.

**B164. (JOHN D. ROCKEFELLER) Latham, Earl (editor), JOHN D. ROCKEFELLER: ROBBER BARON OR INDUSTRIAL STATESMAN?, Boston, MA, D. C. Heath and Co., (1949), vii, 115pp.**
This is an excellent, broad based collection of excerpts from previously published biographical material on John D. Rockefeller by several well-known authors. Included are: Flynn's GOD'S GOLD, Tarbell's THE HISTORY OF THE STANDARD OIL COMPANY, Josephson's THE ROBBER BARONS, John D. Rockefeller's RANDOM REMINISCENCES OF MEN AND EVENTS, Nevins', JOHN D. ROCKEFELLER, and Lloyd's WEALTH AGAINST COMMONWEALTH. Also included are critical reviews of Nevins' and Lloyd's books plus a rebuttal letter from Nevins defending his description of Lloyd as an untrustworthy historian (not convincing). Also included is a list of suggested reading on the subject. All sides are represented in Latham's study and the reader can form his own opinions. A note of caution: Amherst College has a strong liberal history and that might be reflected in the editor's selections. Cited in Swanson on page 163. The softbound first edition is scarce.

**B165. (J. PAUL GETTY) Lenzner, Robert:, THE GREAT GETTY: THE LIFE AND LOVES OF J. PAUL GETTY - RICHEST MAN IN THE WORLD., NY, Crown Publishers, Inc., 1986, xii, 283pp.**
This book was probably stimulated by the incredible 1984 financial fiasco of Texaco's purchase of Getty Oil and subsequent litigation publicity. Although written ten years after Getty's death, Lenzner presents a well documented and systematic view of the life of Getty. Good family photos. The appendixes are very valuable, giving among other things an impressive "Chronology of the Getty Fortune", along with copious notes. From the dustjacket: "Hardbitten oilman, spectacular lover, absent parent, miserly art collector, philandering and bigamous husband, bohemian expatriate, social snob, physical coward, Hitler sympathizer, writer, master businessman, and the richest man in the world - J. Paul Getty was these things and more". Lenzner discusses the five wives, Getty's trust problems, and grandson's ear-loss, along with many of the other family problems. Very good reading about a unique individual. This is the most credible of the many efforts to characterize Getty. The hardbound first edition is common.

**B166. (J. PAUL GETTY) LeVane, Ethel and Getty, J. Paul, COLLECTORS CHOICE, London, The Camelot Press, Ltd., (1956), 360pp.**

No oil and gas in this book, just what big oil and gas income can provide. Part I by LeVane describes European art history while traveling through Europe in a semi-journal style, with and without Getty. Part II contains more art history by Getty. He was an acknowledged expert through experience gained over years of concentrated study. The hardbound first edition is becoming scarce.

**B167. (ARISTOTLE ONASSIS, ET AL) Lilly, Doris, THOSE FABULOUS GREEKS: ONASSIS, NICHAROS, AND LIVANOS, NY, Cowles Book Co., Inc., (1970), ix 373pp.**

The author is a well known society columnist and world-traveled best-selling writer. She did considerable research of history and court records along with countless personal interviews to get this story of the three Greek tycoons. Her story of the three men who revolutionized and devastated the oil tanker industry while amassing incredible wealth is well told. Aristotle Onassis was born in 1906 in Greece. Stavros Niarchos was born there in 1909. The two entrepreneurs became great competitors in the oil tanker business as well as in other pastimes. Stavros Livanos was born in 1891, son of a wealthy Greek ship owner. By the time WW II was over, Livanos was a millionaire shipping magnate and Onassis and Niarchos were generally competing with him for world shipping domination. By the mid-1960's, the three along with other Greek tycoons had the largest merchant fleet in the world. A little sensational, but good reading. The hardbound first edition is common.

**B168. (LEWIS WEEKS) Lloyd, John, THE WEEKS ROYALTY, Sydney, Australia, Fortis Publishing, (1993), xiii, 311pp.**

Lloyd's book concerns Lewis G. Weeks, a famous U.S. oil geologist, and the consulting-fee overriding royalty he received in 1960 for recommending the Bass Straits of Australia as a potential offshore oil play to Broken Hill Proprietary (an Australian mining giant). Prophetic. Lloyd does a fine job of background on Weeks and his impressive career at Standard. Weeks had retired from Jersey Standard and was on a $20,000 annual retainer with BHP. Weeks was an instinctive oil explorationist on the world scene and had been responsible for directing several large oil plays for the company. Much of the book reviews oil exploration history around the world and more particularly the unsuccessful efforts prior to 1960 in Australia. After detailing several major well-known overrides and working interest agreements as paid to consultants such as Gulbenkian, Lloyd concludes that the Weeks royalty of 2.5% was far too high and out of proportion on value basis as well as its historic equity. His approach and conclusion is not surprising as he is a career appraiser and prior to retirement in 1979 was head of the Australia Public Lands Valuations. Usual appraisal methods plus history and hindsight are used to answer the question: How much is it worth to have someone tell you where to drill a rank wildcat well which will eventually yield you three or four billion barrels of oil? Somehow the 2.5% consulting fee does not seem relevant when looking at the big picture. Excellent book about one of America's premier oil geologists. The softbound first edition is common.

**B169. (BOOTS ADAMS) Lobsenz, Norman M., THE BOOTS ADAMS STORY, Bartlesville, OK, Phillips Petroleum Company, (1965), 126pp.**

Born in 1899 Adams had moderate education and went to work for the fledgling Phillips Petroleum Company in 1920 as a warehouse clerk. Frank Phillips spotted his abilities early and saw to it that Adams was trained in every department of the new oil company. Lots of good photos and stories are included, along with a detailed description of Adams' rise to become Frank Phillips' personal assistant, and eventually president of the company. A major career accomplishment for Adams was his handling of Phillips' complicated leasing and development drilling program in a residential area near the capitol building at the Oklahoma City field during the early 1930's. The hardbound first edition is common.

**B170. (CHARLES LOCKHART) Lockhart, Charles, SHORT AUTOBIOGRAPHICAL SKETCH: CHARLES LOCKHART 1818-1905, U.S., Privately Published, Ca 1964, 9pp.**

From the preface: "One day years ago, some of Mr. Charles Lockhart's friends arranged to have a stenographer concealed outside his office while they questioned him about the historical aspects of his career. It is from these notes taken by the stenographer that were later edited by Mr. Lockhart..." Lockhart was born in Scotland in 1818 and came to America in 1836. He had some schooling and experience in the grocery business by that time. He found a grocery-market job in Pittsburgh with Mr. McCully, who would later become an important partner. He worked in the store and managed to do some traveling with his good friend William Frew. In 1846 Lockhart speculated in the purchase of a canal boat. He made some money with it and sold it at a profit to Mr. Jones of Keir & Jones. Starting in 1852 a fascinating story unfolds as a salt well began making oil and Lockhart was brought in to the almost non-existent oil business to try and sell the oil. In early 1860, within months of Drake's discovery at Titusville, Lockhart and his partners had completed a forty-five barrel a day wildcat well on Oil Creek. They sold the oil at Pittsburgh for thirty-four cents a gallon. Incredible drilling success followed as Lockhart & Frew quickly moved into the refining and transportation business, being the first to ship and sell oil overseas. In 1874 Lockhart met with Rockefeller, Flagler and Warden. They decided to pool their barrel factories, refineries and money to form a stock company called Standard Oil. This very short autobiography is the ultimate in understatement. The hardbound first edition is becoming scarce.

**B171. (CALOUSTE GULBENKIAN) Lodwick, John, in collaboration with D.H. Young, GULBENKIAN: AN INTERPRETATION OF CALOUSTE SARKIS GULBENKIAN, London, Heinemann, (1958), vii, 289pp.**

This book was published at almost the same time as Hewins' MR. FIVE PERCENT, coming about three years after Gulbenkian's death. The author had access to Gulbenkian's personal secretary of thirty years, (from 1908), D.H. Young, who provided much previously unpublished information. Gulbenkian was probably the richest man in the world because of the five percent commission royalty he had negotiated on the gigantic Persian oil concession. He actually brokered the concession to Knox D'Arcy at Anglo Persian. There is a good deal of interpretation involved in this biography as to intentions and motivations of the characters. Much of the dialogue Young provides helps us to understand this strange genius. You get the impression that he was destined to be famous for something, even if the Middle East oil had not materialized. The hardbound first edition is common.

**B172. (ANDREW MELLON) Love, Philip H., ANDREW W. MELLON: THE MAN AND HIS WORK, Baltimore, MD, F. Heath Coggins & Co., 1929, 319pp.**
Andrew was seventy-four years old at the writing of this, the first biography specifically about him. Andrew was born in Pittsburgh in 1855 to Thomas and Sarah Mellon. Thomas was a lawyer, then judge, then banker, then millionaire banker. He eventually set the sons up in the banking business and they expanded beyond anyone's imagination. Andrew had a net worth of around a billion dollars when Harding appointed him as treasury secretary. Harding rightly felt that Mellon knew more about money than anyone else in the world. Why not get the best? The bulk of this favorable book is laudatory and speaks to Andrew's general policies, his philosophies, his awards and general doings as the country was in unprecedented prosperity. The timing of publication is interesting as it almost coincides with the beginning of his rapid fall from favor with the government and the public. As Secretary of the Treasury, he later received much of the blame for the Depression. Not cited in Swanson. The hardbound first edition is becoming scarce.

**B173. (J. PAUL GETTY) Lund, Robina, THE GETTY I KNEW, Kansas City, Sheed Andrews and McMeel, Inc., (1977), 183pp.**
Lund was a lawyer and Getty's public relations and legal advisor for sixteen years. She paints an entirely different picture of Getty than the public image which later became exposed. She only relates incidents about which she had direct knowledge and sticks to the warm, humorous anecdotes. They got along very well, although when grandson Paul III was kidnaped, Getty required that she handle all media...an enormous job. She portrays Getty as a human being and steers clear of the miser-womanizer image. The hardbound first edition is common.

**B174. (VARIOUS OILMEN) Lundberg, Ferdinand, AMERICA'S 60 FAMILIES, NY, The Vanguard Press, (1938), xxiii, 587pp.**
The second of Lundberg's major socialist-leaning works, this book was a sensation in 1938 and widely circulated. The author presents a considerable study of the sixty most powerful (mainly the wealthiest) families in America. Morgan, Mellon, Rockefeller and DuPont of the Robber Baron era are his typical targets. He shows how these individuals gained money and power and how they have used it to manipulate every phase of average Americans lives, particularly in education, politics,

industry, science, literature, social life and philanthropy. He admits there are another ninety families of secondary wealth which are the back-up support team. A thorough attack on capitalism. If Lundberg was writing the book today, he could easily make the same case and it would be interesting to see how many of the one hundred fifty families still retain the power they had in the 1930's. Excellent bibliography with reference notes and index. Fun reading. The hardbound first edition is common.

**B175. (ROCKEFELLER FAMILY) Lundberg, Ferdinand, THE ROCKEFELLER SYNDROME, Secaucus, NJ, Lyle Stuart, Inc., (1975), 404pp.**
Lundberg loves to stir up the public with his sensational revelations in the world of the wealthy and powerful. He is cynical and sarcastic and does a tremendous amount of research. His book "America's Sixty Families" published in 1938 influenced public and government opinion against the wealthy, as did his "The Rich and the Super-Rich" published in 1968. The Rockefeller Syndrome specifically studies the way the Rockefellers have operated in the past and how they operate now. He tells precisely how they use their money for power and influence. The stimulus for this educational work was the congressional hearings conducted to see if Nelson Rockefeller (a four term governor of New York) was fit to be appointed Vice President under Gerald Ford. This is a good review of the family's influence on public affairs, however, Lundberg's grandiloquent-sarcastic way with words will wear out all but the dedicated anti-mogul. Great book if you like Lundberg. The hardbound first edition is common.

**B176. (MICHAEL ROCKEFELLER) Machlin, Milt, THE SEARCH FOR MICHAEL ROCKEFELLER, NY, G.P. Putnam's Sons, (1972), 256pp.**
The author is editor of Argosy Magazine, a member of Explorers Club and of the Mystery Writers of America. Well qualified for writing this book. From the cover: "The dramatic story of the young anthropologist who disappeared in the Stone Age world of primitive New Guinea". In November, 1961, twenty-three year old Michael, son of Nelson Rockefeller vanished in waters off New Guinea. Probably a revenge killing by local tribesmen, this is a real life adventure-detective story of a Rockefeller heir. A strong conclusion. The hardbound first edition is common.

**B177. (MIKE BENEDUM) Mallison, Sam T., THE GREAT WILDCATTER, Charleston, WV, Education Foundation of West Virginia, Inc., (1953), 528pp.**

The book details the adventurous life of a real wildcatter. Benedum probably found more oil and gas than anyone in history. He was born poor in West Virginia in 1869. As a young man he luckily fell into a job as leasing agent for one of Rockefeller's companies. He soon became a wildcatter, drilling and finding oil all over the world, much with his partner Joe Trees. Benedum was the antithesis of Rockefeller in that he found oil and gas where Rockefeller only exploited the oil and gas industry and its people. Much discussion of Benedum's fifty million dollar tax case which was brought by the government during the depression. Benedum essentially won the case, but paid a stiff price for standing up for his rights. He had significant oil discoveries in Mexico, The Permian Basin and the Mid-Continent areas. Also many worldwide oil field discoveries were made by other companies he owned. This is one of the very best oilman biographies. Cited in Swanson on page 175. The hardbound first edition is becoming scarce.

**B178. (LONE WOLF GONZAULLAS) Malsch, Brownson, CAPTAIN M.T. GONZAULLAS: LONE WOLF, Austin, TX, Shoal Creek Publishers, Inc., (1980), xvi, 224pp.**

Young Gonzaullas, the orphan, found a father figure in the Texas Rangers of the El Paso area and from an early age was determined to become one. Further impetus to this career was said to have been given when he was forced to observe the brutal murder of his two brothers as well as the wounding of his parents at the hand of bandits. He was the most colorful of the Rangers over his thirty-one year career and became the only Hispanic to obtain the rank of Captain. His interest in bringing law to the oil boom towns and arresting prohibition offenders occupied most of his early career and established his reputation as courageous if not fearless. Expertise in handling border problems also helped gain stature for him. Bringing law to the East Texas oil towns was his most remembered feat. He was in charge while military law was being enforced for proration purposes during the early Kilgore days. Illegal (hot oil) production was rampant and causing great waste in the field as well as economic disaster for the oil industry. As State Chief of Intelligence in 1935, Gonzaullas was noted for successfully using many of the latest crime detection and deterrent techniques. Gonzaullas' own voluminous scrapbook was used to prepare much of this book which gives details of Ranger life and incidents. News reports and commentaries of various papers were meticulously kept in chronological sequence by him. A unique person in Texas and law enforcement history. Well worth reading. The hardbound first edition is becoming scarce.

**B179. (WILLIAM BUCKLEY, SR.) Markmann, Charles Lam, THE BUCKLEYS: A FAMILY EXAMINED, NY, William Morrow & Company, Inc., 1973, xiv, 364pp.**
William F. Buckley, Sr. was born in Texas in 1881. With a law degree he began practice in Mexico and became heavily involved in oil and politics. Evicted from Mexico for political reasons in 1921, he pursued oil interests successfully in Venezuela, the U.S. and Canada. Most of the book is good anecdotal family history, describing their right-wing philosophy. The author believes (1973) that the Buckleys, led by William F., Jr. and Senator James will have a serious impact on the U. S. political scene. This book is a more balanced picture of the family than in the book "W.F.B., An Appreciation". Buckley's estate was estimated at over one hundred million dollars at his death in 1958. Little detail is given here as to exactly how the money was acquired, although it was largely from his many successful oil ventures. The hardbound first edition is common.

**B180. (HOWARD MARSHALL) Marshall, J. Howard II, DONE IN OIL: AN AUTOBIOGRAPHY BY J. HOWARD MARSHALL II, College Station, Texas A&M University Press, (1994), xxi, 282pp.**
Marshall was born in Pennsylvania in 1905 and after graduating from Harvard in 1926 was drawn almost immediately into the oil industry. A solid career was begun in oil politics as he worked for Harold Ickes. He became heavily involved in conservation practice, drafting legislation and working for PAW. He later worked for Standard of California, Ashland, Signal and Union Texas Petroleum prior to going on his own where he became involved in exploration and production. The hardbound first edition is common.

**B181. (HENRY FLAGLER) Martin, Sidney Walter, FLORIDA'S FLAGLER, Athens, GA, University of Georgia Press, 1979, xi, 280pp.**
Henry Flagler was born in 1830, the only son of poor parents in New York. He finished eighth grade and left home at fourteen for Ohio where he worked for his half-brother in a general store. By his early twenties he

was a moderately successful businessman in the family owned stores and had some success and failures while expanding into other ventures in Cleveland. John D. Rockefeller handled most of Flagler's grain shipments and the two aggressive young men got to know each other well. In 1866, Flagler took a job with Rockefeller's ex-partner, Clark, essentially taking his place. Rockefeller bought a large oil refinery and in 1867 the Rockefeller, Andrews and Flagler partnership was formed. Flagler fairly contributed $100,000 as his share. They were perfect partners and the business boomed as they became the most efficient and aggressive refiners in the new industry. They eventually controlled the entire market through shrewd merger and acquisition. It was Flagler's idea to form Standard Oil in 1870. He remained vice president until 1908, although retiring from active participation in the early 1880's. John D. Archbold took Flagler's place in the inner circle at Standard starting in 1882. The major part of this book looks at Flagler's role as a promoter of Florida and the ups and downs he encountered while developing the east coast, from Jacksonville to Key West. Flagler died in 1913, much better known for his work in Florida than with Standard. The hardbound first edition is common.

**B182. (PAUL BLAZER) Massie, Joseph L., BLAZER AND ASHLAND OIL: A STUDY IN MANAGEMENT, Kentucky, University of Kentucky Press, (1960), xiv, 253 pp.**
Born in Illinois in 1890, Paul G. Blazer's career is followed from 1924 through his retirement in 1957. His early work was in crude oil sales which led first to management and then to involvement in the refining business. In 1924 he was hired by Swiss Oil to organize Ashland Refining Company, which he did. Through years of mergers and acquisitions largely directed by Blazer, Ashland became one of the largest integrated independent oil companies. Ashland & Blazer were noted for having a flexible management style which allowed them to adjust to changing industry conditions faster than their competition. Although Ashland occasionally did exploratory drilling with success, they never had much production of their own, specializing more in refining and transportation, particularly east of the Mississippi. The hardbound first edition is common.

**B183. (E. W. MARLAND) Mathews, John Joseph, LIFE AND DEATH OF AN OILMAN: THE CAREER OF E.W. MARLAND, Norman, OK, University of Oklahoma Press, (1951), xi, 257pp.**
Mathews, a half-Osage Indian and famous writer knew Marland very well. Marland found much of his early oil on the Osage Reservation. This book is a brilliantly written and compassionate study of a man who achieved greatness in the oil industry and in politics, finally managing to lose them both. Born at Pittsburgh in 1874, Marland's parents were wealthy and well educated. As a child he was described as both a dreamer and an intuitive gambler. After graduating from college he began exploring for coal in Ohio and West Virginia in the belief that coal was one of the few minerals that could be exploited by an independent and would not be monopolized by such as Rockefeller, Vanderbilt, Carnegie, etc. While boring for coal in West Virginia, he struck oil. His discovery turned out to be only a small pool, but he was hooked on oil for life. He moved to Oklahoma and discovered several large oil fields there, and then in the Permian Basin. His success was largely due to use of geologists as exploration tools. He was one of the first to do so. The J. P. Morgan companies later squeezed Marland out of his company and forced a merger with Continental Oil Company. Marland was down, but not out. He pursued politics and was elected to the House of Representatives in 1930 and became governor of Oklahoma in 1934. A very good governor in an age of very bad governors. The Interstate Oil Compact Commission was formed largely due to his help. Over the years, the state has received considerable revenue from the oil produced on leases around the state capitol which he initiated. This is one of the top oilman biographies. Cited in Swanson on page 175. The hardbound first edition is becoming scarce.

**B184. (HOWARD HUGHES) Mathison, Richard, HIS WEIRD AND WANTON WAYS: THE SECRET LIFE OF HOWARD HUGHES, NY, William Morrow and Co., 1977, 247pp.**
The author is a well known biographer and California magazine editor. He also worked for the L.A. Times and later was bureau chief for Newsweek. Over the years he had collected Hughes stories. His book is particularly concerned with Hughes' girl friends, of which Mathison counts one hundred eight. Generally, it is based on interviews with many of the girls and other Hughes associates who would talk. Jeff Chouinard was one of Hughes' security guards and was probably the main

contributor, although he does not appear to be too credible in the book. The author goes against the general opinion of Hughes' genius and claims that his great wealth was acquired through the work of his employees and inflation rather than through brilliant maneuverings. Lots of strange and funny stories not published elsewhere. The hardbound first edition is common.

**B185. (JACKSON MATTHEWS) Matthews, Elizabeth Turner and John Wood, JACKSON MATTHEWS AND THE MANUELLA CORDOVA LEGACY, U.S., John Wood, (1975), 132pp.**
Born in 1897, Jackson Matthews' father was a traveling salesman in North Carolina and died early. Working his way through college, Matthews began a career in public relations. The story revolves around 30,000 acres of Central Texas land originally patented by General McClelland (of Civil War fame) for a copper mining venture and subsequently passed down through generations to Mathews' wife and several others. The title was continually questioned over the years, but most problems came after Katz and Humble drilled and found oil in the Stonewall County portion during 1951. Lots of oil and money and litigation as the good guys prevail. The hardbound first edition is becoming scarce.

**B186. (ANTHONY LUCAS) McBeth, Reid Sayers, PIONEERING THE GULF COAST: A STORY OF THE LIFE AND ACCOMPLISHMENTS OF CAPT. ANTHONY F. LUCAS, NY, The Author, 1918, 80pp.**
This book was not available for review by Neff, but is regarded as important. Quoting from Swanson: "Life of Capt. Lucas; discovery and development of Spindletop oilfield, salt and sulphur exploration and production; use of rotary drill at Spindletop, development and use of valve to check back-pressure and addition of mud to drilling fluid". Cited in Swanson on page 175; Hardwicke page 45. The first edition is rare.

**B187. (PATILLO HIGGINS) McDaniel, Robert W. with Henry C. Dethloff, PATILLO HIGGINS AND THE SEARCH FOR TEXAS OIL, College Station, TX, Texas A & M University Press, (1989), xii, 174pp.**
Higgins was born at Sabine Pass, Texas in 1863, where his father was a gunsmith. The family moved to Beaumont in 1869 where Patillo had little

schooling, but learned his father's craft. In his teens he was described as dangerous and rowdy. At eighteen he lost his left arm in a Beaumont gunfight with a law officer. The local jury let him off on his plea of self-defense. He became a lumberman and by 1885 had a religious conversion, giving up his wild ways. Patillo read everything he could find on petroleum and geology and was convinced that the anticlinal theory of oil accumulation which was being successfully applied in Pennsylvania could also be applied to a hill outside of town called Spindletop. In 1892 he organized Gladys City Oil, Gas & Manufacturing Company. His plan was for the refining and transportation of oil with pipelines to the nearby ports which would be points of distribution to the world. In 1893 he drilled a small gas producer on the salt dome and completed it at sixty feet. Trying to raise money to drill deeper, he was hindered by skepticism and out- right ridicule over several years. He caused a second dry hole to be drilled to 300 feet in 1895. Finally he convinced Captain Anthony Lucas to come to Beaumont. Lucas was a driller and had financial backing. Lucas leased the top of the dome and drilled a 575 foot dry hole. At this point almost all experts believed Spindletop was done for. Lucas joined with the highly successful exploration team of Guffey & Galey along with the Mellon family and leased most of the remainder of the dome. Lucas then drilled the deep test which blew in at 1,020 feet, a monster discovery. By then they had cut Higgins out of the deal, but fortunately he owned a 33 acre lease within 100 feet of the discovery which he quickly began drilling for himself. The major oil companies of Gulf, Sun, Texaco, Humble and Magnolia essentially came out of the Spindletop boom, spelling disaster for the Standard Oil monopoly. Higgins made good money at Spindletop and went on wildcatting across Texas, finding oil here and drilling dry holes there. He died in 1955, a legend. Very good story of a persistent man who kept after his dream. The hardbound first edition is common.

**B188. (ROY McLAUGHLIN) McLaughlin, Roy P., THE TENDERFOOT COMES WEST: HALF A CENTURY OF PROGRESS IN CALIFORNIA, AND THE PETROLEUM INDUSTRY, NY, Exposition Press, (1968), 124pp.**

Born in 1879, McLaughlin was raised in the mining towns of Colorado and later California. He received a degree from Stanford in 1902 and worked as a draftsman, surveyor and mining engineer. He did general exploration work for Associated Oil for four years and decided to work for the California Mining Bureau to help stabilize their oil and gas problems, which were extensive in the early days of oil there. He later had several successful years in business with Dominguez Oil Field Company. McLaughlin is very weak in providing dates, however sometime after 1916 he began an impressive career in regulation and conservation as Supervisor for Oil and Gas in California. His pioneering legislation was enacted and is recognized as precedent setting as it became adopted by other states as well. The hardbound first edition is common.

**B189. (DAVID LOVE) McPhee, John, RISING FROM THE PLAINS, NY, Farrar Straus Giroux, (1986), 213pp.**
McPhee writes of geology and David Love with great feeling for the science as well as for this historically important geologist. Love was born in 1913 and grew up on an isolated ranch in Wyoming where he was exposed to rocks on a daily basis. He was largely educated by his exceptional, pioneer mother. As in McPhee's other three books of the series titled "Annals of a Former World", he reveals the geology of a region, this time it is the Rocky Mountains. He integrates the structure, stratigraphy and paleontology of the region with the ancient geologic history to make the whole scene come alive and meaningful to laymen and scientists alike. His guide through the Rockies is David Love. Love has a Ph.D. from Yale and has done every kind of geological study over much of the U.S. However, he is generally regarded as "The Grand Old Man of Rocky Mountain Geology". Now in his seventies, Love still does field geologic mapping and admits to having spent a quarter of his nights camped out amidst his outcrops. He was involved in oil exploration in the Appalachian Basin, and the mid-continent, but more intimately, around his own family ranch in Wyoming. Love takes McPhee on a fascinating tour of the Rockies, explaining the regional geology, from theoretical problems to practical applications. He tells of the background and history of various techniques which have been used over the years to unravel the mysteries of this complex area. The hardbound first edition is common.

**B190. (LEE McWORKMAN) McWorkman, Lee G., OILFIELD.DOC, Seagraves, TX, Pioneer Book Publishers, (1988), viii, 176pp.**
From the Forward: "This book is about the oil field as seen through the eyes of an oil field supply man, from 1946 to the present day, in Texas,

New Mexico...." McWorkman is not a doctor, the book title comes from his word processor which assigns a .DOC (short for document) to all files and did so to his book which was to be titled Oilfield. McWorkman was born in 1916 and in 1946 took a job in the oil fields of West Texas, in the employ of Republic Supply Company. He relates business stories in chronological sequence and describes his job with one of the biggest and best oil field supply firms of the day. An oil field supply company serves as middleman or broker between manufacturers and end users. The field representatives put in long hours, being on twenty-four hour call to provide virtually any material or equipment required by the oil drilling and producing companies. He retired in 1979 but continued working as a commission salesman for another oil equipment firm. After the oil crash in 1986 he sadly watched Republic merged and squeezed out of existence along with about two-thirds of the industry. Good oil history with humor and philosophy from another side of the industry. The softbound first edition is becoming scarce.

**B191. (MELLON FAMILY) Mellon, William Larimer and Boyden Sparkes, JUDGE MELLON'S SONS, Pittsburgh, Privately Published, 1948, 570pp.**
This book was not available for review by Neff, but is regarded as important. Quoting from Swanson: "Story of 5 sons of Judge Thomas Mellon; pipe line and refinery activity Pennsylvania; participation in Spindletop development; establishment and development of Gulf Oil Corp. Cited in Swanson on page 175. Hardbound first edition is very scarce.

**B192. (MAX MEYER) Meyer, Lewis, PREPOSTEROUS PAPA, Cleveland, OH, The World Publishing Co., (1959), 214pp.**
Lewis Meyer is a well-known Tulsa attorney and bookseller. He has a radio and TV show about the world of books. In this book, he writes of his father, Max Meyer, with great admiration and affection and most of all with humor. Papa is an unusual person and is described by the author as a comic curmudgeon. Meyer felt he had to write this book about Papa to share this unique personality with the world. Papa settled in Oklahoma Territory in 1906, near Sapulpa, where he became the only Jewish farmer in the Territory. As the stories of his life there unfold, it is obvious that his only successful crop consisted of the thirty-one oil producers drilled on his 160 acre homestead. Papa would probably have been preposterous

without the oil, but less so. Fun reading about a good man. The hardbound first edition is common.

**B193. (LEO MILES) Miles, Leo. HIGH PRESSURE: A PIPELINER'S STORY, Shreveport, LA. Raton Books Ltd., (1988), vii, 179pp.**
Miles was born in Louisiana around 1930. He completed some college and served in the Korean war. After the war his older brother got him a pipeline job with Brown and Root, a large construction company. Much humorous dialogue using pipeline colloquial language as he relates tale after tale, working his way up through to ownership of a pipeline testing company. Pipelining is a critical part of the oil industry, particularly for natural gas, but also for the transportation of crude oil and refined products. Several pipeline histories have been published, but this biography of a work-a-day pipeliner is unusually interesting. The hardbound first edition is common.

**B194. (JOHN WILKES BOOTH) Miller, Ernest C., JOHN WILKES BOOTH IN THE OIL REGION, NY, The Exposition Press, (1947), 78pp.**
Dr. Miller's extensive research on this book and on Booth is aided by the first-hand 1894 interview notes of Louis J. Mackey, which are presented as Part II of the book. Booth, the actor, oilman and assassin, is a fascinating subject. Dr. Miller is Pennsylvania's premier oil historian. From the Dust Jacket of the 1987 revised edition: "A carefully crafted monograph on the progress of Booth's oil speculation, this new book traces the acquisition and disposition of Booth's Homestead and Pithole leases, the activities of his business manager, Joseph H. Simonds, Booth's friendships and associations in the boom towns of Western Pennsylvania". Booth turns out to be a very light-weight speculator and had about as much luck in the oil fields as most of that type. A revised version of Dr. Miller's book, retitled JOHN WILKES BOOTH IN THE PENNSYLVANIA OIL REGION was published by Crawford County Historical in 1987 with 89pp. Cited in Swanson on page 175. The hardbound first edition is scarce.

**B195. (J. PAUL GETTY) Miller, Russell, THE HOUSE OF GETTY, NY, Henry Holt and Company, (1986), 362pp.**
Another biographical work created ten years after Getty's death, this one reads easier than Lenzner's book and contains many Getty quotes along

with intimate family photos of the later years. No notes or appendix. From the dust jacket: "The tormented saga of the Getty family reads like the script for Dynasty, interweaving boardroom battles, sex, money, drugs, power, crime, tragedy, and family intrigue. At the center stands John Paul Getty, the grandfather, an eccentric oil billionaire..." The old adage, "Truth is stranger than fiction", surely holds for Getty, as it would take an exceptional author to create such a bizarre saga from imagination alone. Enjoyable reading. The hardbound first edition is common.

**B196. (ELIZABETH MILLS) Mills, N. Elizabeth, THE LADY DRILLER: THE AUTOBIOGRAPHY OF N. ELIZABETH MILLS, NY, Exposition Press, (1955), 176pp.**

Mills' autobiography is a tribute to entrepreneurship and persistence. Her story lacks the extreme highs and lows of the Sue Sanders saga (Our Common Herd), but the same result of successfully operating in the "man's world" of the oil fields is gained. Elizabeth Mills was born in Illinois around 1880. The family moved to Nebraska where they were not too successful in farming. She graduated from a business college and finally married a young doctor who turned out to be a compulsive gambler and heavy drinker. She found herself to be a successful businesswoman in that she was able to run a local cafe and other interests with ease while supporting her worthless husband. After a divorce, she caught oil fever from two of her brothers who had been working the Wyoming oil fields and were on the way to Oklahoma as it was then booming. The three of them formed an oil field supply company and began investing in leases and drilling wells. Pretty exciting times were had as they drilled an expiring lease near Muskogee and got a real teaser. The near miss encouraged two offset wells which discovered significant natural gas. Protracted litigation with a large gas company eventually netted her significant income and a gas contract with the city of Muskogee. Always relying on solid geological advice, her many successful drilling campaigns gained national recognition. Her later success extended to Texas. Cited in Swanson on page 175. The hardbound first edition is scarce.

**B197. (THOMAS GILCREASE) Milsten, David Randolph, THOMAS GILCREASE, San Antonio, TX, The Naylor Co., (1969), xxii, 468pp.**

Gilcrease was born in Louisiana in 1890, one-eighth Creek Indian and the first of fourteen children. His father ran a cotton gin. Tom had little education in eastern Oklahoma where he was raised, but at age sixteen he gained an Indian headright of 160 acres within the Creek reservation. He continued some schooling and married at eighteen, being a farmer and rancher at the time. Tom's father, as guardian, had leased the 160 acres for oil exploration to William Milliken, who proceeded to drill forty-one producing oil wells (in the middle of Glenn Pool Field). Tom, at nineteen, sued his father for mishandling the trade and sought his majority which was granted by the court. Tom made a separate lease to another party and spent the next ten years in and out of court trying to get more than his fair share of the proceeds. The U.S. Supreme Court finally ruled against him. His Gilcrease Oil Company, formed in 1922 had exceptional success in wildcat and development drilling over the next several years. Two cases of inattention to business are related. Tom had authorized the purchase of the Signal Hill property in California, prior to its discovery, but his people did not follow through on the acquisition. Also, Tom had authorized the purchase of what turned out to be a big portion of the East Texas field and again his people got the blame as they did not follow through. Nobody's perfect. The remainder of the book documents his fabulous art collection and the Gilcrease Institute of American History and Art at Tulsa, Oklahoma. Amazing story. The hardbound first edition is becoming scarce.

**B198. (JOHNNY MITCHELL) Mitchell, Johnny, THE SECRET WAR OF CAPTAIN JOHNNY MITCHELL, Houston, TX, Pacesetter Press (1976), vii, 96pp.**

Mitchell was born around 1913, graduated from Texas A & M in 1934 and became an independent oil producer and drilling contractor on the Texas gulf coast. He entered the service as a lieutenant in August 1942. Mitchell kept good notes of his three year wartime saga and thirty years later decided to write the book. Prior to the war, he had established some oil production and was surviving, although not getting rich. His first war assignment was near Reykjavik, Iceland for about a year. He had a delightful time. He was then assigned to Northern Ireland and became part of the Normandy invasion. He fought through France, Luxembourg and Germany and ended up in Czechoslovakia as a field artillery commander under George Patton. Mitchell is a practical minded soul who pushed the system to its limits and managed to make the best of almost every situation. Good reading. The hardbound first edition is common.

**B199. (ISADOR MOLK) Molk, Isador, THE MAKING OF AN OILMAN, NY, The Citadel Press, (1958), 256pp.**
Secondary title: THE AUTOBIOGRAPHY OF A TALMUDIC SCHOLAR WHO MIGRATED TO THE UNITED STATES FROM LITHUANIA AND BECAME A PIONEER IN THE OIL FIELDS OF THE MIDWEST. This is a well-told autobiography. Molk was a child prodigy in Lithuania who sought more freedom. At thirteen he came to America where he attended college and received several degrees. He studied geology along the way and taught school for a while, but seeking still more freedom, ended up in the Kansas oil fields where he built a small oilfield pipe business into a thriving oil company. No wildcatting for Molk; he bought producing leases, meticulously repairing and maintaining them to great profit. He bought undeveloped leases and then drilled them in an efficient manner. This is one of the most unusual oil biographies which leaves the reader with the impression that if one is smart and focused, he can be a success at whatever he chooses to do, no matter how incongruous to his background. Cited in Swanson on page 175. The hardbound first edition is common.

**B200. (JOHN D. ARCHBOLD) Moore, Austin Leigh, JOHN D. ARCHBOLD AND THE EARLY DEVELOPMENT OF STANDARD OIL, NY, Macmillan, Nodate, 349pp.**
This book was not available for review by Neff, but it is regarded as important. Quoting from Swanson: "Archbold was leading figure with Standard Oil for half a century". Cited in Swanson on page 163. The hardbound first edition is very scarce.

**B201. (ROBERT S. KERR) Morgan, Anne Hodges, ROBERT S. KERR: THE SENATE YEARS, Norman, OK, University of Oklahoma Press, (1977), xiv, 337pp.**
Kerr was born at Ada, Indian Territory in 1896. He attended the University of Oklahoma and WW I, later becoming a lawyer. He then became a drilling contractor and oil producer by 1926. With Dean McGee he formed Kerr-McGee Corporation which became a large independent drilling, producing and exploration company. Kerr was elected governor of Oklahoma and served from 1943 to 1947. He was elected to the U.S. Senate in 1948, 1954 and 1960, serving until his death in 1963. This book concentrates on his Senate years where he quickly gained national recognition and proceeded to become a powerful force in Washington.

Large notes and bibliography section is included. Kerr was one of the most influential Oklahomans of all time. The hardbound first edition is common.

**B202. (ROCKEFELLER FAMILY) Morris, Joe Alex, THOSE ROCKEFELLER BROTHERS: AN INFORMAL BIOGRAPHY OF FIVE EXTRAORDINARY YOUNG MEN, NY, Harper & Brothers, (1953), x, 275pp.**

Morris had the idea of writing a book on the five sons of John D. Rockefeller, Jr. The sons were in a transitional stage of sorting out their careers in 1950 and the Rockefellers had virtually dropped from the public eye. After two years of trying to get the brothers to agree to timing and logistics, Morris finally cornered all of them and received good cooperation. John D. III, born 1906 was trustee of several family foundations and trusts; Nelson, born 1908 was involved in international affairs and domestic social issues; Laurence, born 1910 held the family seat at NYSE; Winthrop, born 1912 was involved in restoring Williamsburg and in oil; David, born 1915 was already senior vice president of Chase National Bank. This is a positive non-investigative book. A father or grandfather would be proud. The hardbound first edition is common.

**B203. (ROTHSCHILD FAMILY) Morton, Frederick, THE ROTHSCHILDS: A FAMILY PORTRAIT, NY, Atheneum, 1962, xii, 305pp.**

A two-hundred year history of the Rothschilds is so very compressed so that it is difficult to comprehend the real magnitude of their influence on the world scene. A serious study of this family biography will, however give an idea of how pervasive they are in everyone's life. From Mayer Rothschild who was born in the early 1700's and became a money changer in Frankfurt's Jewish ghetto, to the unimaginable family wealth generated in the 1800's is a remarkable saga. The family investment in oil is briefly mentioned in several places throughout the book. Around the turn of the century, they virtually controlled what the Nobel family did not control of the Russian oil fields at Baku. They made the decisions as to who got how much oil, between major powers Standard and Shell. Largely a banking family, their influence is felt in about every country and industry in Europe. The book is well illustrated and includes a family genealogy and index. The hardbound first edition is common.

**B204. (ROCKEFELLER FAMILY) Moscow, Alvin, THE ROCKEFELLER INHERITANCE, Garden City, NY. Doubleday & Co., Inc., 1977, xii, 462pp.**

The author took eight years to write this story of the five sons of John D. Rockefeller, Jr. He claims they agreed to give him personal interviews and unprecedented access to Rockefeller family personal papers. Still, the biography is called unauthorized and unsupported by the Rockefeller family. It also includes no source notes and only a meager bibliography. The story begins as the five brothers, John D. III, Nelson, Laurence, Winthrop and David return from WW II and establish their respective offices at Rockefeller Center. Their big problem was deciding what they wanted to do in life. John D., Jr. still controlled the family fortune and was giving it to worthy causes as fast as he could. The five sons, plus one daughter, had only the income from their several trust funds, not the assets. They all had a strict money-oriented up-bringing along with history lessons on the Standard Oil and were well versed in the tradition of philanthropy. A good chapter on each son relates what he decided to do along with the how and why. John D. III's interest ran to the Rockefeller Foundation which centered on medicine, public health and world food supply. Nelson's interest ran to politics (all the way to Vice President) which was in opposition to the family policy, discouraging public life. Laurence became a very successful investor. Winthrop went back to the oil business for several years but was never happy there. The family put him in charge of the on-going reconstruction of Colonial Williamsburg and he became a champion of social causes as a two-term governor of Arkansas. David, the baby of the family became president of the Chase Manhattan Bank. A further chapter on each son discusses their accomplishments to date. Very good analysis of the family of John D. Rockefeller, Junior. The hardbound first edition is common.

**B205. (DEVITT FAMILY) Murrah, David J., OIL, TAXES, AND CATS: A HISTORY OF THE DEVITT FAMILY AND THE MALLET RANCH, Lubbock, Texas Tech University Press, (1994), xii, 247pp.**

The Mallet Ranch in West Texas was formed by David DeVitt in 1895. He died in 1934 and his daughter Christine came to run the large spread. Better known as the eccentric "cat lady", she dominated other owners and partners through manipulation and became well respected in a man's world. Her influence was particularly felt after oil was discovered in 1937 and the Mallet became a significant part of the two billion barrel Levelland- Slaughter oil field. Fascinating story. The hardbound first edition is becoming scarce.

**B206. (ALFALFA BILL MURRAY) Murray, William H., MEMOIRS OF GOVERNOR MURRAY AND TRUE HISTORY OF OKLAHOMA, Boston, MA, The Meador Press, 1945, 1685pp.**
Presented in three volumes. It is appropriate that Alfalfa Bill would write his memoirs and Oklahoma history together, as he was an integral part of early Oklahoma history. He was seventy-two at the time his book was published. Among his many endeavors, he had been president of the Oklahoma Constitution Convention, A U.S. congressman, and later governor of the state. He was constantly involved in dealings with the oil industry while in public office and received national attention for those conflicts along with many others. In the early 1930's when Oklahoma City field and East Texas field flush production was collapsing the price of crude, he had the state militia shut in 3,108 producing wells. They would remain shut in until the price rose to the arbitrary $1.00 he believed necessary to restore economic stability and order to the beleaguered industry. Quoting a 1931 Time Magazine article: "William H. Murray leaves no man neutral. To his friends he is a second Andrew Jackson sent to lead the plain people out of economic bondage. To his foes he is another William Jennings Bryan, threatening the foundation of U.S. economic life." He was definitely for the common man. His narrative is entertaining and enlightening as it is rendered in his own unique style. The Hardbound first edition is scarce.

**B207. (JOHN D. ROCKEFELLER) Myers, Elisabeth P., JOHN D. ROCKEFELLER: BOY FINANCIER, NY, The Bobbs-Merrill Co., (1973), 200pp.**
A juvenile biography of John D. Rockefeller with large print, concentrating on his youth and inquisitive, industrious nature. There is much made-up dialogue to stress "Johnny's" attributes. The entire family is portrayed as normal, including John. This is one of the Bobbs-Merrill series entitled Childhood of Famous Americans, and consisting of about two hundred biographies. The hardbound first edition is common.

**B208. (CULLEN DAVIS) Naifeh, Steven and Gregory White Smith, FINAL JUSTICE; THE TRUE STORY OF THE RICHEST MAN EVER TRIED FOR MURDER, NY, Dutton, (1993), ix, 466pp.**
The story of Cullen Davis and his sensational murder trials has been re-hashed several times, by Cartwright, 1979, Phillips, 1979, and Cochran, 1980. Davis was tried and found not guilty on charges of killing his twelve year old step-daughter, despite overwhelming evidence to the contrary. Davis was released and this book tells what happened to him and his family and his oil empire after that. Truth is stranger than fiction and justice prevails. The hardbound first edition is common.

**B209. (CLAUDE BROWN) Nash, Robert, and Peggy Nichols Nash, "MR. McCAMEY" CLAUDE W. BROWN: THE LIFE OF A TEXAS OIL MAN, Austin, TX, Eakin Press, 1994, x, 98pp.**
Brown was born in 1904 on a Texas farm shortly after the Spindletop discovery. After high school he roustabouted at Mexia. For the next seventy-one years he was an oil man. He endured many ups and downs with oil operations and with the economic and logistical problems largely in the Permian Basin town of McCamey. He was a pragmatic man who helped develop the West Texas area, both in oil and in community. The hardbound first edition is common.

**B210. (HAROLD NEELY) Neely, Harold G., WILDCATTING FROM PENNSYLVANIA TO TEXAS, Fort Worth, TX, Privately Published, (1976), Ltd Ed. 54pp.**
The author was born into an oil family in 1895. His uncles had found work in the oil fields immediately after the Drake discovery in 1859. His grandfather also worked in the oil fields. The relatives got together and drilled a well on the old Neely family place near the Clarion river. The well was started in 1870 and during the summer of 1871 hit gas in the "third sand", but later turned into a fiasco. The well finally plugged itself off. Neely's Dad took an oil well pumping job at age eleven and held that for several years. By eighteen he was a driller in the family owned drilling firm and a part owner. They drilled a well under contract for Senator Cal Brice of Ohio and discovered the Lima oil field while looking for gas. The Neely-Clover company had established its own production by the late 1890's while continuing to drill for others. The Spindletop discovery in Texas, which wrecked the oil market, was just one of the many boom to bust events which had a serious effect on this early oil

family. They expanded into Indiana where Harold's Dad formed St. Mary's Drilling Company. By 1907, the family started Iron Mountain Oil Company and was drilling successfully in Kansas and Oklahoma. The author left college to get married and began working full-time for the family business. He speculated successfully in oil deals on the side. In 1918 he caused a second well to be drilled on the old family place and discovered a good oil field. He made a lot of money and lost it all in a bank failure as the depression hit. By 1931 he was almost destitute, moving the family across country to avoid the Pennsylvania creditors who were hounding him. In 1938 he took a drilling deal from Jones and Stasney of Albany, Texas to test one of their wildcat prospects. Humble Oil joined Neely as a partner and after a considerable saga, they discovered a large oil field. A lot of development drilling ensued. Creditors were finally paid off and he moved back to Pennsylvania with the family and lived happily ever after. Lots of good photos and history. The limited softbound first edition is scarce.

**B211. (RICHARD NEFF) Neff, E. Richard. HYDROCARBON QUEST: ONE MAN'S SEARCH FOR OIL AND GAS., Santa Fe, Camelrock Publishing, (1990), v, 221pp.**
From the back cover: "Neff gives a chronological overview of his three decade oil career from an exploration geologist's perspective. His recollections of southeast New Mexico and West Texas events range from mundane to bizarre and from private to public. If you have ever invested in an oil drilling deal or known a geologist, geophysicist, engineer, landman or promoter who has participated in the search for oil and gas, you will enjoy reading this autobiographical effort". The softbound first edition is becoming scarce.

**B212. (ED NELSON) Nelson, Ed, TELLIN' THE TALE, Sterling, CO.: Privately Published, (1984), i, 159pp.**
The first part of this autobiography is dedicated to oilfield characters. Fourteen of Nelson's friends and business associates are given about a page each, where he relates a story or series of stories about his relationship with them. Nelson is a humorous writer and makes you want to be his friend, particularly when you read about how he regards his friends. These characters are geologists, engineers, operators, promoters and other oil field sales people like himself. He was born around 1920 and worked for several oilfield tank companies before starting his own

business. His short stories continue through the book and along with several letters are easy reading and enjoyable prose. He also includes some poetry. Many of the stories concern hunting and fishing and philosophy. Nice man. The softbound first edition is becoming scarce.

**B213. (JOHN D. ROCKEFELLER) Nevins, Allan, JOHN D. ROCKEFELLER: THE HEROIC AGE OF AMERICAN ENTERPRISE, NY, Charles Scribner's Sons, 1940, 2 Vols., xiii, 683 & x, 747pp.**

Many potential Rockefeller biographers had been considered for this job over the years, but finally in 1934 Nevins was chosen by John D., Jr. Nevins had previously won a Pulitzer Prize for his biography of Grover Cleveland. Nevins writing was thought to be somewhat dry, but he was highly respected. This two volume book became the most popular literary work on Rockefeller, first because of Nevins' technical ability and second for the personal Rockefeller information he became privy to. Basically Nevins defends Rockefeller throughout the book and tries to humanize him. Volume one is about one-third childhood and early business ventures. The remainder carries through the formation of Standard Oil Trust in 1885. The author defends almost every move Rockefeller made from the vicious South Improvement Company formed to squelch the small producer-refiners of the oil regions, to the railroad rebate scheme designed to squelch everyone else in the industry. The trust was formed to get around government anti- monopoly legislation and worked for several years in one form or another. Standard's first serious challenge was in 1895 when The Pure Oil Company was formed by fifty independent refiners and producers to combat the Standard strangle hold over the oil business. Standard tried everything to bring Pure into the fold, but they were unsuccessful. In 1899, Rockefeller was sixty years old and beaten down from years of government litigation. John Archbold officially took over the reins at that time although he had already been doing the job for several years. Much in volume two of the Rockefeller Foundation and charities describing John D., Jr.'s part in them. Of John D., from the epilogue: "...by virtue of his organizing genius, his tenacity of purpose, his keenness of mind, and his firmness of character, he looms up as one of the most impressive figures of the century..." This is a defensive and apologetic work which still has considerable merit from its historical value. Another two volume version of the same book was published in 1953 and retitled JOHN D. ROCKEFELLER: A STUDY IN

POWER. It is similar to the first, but more comprehensive and less defensive of the Rockefellers. A one-volume abridgement was published by Greenleaf in 1959. Cited in Swanson on page 164. The slipcased first edition of both is common.

**B214. (JIM ABERCROMBIE) Nicholson, Patrick J., MR. JIM: THE BIOGRAPHY OF JAMES SMITHER ABERCROMBIE, Houston, Gulf Publishing Co., (1983), vii, 364pp.**
The first half of Nicholson's book is a compilation of Abercrombie family history from the Scottish roots to Jim (1891-1975). Born on a dairy farm in Houston and enduring much family tragedy while growing up, Jim entered the oil business at eighteen, rapidly progressing from roustabout to independent drilling contractor. During the 1920's he formed a friendship and association with Harry Cameron, who was a mechanical genius. Together they started Cameron Iron Works which became a billion dollar oil equipment manufacturing and supply firm. One of their main claims to fame was Abercrombie's invention of the first practical oil well blow-out preventer. The author relates Abercrombie's industry stories, his politics and his philanthropy. The hardbound first edition is common.

**B215. (ANDREW MELLON) O'Connor, Harvey. MELLON'S MILLIONS: THE BIOGRAPHY OF A FORTUNE, THE LIFE AND TIMES OF ANDREW MELLON., NY: The John Day company, 1933. xv, 443pp.**
This book was the first full family biography of the famous Mellon family which was widely circulated. There had been a few earlier efforts which were privately printed and meant for family and company people only. O'Connor has been accused of muckraking but today he would just be another investigative journalist doing his thing. The book is not very favorable to the Mellons, however they were pretty easy pickings. Worth over two billion dollars in the depths of the depression, the Mellons held a vertically integrated aluminum monopoly with Alcoa, one of the major U.S. banking empires with Mellon Bank and one of the top vertically integrated oil companies with Gulf Oil. Many more smaller companies were owned outright or controlled by the family. Thomas Mellon (1813-1908) was born near Pittsburgh to a hard working and successful farm family. He was inspired by Ben Franklin's autobiography (as most who read it), and left the farm to become a lawyer. He speculated in

mortgages, liens, notes, real estate and became a foreclosure fanatic, making a fortune. Elected to a judgeship in 1859 he served ten years and served well. In 1870 he established T. Mellon & Sons bank. Andrew (1855-1937), then nineteen began his rise to power and greatness. Just one of his several major victories was the decision to provide funds to Lucas and Guffey & Galey at the Spindletop discovery, gaining him control of what turned out to be a major oil company. President Harding appointed Andrew to his cabinet as Secretary of the Treasury in 1921 because the Mellons knew more about money than anyone else. Some historians think Mellon was responsible for the financial collapse in 1929 and many of those are convinced he could have prevented it. We will never know. Before long, public pressure caused then president Hoover to get Andrew out of the Treasury so he sent him as ambassador to England in 1932. As the book ends in 1933, the feds are beginning to press their monster tax suits against Andrew. Excellent biography of an impressive mogul. The book was a sensation when it came out and is still worth reading. Cited in Swanson on page 175. The hardbound first edition is becoming scarce.

**B216. (WILLIAM ORCUTT) Orcutt, Mary Logan, MEMORABILIA OF WILLIAM WARREN ORCUTT, 1869-1942, Los Angeles, Fred S. Lang Press, 1945, xxi, 93pp.**

This book was not available for review by Neff, but is regarded as important. Quoting from Swanson: "Orcutt was early practitioner applying geology in field of oil exploration; was famous head of geological department of Union Oil Company of California". Cited in Swanson page 176. The first edition is very scarce.

**B217. (RODERIC OWEN) Owen, Roderic, THE GOLDEN BUBBLE: ARABIAN GULF DOCUMENTARY, London, Collins, 1957, 255pp.**

This book is essentially a journal of the author's activities and thoughts during his Arabian Gulf adventure. Overwhelming oil wealth imposed on some of the most primitive culture in the world created some interesting situations. He became involved with sheiks, politicians and oilmen and provides a fascinating commentary on Arab life with oil. The hardbound first edition is common.

**B218. (SHAH OF IRAN) Pahlavi, Mohammed Reza Shah, AN ANSWER TO HISTORY, NY, Stein and Day, (1980), 204pp.**
As this autobiography is written, the Shah is in exile, fighting public opinion which had turned against him, moving from one country to another, trying to find a place he is welcome after the Khomeini takeover. This is a well written story including much Iranian history, biographical information on his father, and notes on his own growing up. Shah relates the early, British dominated oil dealing with Anglo-Persian and Knox D'Arcy which resulted in the discovery of oil in 1908. Constant turmoil existed between Persian and British governments, each trying to control the country as its oil wealth became increasingly important. Exiled in January 1979, the Shah had ruled Iran for thirty-seven years. He makes a strong point that he had been a friend to the West for that entire time. It was an unsuccessful plea for a little reciprocity. The hardbound first edition is common.

**B219. (SHAH OF IRAN) Pahlavi, Mohammed Reza Shah, MISSION FOR MY COUNTRY., NY, McGraw-Hill Book Co, (1961), 336pp.**
The Shah was born in 1919 in Teheran to a privileged family and his father became Shah of Iran in 1926 with a vow to Westernize the country. Shah was well educated there and married an Egyptian princess, Fawzia in 1938. Young Pahlavi narrowly escaped an assassination attempt in 1949. His father had abdicated in 1941 in favor of his son, aged twenty-one. The Shah briefly gives Middle East oil history. Anglo-Persian had found and developed several large oil fields and was in continual disagreement from the 1930's until 1950 when a 50/50 profit sharing plan was developed, quickly followed by nationalization. The Shah was exiled for a period as Mossadegh took power. Within two years the Shah was back and having success with his Iranian Oil Company as they did their own exploratory drilling. At the time the Shah wrote this book, he had recently granted concessions to Standard Oil, Pan American and AGIP who were all commencing drilling operations under stringent contracts. The hardbound first edition is common.

**B220. (DR. GEORGE STOKER) Parker, George (pseudonym), OIL FIELD MEDICO, Dallas, Banks Upshaw and Company, (1948), 139pp.**
Stoker, under this alias wrote of his adventures during several years in the Texas Gulf Coast boomtowns of the early 1900's. This book is on par

with Sue Sanders', "Our Common Herd". The non- fiction accounts are so vivid, bizarre and intriguing that they are hard to believe. He was born about 1880 into a wealthy family and breezed through college and medical school to a lucrative practice somewhere in the East. An adventurous soul, he gave all of this up to go to Spindletop (he uses the uncommon spelling "Spindle Top") immediately on hearing of the Lucas gusher where he became the only doctor in town. From Spindletop to Batson Prairie and then on to Saratoga he followed the boomtowns and practiced some tough medicine. These oil boomtowns were as rough as any frontier towns with rampant disease, gunfighting and accidents, all of which keep a doctor busy full time. Bone setting and baby delivery were about the only normal business. First-hand accounts of mud and blood and oilfield disasters are told with humor and understanding of the human condition. His descriptions of the oil fields and the people who worked them are such that the reader feels like he is reliving these events of almost one hundred years ago. Excellent reading. Cited in Swanson on page 40. The hardbound first edition is common.

**B221. (NELSON ROCKEFELLER) Persico, Joseph E., THE IMPERIAL ROCKEFELLER: A BIOGRAPHY OF NELSON A. ROCKEFELLER, NY, Simon and Schuster, (1982), 314pp.**
Persico was chief speech writer for Nelson Rockefeller for eleven years (1966-1977), through part of his New York governorship, and the vice presidency. Although he recognizes Rockefeller's faults, he is a solid supporter and friend. The biography is not very revealing or controversial as it is basically a chronological recollection along with some family history. No source notes or bibliography are included. From the dust jacket: "It is also a deeply human portrait of a fascinating man, at once powerful and vulnerable, ambitious and self-sacrificing, democratic yet deeply aristocratic, a man who preferred labor leaders over Republican fat cats, a family man whose life ended in a front-page sensation". Nelson was born in 1908, the second son of John D. Rockefeller, Jr. From an early age he apparently thought he would become president of the U.S. He accomplished much in business and philanthropy, but much more in politics, particularly as governor of New York. He lost his bids for the Republican presidential nomination against Goldwater and Nixon, finally getting close to the office when appointed Vice President by Gerald Ford after Nixon retired from office. He tried to make something consequential of the vice presidency, but like all of those before him failed. Genetic

longevity also failed him as he died at age seventy. His great grandfather died at ninety-four, his grandfather at ninety-eight and his father at eighty-six. The unfortunate circumstances of his death, in a compromising situation, are probably better remembered than his many other accomplishments. A very good book. The hardbound first edition is common.

**B222. (HOWARD HUGHES) Phelan, James, HOWARD HUGHES: THE HIDDEN YEARS, NY, Random House, (1976), xvi, 201pp.**
The author was helped in preparation of this book by Stewart and Morgulis, two of Hughes' closest aids up to the time of his death. He uses the Hughes Secrecy Machine as his theme for the book. Phelan takes a very insightful look at Hughes by examining only his most interesting deals, associates and quirks and presenting them in chronological order. This is a popular and widely circulated biography. The hardbound first edition is common.

**B223. (CULLEN DAVIS) Phillips, David Atlee, THE GREAT TEXAS MURDER TRIALS: A COMPELLING ACCOUNT OF THE SENSATIONAL T. CULLEN DAVIS CASE, NY, Macmillan Publishing Co., 1979, 270pp.**
Philips was a newspaper editor and writer before serving twenty- five years with the CIA. His book is probably the first of the three early versions of the T. Cullen Davis trials, although they were all released about the same time. Davis had been acquitted of the murder of his step-daughter and the boyfriend of his wife Priscilla. He had also survived a later trial for soliciting the murder of Judge Eidson, a long time antagonist, as well as several other people on a hit list. The fabulous "Racehorse Haynes", of course, got him off, even after three eye witnesses identified Davis as the killer. The most expensive murder trial in Texas history at $3,000,000. From the dust jacket: "This is a modern frontier saga of sex, violence, and the degree to which some people believe money talks". The hardbound first edition is common.

**B224. (BOONE PICKENS) Pickens, T. Boone Jr., BOONE, Boston, Houghton Mifflin Company, 1987, xii, 304pp.**
An investigative journalist needs to take this whitewash autobiography of one of the most disliked men in the oil industry (as of 1994) and expand on it. The average layman believes Pickens to be a folk hero, however the same is not true of those still left in the oil business. Pickens was born in

1923 at Holdenville in eastern Oklahoma just one year before the big Seminole field discovery. His father was an independent landman who bought and sold oil and gas leases. Boone played ball and had a paper route while growing up, later moving to Amarillo for high school. He received a geology degree from Texas A & M and went to work for Phillips Petroleum in Bartlesville in 1951. A couple of years of big company bureaucracy and politics frustrated him into leaving in 1954. He had early success on a farmout of leases on one of his old Phillips prospects which encouraged him to continue consulting and selling drilling deals. By 1957 he had formed a small company and was selling his own drilling partnerships. Good success was had in Canada and in the Panhandle field. He changed the company name to Mesa in 1964 and was listed on the American Stock Exchange in 1968. Success continued as he merged with Houghton Company for their gas reserves and with Pubco for their oil reserves. The last forty percent of the book deals with his raids and takeover attempts on Cities Service, Gulf Oil, Phillips and Unocal. Not pretty, but still an important part of oil history. The hardbound first edition is very common.

**B225. (NELSON ROCKEFELLER) Poling, James (editor), THE ROCKEFELLER RECORD: A POLITICAL SELF-PORTRAIT, NY, Thomas Y. Crowell Co., (1960), xi, 177pp.**
In 1960, Poling is a free lance writer who had previously been managing editor of Doubleday. He gives a brief background and career of Nelson Rockefeller. Then as the title indicates, lets Nelson speak for himself. The later chapters are selected speeches delivered by Nelson and organized in categories of important issues of the day. In the final chapter titled "Prognosis", Poling states that if elected president, Nelson would have an administration characterized by social liberalism and economic conservatism. Evidently the public felt that way also. The hardbound first edition is common.

**B226. (CLARENCE POPE) Pope, Clarence, AN OIL SCOUT IN THE PERMIAN BASIN, El Paso, Permian Press, 1972, xi, 150pp.**
Pope saw it all in the Permian Basin as an oil scout and landman. From the Big Lake field development through Hendricks and the Yates discoveries to the industry recession caused largely by cheap imported oil in the late 1950's and early 1960's. Born in 1895 to poor Tennessee parents, he was an industrious child and eventually attended business

college where he learned typing, steno and speed writing. He served in WW I and after release went to work for Sun, first in Tulsa, then Kansas, before transferring to San Angelo, Texas, just as the Permian Basin opened up as a major oil province in 1924. He scouted most of the early wildcats and discoveries in the basin and was promoted to landman in 1930, serving in that capacity with Sun until his retirement in 1960. His chapters on the oil fraternity and wildcatters have great descriptions of the Permian Basin movers and shakers. The chapter titled Black Gold in the Permian Basin provides history of the specific field discoveries and their relative importance. Much knowledge is concentrated in this fine book and the form is not too technical for the layman. Very good reading. The hardbound first edition is becoming scarce.

**B227. (VARIOUS) Porterfield, Bill, A LOOSE HERD OF TEXANS, College Station, Texas A&M University Press, (1978), xiv, 198pp.**
Some of Porterfield's friends and acquaintances are profiled in this well written rambling biographical work. Some of the characters are famous and others obscure. All are interesting. Of particular interest is Porterfield's trip to the oil fields with Loyd Powell, Jr. He gets a full dose of history and philosophy from a modern independent oil operator. Also Porterfield does a nice job of dissecting H.L. Hunt's unique character. During 1978, Hunt was probably the richest man in the world although Libya had just absconded with a major oil field he had developed there. The hardbound first edition is becoming scarce.

**B228. (BILL PORTERFIELD) Porterfield, Bill, TEXAS RHAPSODY: MEMOIRS OF A NATIVE SON, NY, Holt, Rinehart and Winston, (1981), xvii, 266pp.**
Porterfield grew up during the depression and moved from town to town in the Texas oil fields where his father was a roughneck and sometimes driller. He attended college and became a journalist- reporter of some note. He was a well respected columnist for the Dallas Times-Herald. He writes largely of the rural Texas of his youth. In six loosely related parts he give us a fine series of short stories. The first section is family: "My father, being a roughneck and a driller, wanted me to become a geologist. He saw me following his footsteps in the oil field, but wearing soft, expensive loafers instead of steel-toed boots". Not much oil in the book, but Porterfield is definitely an oil field child. The hardbound first edition is common.

**B229. (BILL PORTERFIELD) Porterfield, Billy, DIDDY WAW DIDDY: PASSAGE OF AN AMERICAN SON, NY, Harper Collins, (1994), xviii, 420pp.**

Autobiographical work by the much published journalist-author. He was born about the time the Depression began and this is the story of his family and growing up through the oil boom towns of Oklahoma and Texas. Largely about his somewhat difficult father who decided to up root his farm family and follow the oil rigs. Something like twenty-two moves were involved while working as a high paid driller. However, seeking more stability for the family, he eventually took lower pay as roustabout and pumper. Poignant reminiscences flow through this saga which Porterfield's fans think was long overdue. Diddy Waw Diddy is the mythical town where everything is better. He may have found it. An expanded and finer version of the author's earlier book, Texas Rhapsody. This is an easy reading and generally uplifting book. The hardbound first edition is common.

**B230. (EVERETTE DeGOLYER) Pratt, Wallace E., MEMORIAL TO EVERETTE LEE DeGOLYER (1886-1956), NY, Geological Society of America, 1958, Reprinted from GSA Bulletin 1957 Annual Report, p95- 104.**

A memorial written by another legend. Throughout their careers, Pratt and DeGolyer had worked together and developed deep mutual respect. Pratt believed that DeGolyer's primary contribution to geology was his role of developing the use of geophysical techniques which allowed geologists to determine structure and form of rock units far below the surface prior to drilling. DeGolyer was a founder and president of the American Association of Petroleum Geologists and had gained honorary membership in many of the industry organizations. An impressive list of DeGolyer's writings from 1912 to 1952 is given at the end of this pamphlet. The softbound first edition is becoming scarce.

**B231. (W. C. PRICE) Price, W.C., MEMOIRS, U.S., Privately Published, Nodate, 17pp.**

This short autobiographical sketch is well written literary prose, done in the third person. Most of it is memories of growing up and hunting and fishing. The center section reveals his life as a pipeline construction man. A sensitive one. No times or dates are mentioned, along with few places. Well worth reading. The softbound first edition is becoming scarce.

**B232. (WILLIE & MURPH PURCELL) Purcell, Ann (Willie), DON'T CALL ME DIRTY NAMES, Dallas, TX, Px Press, (1975), x, 164pp.**

Mrs. Purcell wrote this autobiography of herself and husband, Murph, after becoming enraged at the poor public image the oil industry acquired as a result of the 1973 Arab oil embargo and subsequent gasoline price rise (and long gas station lines). The title comes from her frustration with the dirty names which oil people were being called at the time. Her story portrays oil industry people as real and hard-working just as those in any other American industry. She married Murph in 1930 while he was in the employ of a major oil company doing geological field work. From the honeymoon on, she maintained a deep interest in his geological-geophysical career, helping and supporting him at every chance. She portrays herself as being slightly uninformed in the ways of the world and oil business, but has perceptions which go far beyond this ploy. During the depths of the depression she learned oil field lingo and procedures the hard way. She was present when Daisy Bradford #3 came gushing in as the discovery of the giant East Texas Field, not knowing that a high percentage of geologists would lose their jobs as the crude oil of East Texas flooded the market and the search for new crude almost stopped. Murph lost his job shortly, but by 1935 was back at work in Wichita Falls. The next twenty-five years are a series of transfers and job changes, one after another, as they barely scraped by in Oklahoma, Wyoming and North Texas. Murph underwent a conversion from geologist to geophysicist in 1945 and continued his odyssey. Purcell Exploration Services was formed by him in 1950 with one seismic crew and a moderately booming oil business. Good times were enjoyed until 1960 when imported crude depressed the price such that drilling and exploration activity quieted, and his crew went idle. Much of her story is told in dialogue and is more than just a glimpse of oil history from 1930 to 1960, it is a poignant look at the lives of people in an extremely cyclical industry, trying to survive. Excellent reading. The softbound first edition is scarce.

**B233. (ROCKEFELLER FAMILY) Pyle, Tom and Beth Day, POCANTICO: FIFTY YEARS ON THE ROCKEFELLER DOMAIN, NY, Duell, Sloan and Pearce, (1964), 240pp.**

Tom Pyle worked for the Rockefellers for fifty years and according to the writeup was their official game warden and guardian at Pocantico, the family estate. From the dust jacket: "It took a resident staff of nearly one hundred to maintain the home place of the Rockefellers, Pocantico, whose twenty-five hundred acres of woods, bridle trails, and gardens are within sight of the New York skyline". Beth Day did the actual writing. She is a reporter for the Readers Digest, and a good one, providing many interesting stories about Rockefeller which have not been published elsewhere. The hardbound first edition is becoming scarce.

**B234. (HUEY LONG) Reed, Ed, REQUIEM FOR A KINGFISH: THE STRANGE AND UNEXPLAINED DEATH OF HUEY LONG, Baton Rouge, LA, Award Publications, 1986, iv, 250pp.**
The author has a well supported but circumstantial theory of how Long was killed in 1935, different from the generally accepted version of assassination by Dr. Weiss. His case has some merit and involves Long's own people. A good biography with Long's constant battle against Standard Oil carried throughout. The hardbound first edition is common.

**B235. (HOWARD HUGHES) Rhoden, Harold, HIGH STAKES: THE GAMBLE FOR THE HOWARD HUGHES WILL, NY, Crown Publishers, (1980), 404pp.**
This is a lightly embellished non-fiction account of the so called Mormon Will, purported to have been hand written by Howard Hughes in 1968. Hughes died in 1976 and the will mysteriously appeared later that year at the Hughes offices and later was filed by the LDS church. The will gives his estate to: his medical institute, various universities, the Boy Scouts, orphan children and the Mormon Church with a little to individuals such as Jean Peters and Melvin DuMar....Melvin who? Noah Deitrich, who Hughes had not spoken to for twenty years was to be the executor. The author of this book was Deitrich's attorney, who stood to gain massive fees if the will could be proved. They made an excellent case but in 1978, the jury did not buy it. This fight for one of the world's largest fortunes reads more like a mystery novel than the real thing. Fascinating. The hardbound first edition is common.

**B236. (HENRI DETERDING) Roberts, Glyn, THE MOST POWERFUL MAN IN THE WORLD: THE LIFE OF SIR HENRI DETERDING, NY, Covici- Friede, (1938), 448pp.**
This biography-attack on Deterding portrays him as Rockefeller's successor on the worldwide oil scene, as he had gained incredible

political and economic clout through his firm leadership of Royal Dutch-Shell. Book subtitle: "The amazing story of the dictator nobody knows". The book is very entertaining as Roberts slants his anti-capitalist version of Deterding's rise from obscurity to power. Deterding was born in Amsterdam in 1866. His father was a seaman and died when Henri was only six, leaving a wife and five children in poor shape. A good student and hard worker, at sixteen Henri got a job as clerk in a bank. Six years there trained him for a job with the government connected banking enterprise which sent him to the colonies in the Dutch East Indies. He became the head of the gigantic international trust and gained reputation in his country. He decided he wanted a piece of the action and the trust would not give it. Mr. J.B.A. Kessler offered Deterding a job with his small oil company, Royal Dutch, and they came to terms in 1896. Royal Dutch operated small wells in the Dutch East Indies. Roberts' story takes Deterding's career from this inauspicious beginning to where he proves that Rockefeller lost the title of The Most Powerful Man in the World to him. One of the oil industry's most important characters. Cited in Swanson on page 176. The hardbound first edition is scarce.

**B237. (AHMED YAMANI) Robinson, Jeffrey, YAMANI: THE INSIDE STORY, NY, The Atlantic Monthly Press, (1988), xix, 302pp.** Biographical account of one of the most influential oil men of all times. Ahmed Yamani was born at Mecca, Saudi Arabia in 1930, shortly before oil was discovered in the country. He obtained Law Degrees at Cairo, New York University and Harvard prior to becoming legal advisor to Prince Faisal in 1957. He was named Minister of Petroleum by Faisal in 1962. Yamani led an increasingly public life which included standing next to Faisal as he was murdered by international terrorist, Carlos. Yamani was responsible for many accomplishments within OPEC, largely dictating the world price of crude oil. Over his twenty-four year career on the ministry, Yamani continually fought OPEC members (primarily Iran and Iraq, the petulant ones) who were demanding higher prices and higher production levels. Saudi Arabia was almost always the moderate on producing rates and prices. He also negotiated the Tehran Participation Agreements with major oil companies for the Persian Gulf producers. Yamani's fabulous career at the Ministry ended in 1986 after he blatantly disobeyed King Fahd's orders at an OPEC clash with Iran over prices. Now a consultant and speaker around the world, he is enjoying this early retirement at his English estate. The hardbound first edition is common.

**B238. (JOHN D. ROCKEFELLER) Rockefeller, John D., RANDOM REMINISCENCES OF MEN AND EVENTS, Garden City, NY, Doubleday, Page & Co., 1909, ix, 188pp.**
This is Rockefeller's response to the Lloyd and Tarbell books which attacked his companies and him personally. It was written in the heat of the Standard anti-trust controversy, just two years before the court-ordered breakup of 1911. He had retired years before and his son John, Jr. was making progress in damage control by creating the charitable foundations with the help of Reverend Frederic Gates, Rockefeller confidant. Most of the book concerns John D.'s charitable giving. He was seventy at the time and discusses his philosophy of business and wealth, also gives his thoughts on close friends and associates, many of whom were deceased by then. Rockefeller presented the facts as he remembered them, in this the only book he wrote for public consumption. Reprinted in 1916. Reprinted in 1933 w/199pp. Cited in Swanson on page 165; Hardwicke page 45. All hardbound editions are scarce.

**B239. (NELSON ROCKEFELLER) Rockefeller, Nelson A., THE FUTURE OF FEDERALISM, Cambridge, MA, Harvard University Press, 1962, 83pp.**
This is the published version of three lectures given by Nelson in February, 1962 at Harvard University, under the sponsorship of "The Godkin Lectures on the essentials of free government and the duties of the citizen". Lecture titles were Freedom and Federalism; Federalism and National life; and Federalism and Free World Order. His speeches explain and tout federalism. He believes that our system should be applied to development of all nations of the free world. Good Rockefeller propaganda. The hardbound first edition is common.

**R240. (NELSON ROCKEFELLER) Rodgers, William, ROCKEFELLER FOLLIES: AN UNAUTHORIZED VIEW OF NELSON A. ROCKEFELLER, NY, Stein and Day, (1966), 224pp.**
Another timely political attack on John D. Rockefeller's grandson who dedicated his career to politics and suffered a similar fate as others not so endowed. From the DJ: "His adventures in living like a corporation,

governing like a baron, and outfoxing his constituents, seems to give him a chance of becoming an American legend as memorable as that of his grandfather, one the well- concealed facts are known". The hardbound first edition is common.

**B241. (CHARLES ROFF) Roff, Charles L. A BOOM TOWN LAWYER IN THE OSAGE 1919-1927., Quanah, TX, Nortex Press, 1975, xii, 122pp.**
Roff was born in Indian Territory in 1894. His family moved from place to place in Oklahoma and Texas, finally settling back in Oklahoma in 1905. He graduated from University of Oklahoma Law School and served in WW I. This book is the story of his eight- year exposure to the Osage oil boom. After WW I he saw opportunity in the oil boom which was taking place on the great Osage Indian lands. He began practice at Pawhuska, the county seat and Osage Nation capital. He was soon elected county attorney. His stories of prosecuting the criminals of Osage County are fascinating. Many include murder and mayhem. Some are humorous and some are not. The crime rate was higher in the county than in larger cities because of its unique situation. Crimes against the Indians were varied and constant because of the incredible wealth. Unlike other Indians, the Osage leaders had seen to it that their minerals were severed from the surface estate and left in control of the Tribe. The surface was then allotted to the Osages and later sold to others. Thus the Tribe owned nearly 1.5 million acres of mineral rights, much of which would become oil productive. During the 1920's, each headright was receiving around $14,000 annually. Oil does not play much of a part in Roff's stories, however the boom towns and the oil money were the cause of almost all of the crimes he dealt with. Roff had his hands full. The hardbound first edition is becoming scarce.

**B242. (JOSEPH JONES) Rowe, Melodia B., CAPTAIN JONES: THE BIOGRAPHY OF A BUILDER, Hamilton, OH, Hill-Brown, 1942, 262pp.**
Joseph Thomas Jones was born in Philadelphia in 1842. His father who was a building contractor died the next year. Still, young Jones grew up with an aptitude for building and engineering. He fought in the Civil War at nineteen and gained a commission after the bloody Fredericksburg battle. His regiment fought at Chancellorsville and Gettysburg as well. He was wounded and promoted to Captain prior to release in 1864. Captain Jones recovered and got caught up in oil fever the next year. The story of his incredible oil success comprises most of part one of the book. His first

twelve drilling ventures were dry holes, gaining him the nickname "Dry Hole Jones". That title did not stick. At one time he was the largest producer in the U.S., operating 586 wells. He knew all of the oil characters and was involved in many of the historic events. Part two begins about 1896 where he almost single handedly creates the city of Gulf Port, Mississippi by building a major railroad, a deep water harbor and developing the city. Cited in Swanson on page 176. The hardbound first edition is scarce.

**B243. (JOHN CADMAN) Rowland, John and Basil, Second Baron Cadman, AMBASSADOR FOR OIL: THE LIFE OF JOHN, FIRST BARON CADMAN, London, Herbert Jenkins, (1960), 191pp.**
Lord Cadman was born in 1877. His father and grandfather were mining engineers so he studied that subject in school. He later became a professor of mining at Birmingham University where he developed a course in oil technology which soon became a world standard. He was knowledgeable in geology and became an early developer of oil at Trinidad. Well respected by the British government, Cadman persuaded Churchill to convert the Navy from coal to oil-burning fuel. After WW I, he formed Anglo-Persian and quickly rose to chairman. He was an excellent negotiator and was frequently involved in international decisions of great magnitude. During the 1920's, Middle-East oil grew very rapidly with gigantic discoveries, development drilling and construction of pipelines and refineries. Cadman died in 1941, after an incredible career which was largely hidden from the public eye. Important figure in world oil. Cited in Swanson on page 176. The hardbound first edition is becoming scarce.

**B244. (JERRY SADLER) Sadler, Jerry with James Neyland, POLITICS, FAT-CATS & HONEY-MONEY BOYS: THE MEM-WARS OF JERRY SADLER, Santa Monica, CA, Roundtable Publishing, (1984), xii, 316pp.**
Sadler was born in Anderson County, Texas in 1907. He was destined to a life of politics and unpopular causes. His grandfather W.T. Sadler fought in the battle of San Jacinto and his father kept his hand in politics. Sadler was extremely bright and precocious, skipping several grades in school. As indicated in the introduction, several of his stories are contrary to the accepted historical versions and others shed light on motivations for political happenings which were hardly suspected. This book only covers Sadler's career to 1945 although he was active in politics until his

death in 1982. He met the fat cats and honey money boys early in his career as he was a small refiner during the East Texas hot oil episode where he fought the oil companies, the feds and the state for fair treatment. The pattern for his career was established with the theme: seek out injustice imposed on the little guy and try to make it right. From the dust jacket: "Sadler was many things in his lifetime, FBI agent, oilman, lawyer, Railroad Commissioner of Texas, candidate for governor, friend and advisor to Franklin Roosevelt, army officer, legislator and Texas Land Commissioner. He was greatly loved in the State of Texas, and almost equally as hated". The hardbound first edition is common.

**B245. (MERLIN SAILOR) Sailor, Merlin F., YOU MIGHT STRIKE OIL, Norman, OK, University of Oklahoma Press, (1965), 194pp.**

Sailor, a landman for a major oil company, tells how he goes about his work, some of it fun and some of it frustrating. He gives a continuous stream of stories largely relating to a landman's dealings with recalcitrant or crazy farmers, oil companies, lawyers, heirs and royalty owners. He has a humorous approach to his work, but also does a good job of educating laymen to the critical job of a landman in the oil industry. No drilling is done without their expert work. Not much personal information on Sailor although he was born in Iowa, around 1900, received a law degree from Nebraska, practiced twelve years and then began work in the oil business in 1945. Interesting stories. The hardbound first edition is common.

**B246. (SUE SANDERS) Sanders, Sue, OUR COMMON HERD, NY, Garden City Publishing Co., Inc., 1939, 261pp.**

Cal Sanders, of part American Indian (red-blood) and Jane Weiland of British (blue-blood) were married in 1858 in Texas over their parents loud objections. They began by farming, first in East Texas and then in the central Texas area of Stephenville. Sue was born in 1880. She was crippled with bad ankles from some sort of birth defect and could not move her feet until three years old. She got along fairly well after that but was always embarrassed at her condition. At age five her father died and the family progressively became poorer and poorer with Sue learning to fight her own battles as she grew. She was a large and homely girl which coupled with the humiliation of poverty gave her much to fight about. She married Joe Handley at age fourteen in 1895. They had a daughter that year and tried to farm, first at Stephenville, then moving from farm to farm on an odyssey which carried them across Texas, New Mexico, Arizona and Oklahoma. She finally left Joe and tried, on more than one occasion, to start a boarding house-restaurant. Just as something good was about to happen she ran into another disaster, one right after another. Anyone reading her story will agree that the lady was persistent. She first became interested in the oil business in 1913 while working as a housekeeper at a hotel in Muskogee, Oklahoma. After saving a little money and meeting many oil field people she was advised by her attorney to get into oil. For $125.00 he sold her 1/32 interest in a nearby drilling well. The well was a gusher and the operator offered her $5,000 on the spot for her interest. She sold, and the well quit producing shortly thereafter. This same scenario happened over and over again as she parlayed her initial profit into a fortune, Then she lost it. Then she re-gained it etc. Her story is so bizarre that a fiction writer could not have created it. Remarkable story and one of the very best oil biographies. Not mentioned in Swanson. The hardbound first edition is becoming scarce.

**B247. (HUNT FAMILY) Sarnoff, Paul, SILVER BULLS: THE GREAT SILVER BOOM AND BUST, Westport, CT, Arlington House, (1980), 199pp.**

Sarnoff is a knowledgeable highly experienced commodities trader who specializes in precious metals. His writing of this book provides an educational tool for anyone who is interested in the subject. The story is a slightly fictionalized account of the Hunt brothers' (Bunker, Herbert and Lamar) seven year attempt to corner the silver market through buying silver in all forms, but especially in long term futures contracts. The Hunt brothers along with their friends worked hard to get their position and as they acquired more silver the price slowly rose. Others joined in the frenzy. A trader from New Jersey, another from Saudi Arabia, Louisiana, Brazil and Idaho. These men and their associates were the Silver Longs. Their job was to buy the metal and force prices up. Opposing this group were the Silver Shorts, who were generally dealers and speculators, but more often, silver producers who were hedging their bets that the price would fall. By January 1980 the Hunts controlled about 200 million ounces ($4.7 billion dollars worth) After collapse of the price in March of 1980, margin calls were made on the Hunts for $100 million as options expired and panic set in forcing the price lower. The per ounce price had gone from $6.00 in January 1979 to $50.00 in January 1980, then falling to $10.80 in March of 1980. The government got involved with bail-out loans and congressional investigations. The story was not finished when Sarnoff wrote this book. Later the price dropped to around $3.00 per ounce and the Hunt Brothers were forced into bankruptcy. A study in greed. The hardbound first edition is common.

**B248. (WILLIAM LEE) Schofield, Donald F., INDIANS CATTLE SHIPS & OIL, Austin, TX, University of Texas Press, (1985), x, 205pp.**

Born in Wisconsin in 1841, William McDole Lee grew up on the Indiana frontier, served with Sherman in the Civil War and with the government for a few years after that. At the age of thirty-eight, he and a partner named Reynolds gained the right to trade with Indians at Camp Supply, Indian Territory. He made a fortune trading buffalo hides and even more money in cattle ranching. In 1888 he financed the first Texas deep-water port with a railroad connection to Houston. The last chapter deals with Lee's oil ventures. From 1908 to 1920 he persistently leased and drilled on the West Columbia salt dome, south of Houston, with no real success. He did have some marginal luck around the Spindletop field near Beaumont. His poor luck was due to everything from tornadoes to equipment failure. Finally when Lee was aged eighty, Humble and Texaco drilled monster oil producers on some of his remaining royalty interests at the West Columbia dome. He was a late bloomer in the oil business, but his story makes one more good case for persistence. Good early-days history of the frontier, largely Texas. The hardbound first edition is common.

**B249. (HAROLD SCOTT) Scott, Harold W., THE SUGAR CREEK SAGA: CHRONICLES OF A PETROLEUM GEOLOGIST, Cushing-Malloy Inc., Ann Arbor, MI, (1986), xviii, 308pp.**

Scott is a well known University of Illinois Geologist who has applied geology to the practical world of energy. He was born to an Illinois farming family near Sugar Creek in Illinois in 1906. He left for the University in 1924 and did not return. After graduating, he held a few consulting jobs before taking a professorship at the university. He had forays with oil operations for his own account and consulted for several oil companies over the years. At one point he took a sabbatical to do an extensive exploratory survey of foreign lands for H.L. Hunt. He studied

Libya, Egypt, Pakistan and several other potential areas. Scott gives his philosophy and political views along with a fine rendition of oil history as it unfolded through his career. The epilog contains a chronology of important dates in oil history. The hardbound first edition is becoming scarce.

**B250. (MONTY MONCRIEF) Scott, John David, TRUE LEGACY: THE BIOGRAPHY OF A UNIQUE TEXAS OILMAN, W. A. "MONTY" MONCRIEF, U.S., Privately Published, (1982), 240pp.** W.A. Moncrief was an unusual subject for a biographer to pick from the field of oilmen. He was a steady, honest, religious family man who found lots of oil through good times and bad. As this book was being prepared, he was still working every day at the office at age eighty-seven. Born in 1895 at Sulphur Springs, Texas, he was a hard-working child as he grew up in Checotah, Indian Territory. He attended the University of Oklahoma for a time in 1914, later serving in WW I with the Cavalry Division. A wartime acquaintance helped get him a job with Marland Oil in their accounting department. He quickly moved to the land department where he learned the oil business and over the next ten years became directly involved in many of Marland's significant oil discoveries in the Mid-Continent and Permian Basin. Moncrief opened an office in Fort Worth in 1929 as an independent oilman. His son, Tex was ten years old in 1931 when Dad's #1 Lathrop wildcat (which expanded the East Texas field by many square miles) came in. Tex determined then and there that he wanted to be an oilman also. Oil soon went to ten cents a barrel so Moncrief and partners were prompted to sell out to Yount-Lee for two and a half million dollars which was worth quite a bit during the depths of the depression. Always the persistent wildcatter, Moncrief and his sons Tex and Dick continued successful oil drilling as many others fell by the wayside. Moncrief was a friend of politicians and presidents as well as Hollywood personalities. The only sour business story in the book is the six page explanation of a complex oil deal he had made with Bob Hope and Bing Crosby during the 1950's. Through Moncrief the three of them shared equally in a large reef discovery in Scurry County, Texas where they made over seven million dollars apiece. At the end of the story he calls Hope an ingrate and I can see why. Good biography of a good oilman. The hardbound first edition is becoming scarce.

**B251. (DAVE TRUE) Scott, John David, WYOMING WILDCATTER, Palm Beach, FL, Private Profile Publishing, 1985, 277pp.**
Born in Cheyenne, Wyoming in 1915, Henry Alfonzo True XIII (Dave) was destined to become an oil man. His grandfather explored Salt Creek anticline as a potential oil area while working for a mining company. His father was a civil engineer and was involved in oil field surveying and construction, eventually working for Texaco as a petroleum landman. Dave graduated from Montana State University in 1937 and went to work for Texaco as a roustabout in the Colorado oil fields. He suffered much early hardship in the field with a young family, however by 1946 he had worked his way up to superintendent of drilling and production for Wyoming. Reserve Drilling Company hired him to run their contract drilling rig and gave him an option to acquire a quarter interest in the company in 1948. By 1951 Dave and a partner owned the company. An impressive list of successes followed, as they participated in most phases of the oil industry. He has been much honored by the industry since the 1950's for his work in promoting the industry and for his innovations. Lots of good family and friends photos. Much on his fishing and his large ranch, the Double Four, near Laramie. Good reading although there is not much mention of Dave's considerable wildcat successes. He died in 1994. The hardbound first edition is becoming scarce.

**B252. (J. B. SAUNDERS) Scott, Otto J., THE PROFESSIONAL: A BIOGRAPHY OF J.B. SAUNDERS, NY, Atheneum, 1976, xi, 497pp.**
Saunders was born in 1901 and was raised at his parents ranch in Indian Territory. Early schooling was in Sapulpa in 1912, an oil boom town near Tulsa. He began work on an oilfield pipeline for Gypsy Oil in 1917. Over the next few years he worked as roustabout, then tool dresser while trying to attend school. In 1923 he took a job with Imperial Refining as a shipping clerk in their Ardmore office. He quickly moved up to refinery construction engineer and by 1929 was sales manager for Imperial's four Oklahoma refineries. He worked for Hassett & Sons for several years prior to forming Triangle Refineries in 1937. A broker-jobber, Triangle negotiated for refined products and marketed them to customers. The business has a small profit margin, requiring large volumes. J.B.'s previous business contacts made initial start-up easy. By 1957 Triangle was one of the largest oil product marketing firms in the world and J.B. decided at that time to merge with Kerr-McGee under very favorable

terms. J.B. was elected to the Kerr-McGee board and retained ownership of his other related businesses which were substantial, later selling them at considerable profit. He moved into corporate power alongside Dean McGee and helped guide the company into nuclear energy and many other diverse fields. He retired in 1971 although remaining on the board. This is a well composed history book with J.B.'s biography as a kind of sidelight. Either J.B. was perfect or the author thought so. Too much book for the subject. The hardbound first edition is common.

**B253. (JOHN PEW) Sennholz, Mary (compiler), FAITH AND FREEDOM: A BIOGRAPHICAL SKETCH OF A GREAT AMERICAN, JOHN HOWARD PEW, Grove City, PA, Grove City College, 1975, xi, 179pp.**
This memorial to J. Howard Pew was compiled by Mary Sennholz for the college from which he graduated, and which he supported. J. Howard was the son of Joseph Newton Pew, founder of Sun Oil Company. He was born at Bradford, Pennsylvania in 1882, where his father was developing that giant oil field. J. Howard became President of Sun in 1912 and remained as Chairman until 1970. He died at age 89 in 1971. No spectacular career is presented for this brilliant man, only the steady direction of the growth of his family's company. He was deeply religious and involved in the well-being of his employees. The hardbound first edition is scarce.

**B254. (ROCKEFELLER FAMILY) Shaplen, Robert, TOWARD THE WELL- BEING OF MANKIND: FIFTY YEARS OF THE ROCKEFELLER FOUNDATION, London, Hutchinson of London, (1964), ix, 214pp.**
Commemorating fifty years of Rockefeller Foundation philanthropic work worldwide. Much as in the earlier devised Rockefeller Institute, Frederic Gates was the idea man for this foundation. Federal taxation of income and corporation earnings was a new fact of life and the wealthy were scrambling to protect what they had. In 1909, Rockefeller deeded 72,000 shares of Standard Oil to the foundation. Their main areas of concentration over the years have been 1) Health, where medical and disease research, education and treatment are performed worldwide, 2) Food, where agriculture research, education and production are promoted on a worldwide scale, 3) Arts & Humanities, where developing countries are assisted at the graduate level. The arts are mainly supported in the

U.S. A prime example is the Lincoln Center for the Performing Arts. The hardbound first edition is scarce.

**B255. (SHAH OF IRAN) Shawcross, William;, THE SHAH'S LAST RIDE., NY, Simon and Schuster, (1988), 463pp.**
This is an exciting version of the Shah's journey into exile and death. The Shah's government had collapsed and he was fleeing from his enemies. Dying of cancer, the Shah sought refuge with his former allies and was surprised at the universal rebuff. International politics and old loyalties are as changeable as the weather. Shawcross also covers the good and the bad in the Shah's reign as well as the rise of Kohmeini. The Carter administration handling of the Shah in exile as well as the Iran hostage crisis is covered in its unpleasantness. The Shah died in 1980. Today, it is like he never existed. Curiously, when the Shah was in power and even when he was in exile (more than once), his every move was followed and covered by the media on a daily basis. This was partly because he was a powerful dictator and partly because he had such huge oil reserves at his disposal. Generally speaking he was the only Middle East ruler who was on good terms with the U.S. When he went out of power, we dropped him like a hot potato, which he was. The hardbound first edition is common.

**B256. (VARIOUS) Sheehy, Sandy, TEXAS BIG RICH: EXPLOITS, ECCENTRICITIES, AND FABULOUS FORTUNES WON AND LOST, NY, William Morrow and Co., Inc., (1990), 415pp.**
A somewhat sensational account of the titled subject by Sheehy who is a contributing journalist to several magazines. The oil characters included are The Hunts, Basses, Boone Pickens, Ross Perot, Cullen Davis, Jack Grimm, George Mitchell, Clayton Williams and others. From the DJ: "..the true story of how the Texas Big Rich got where they are...and what happened after the boom was over". Oil, real estate, ranching and computer tycoons are all covered, but Sheehy likes Perot the best.

**B257. (TINY SHIELD) Shield, Charles F., MY MOST MEMORABLE MOMENTS, Eugene, OR, Privately Published, (1972), 324pp.**
Born to a Texas family around 1900, Shield's father was a Preacher. His father died when he was still in school so he went to work as a tool dresser on a cable tool rig at fifteen. His stories are told in colloquial language and are very humorous if you can tolerate the bad spelling and

grammar. He traveled extensively while drilling wells in Canada and Alaska, eventually coming home to drill a well on the family place at Trickham, Texas in 1948. The saga of this drilling venture includes continual money raising and drilling problems coupled with enticing shows of oil and gas. Basically the project ended up being a near miss. Shield was also involved in another interesting drilling venture, a remote wildcat well near Eureka, Alaska. Shield's memory for detail and dialogue are exceptional and he makes his stories come to life. Most unusual book. The hardbound first edition is scarce.

**B258. (GENE SIDWELL) Sidwell, Eugene Chenoweth. as told to Stan Thorne, ONE MAN'S JOURNEY, U.S., Privately Published, (1983), 152pp.**

Born in Kentucky in 1905, his father and uncle owned a few drilling rigs and made a fair living drilling shallow wells in the area. Gene grew up in the oil field town of Blacksville, West Virginia. The family moved to the Oklahoma oil country in 1922 and at seventeen he began oil field work. Within a few years he had changed jobs several times, working his way from warehouse clerk to field foreman and by 1926, tool pusher for his uncle's drilling firm at Borger, Texas. A partnership with Mr. Kates in 1933 was the beginning of success in drilling and producing in the Pampa, Texas area. Good history of the oil business in the 1920's and 1930's with vivid description of the living conditions. The hardbound first edition is scarce.

**B259. (RED ADAIR) Singerman, Philip, AN AMERICAN HERO: THE RED ADAIR STORY, Boston: Little, Brown and Co., (1990), xi, 276pp.**

Singerman uses the first five chapters to describe Adair's involvement in the 1988 Piper Alpha offshore platform disaster in the North Sea. Subsequent chapters flash back to his extremely poor childhood in Houston. Born in 1915, he had little schooling, but was a very hard worker, holding down numerous odd jobs until he entered the service for WW II. His dare-devil style flourished there and he became part of an elite bomb disposal squad. After the war he began working full time for his mentor, Myron McKinley. McKinley was the first to fight oil and gas well fires as a business. Eventually Adair started his own business and made history while training the current group of oil well fire fighters who now have their own companies. On reading the book, it is hard to believe that Adair has survived eighty years (as of 1995) of hard drinking, hard fighting and incredible risk taking. Many good stories of Adair's more famous worldwide fire fighting adventures are related. An excellent book about a unique individual. The hardbound first edition is common.

**B260. (ALFRED SMILEY) Smiley, Alfred Wilson, A FEW SCRAPS, OILY AND OTHERWISE, Oil City, PA, The Derrick Publishing Co., 1907, 220pp.**

Born to a farming family at Union City, PA in 1843, Smiley received a typical farm boy's education there and at age seventeen (1860) along with hundreds of others, moved the twenty-five miles to Titusville to begin working in the new oil fields. All the action of a new industry was taking place between Oil City and Titusville in the valley of Oil Creek. He worked at countless jobs and relates his experiences along with chronological history. He quotes freely from McLaurin's SKETCHES IN CRUDE and from Derrick's HAND BOOK OF PETROLEUM. Both authors were personal friends of his. He met or knew the oil regions characters such as Edwin Drake, Coal Oil Johnny Steele, John Wilkes Booth and many others. He believed that Drake received too much credit for drilling the first successful oil well and thinks that the idea and moneymen (Kier, Bissell and Brewer) deserve a larger place in history. His descriptions of the Pithole boom town are great. He invested in a well there and made a lot of money. He kept investing and made a small fortune, and then like most Pithole investors lost it all within a short time due to over speculation and erratic market price. Pithole went from barren land to a 20,000 population in one year and then back to zero within five more years as the production gave out. The author was manager of construction for the first successful area pipeline to the Miller Farm in 1865. He and partner Coutant were the first to connect pipelines from wells to shipping points, which spelled doom for the teamster's business. It is interesting to note that in Smiley's few comments about Standard Oil he never says a bad word although at the time of this writing Standard still had a virtual monopoly on transportation and refining in the area. He probably did not want to make anyone mad. Excellent history-biography by a man who saw it all. Cited in Swanson on page 176; Hardwicke page 40. The hardbound first edition is scarce.

**B261. (JIMMIE OWEN) Smith, Corita Jean Owen, LET THE EAGLE FLY...AGAIN, CA, Privately Published, (1989), x, 287pp.**
James P. (Jimmie) Owen was born in Waco Texas in 1898. His father was an attorney and died when Jimmie was five years old. His mother had died earlier and he and four siblings were raised by an uncle and his wife. He attended S.M.U in 1915 and was in the Army Air Force from 1917 to 1919. He went to Burkburnett for the boom and worked his way from mule skinner to roustabout to gang pusher with Gilliland Oil Company. He quit and started trading in oil and gas leases where he turned a fair profit. He plunged on several oil drilling deals but to no avail. Lots of near misses. He went to Mexico for an oil company combine, to spy on revolutionary activities there as they were expected to seriously affect U.S. oil interests. He made money in cottonseed trading and during WW II ran several bus lines for the army, very successfully. A major 1949 oil discovery with H. L. Hunt as a partner in Louisiana assured success for his company. Owen was directly involved in the discovery of Holly and Maxie fields there. A chance meeting with Winthrop Rockefeller while he was in the service resulted in later beneficial dealings with that family. Owen sold out to General American in 1962 for fifteen million dollars. Later wheeling and dealing unfortunately resulted in bankruptcy's in 1972 and 1982. The last part of the book explains what the All American Wildcatters honorary organization is, and has photos of all honorees (including him). Chapter nineteen is titled Wildcatters I have known, and contains several good stories of his dealings with oil finders. The book is well done and worth reading. The hardbound first edition is common.

**B262. (WEETMAN PEARSON) Spender, John A., WEETMAN PEARSON, FIRST VISCOUNT COWDRAY, 1856-1927, London, Cassell & Co., 1930, xi, 315pp.**
This book was not available for review by Neff, but is regarded as important. Quoting from Swanson: "Cowdray founded famous Mexican Eagle Oil Company, One of the largest Mexican oil companies". Cited in Swanson on page 176. The hardbound first edition is very scarce.

**B263. (TEX THORNTON) Stanley, F., THE TEX THORNTON STORY, Nazareth, TX, Privately Published, 1975, 19pp.**
This short tribute to Tex Thornton never reveals his real name (Word Thornton), but does a good job of describing the man and his work. He was born at Oxford, Mississippi in 1892. He moved to Ohio in 1913

where he learned how to use nitroglycerine and other explosives. A strong demand for oil well shooters drew him to Electra, Texas where he became the expert for North Texas, the Panhandle as well as Southeast New Mexico. Numerous stories are related where his use of nitroglycerine torpedoes stimulated a poor or seemingly dry well into an economic producer. He expanded the use of nitro to extinguishing oil and gas well fires and achieved even more fame in that field. He was murdered in 1948 in a bizarre incident involving two hitchhikers, booze and guns. Thornton was an oilfield pioneer, legend and innovator who came to an unlikely end. Unusual story worth reading. The softbound limited edition is becoming scarce.

**B264. (COAL OIL JOHNNY STEELE) Steele, John Washington, COAL OIL JOHNNY, NY, Hill Publishing, 1902, 211pp.**
Steele was born near Sheakleville, Pennsylvania in 1843. He was adopted at age one and a half by a wealthy farm family who lived on Oil Creek. He received better than average education for the time as well as strong religious schooling. Steele describes the oil springs which bubbled into the creek on their farm and how they would gather "Seneca Oil" by damming the creek until oil accumulated on the surface. They would then soak it up with flannel and wring it into gallon jars. A gallon sold for seventy-five cents to the medicine dealers. Steele's adoptive father died when he was twelve and left the farm to his wife, with provision to go to Johnny on her death. Drake discovered oil at Titusville and the oilmen came to the farm for leases. His adoptive mother leased the low parts of the land to several different operators (who were using creekology as their oil exploration tool) at around $500 apiece. The local farmers received incredible wealth from their 1/8 royalty and lease bonus payments. Steele admits to being a hard worker, but not a sharp trader. When his adoptive mother died of severe burns from a crude oil kitchen fire in 1864, Steele inherited the newly discovered oil wealth. He was at first content to stay on the farm and collect his royalties, but ventured out to buy a first class team of horses. He soon became known as a rich country boy, an easy mark. A few bad local real estate deals took him for $100,000. Then the Hammond well came in on his farm for 300 barrels daily and his income increased considerably. A sharp character named Slocum befriended him and the two of them became traveling and

drinking buddies. In six months of doing Philadelphia, he blew something over $100,000. Steele had made a deal to sell the farm and oil property to Mr. Wickham, whereby Steele received an initial cash consideration of $30,000 (a loan?) with a closing payment to be $1,200,000 payable in six months if Steele could clear the title of various liens and judgments which had been spuriously filed against him. Steele admits to making one stupid business move after another, all the while his friend Slocum was taking advantage of his naivete and more particularly, his drunkenness. Oil production slowed on the farm and Wickham maneuvered out of his agreement to buy it. Everyone and his dog filed liens and judgments against Steele, a few valid, but most not. He sobered up, took bankruptcy, lost the farm and after traveling around the country for a year or so began working seventeen-hour days at a train depot near his old farm. He re-gained the respect of the community and his wife and son, although he could not escape the incredible publicity of being "Coal Oil Johnny", the most famous oil field dunce. Not until fifteen years later, living and working in Kansas and Nebraska, did he feel like he was actually free of the publicity. Thirty- five years after the famous episode he wrote this autobiography. It is a classic story and says much about what can happen if you don't take care of business. A great reading experience. Reprinted in limited softbound edition in 1994 by Mong. Cited in Swanson on page 176; Hardwicke page 45. The hardbound first edition is scarce.

**B265. (JOHN POWELL) Stegner, Wallace, BEYOND THE HUNDREDTH MERIDIAN: JOHN WESLEY POWELL AND THE SECOND OPENING OF THE WEST, Boston, Houghton Mifflin Co., (1954), xxiii, 438pp.**

Exciting story of a brilliant powerful man who was both loved and hated by his associates and government politicians. From the cover: "Powell's most dramatic achievement was his leadership of the first expedition to descend the Green and Colorado Rivers through all their canyons, including the Grand Canyon, in 1869 and again in 1871-72...Powell's great contribution to the country was his lifelong struggle to organize and encourage the participation of the government in scientific studies of the nation. He can justly be called the father of the Geological Survey". Great man. The hardbound first edition is scarce.

**B266. (LONE WOLF GONZAULLAS) Stephens, Robert W., LONE WOLF: THE STORY OF TEXAS RANGER CAPTAIN M.T. GONZAULLAS, Dallas, TX, Taylor Publishing Company, (ca 1979), 79pp.**

In the introduction to this biography Stephens explains that he was asked to be the biographer only three weeks before Gonzaullas' death in 1977. Stephens further states that the book contains only the stories which Gonzaullas previously discussed with him and felt were of particular significance. Born in 1891 in Spain to American parents of Spanish and German decent, he shortly came to the U.S. His parents were killed in the Galveston tidal wave of 1900 and his childhood was an orphan odyssey of being passed from family to family. He grew up admiring the Texas Rangers and joined them in El Paso in 1920. Prohibition was under way at the time and he fought border smuggling with a vengeance. Preferring to work alone, he was aptly nicknamed Lone Wolf. His first oil duty was to restore law and order to the roaring boomtowns of Ranger and then Desdemona. He later proceeded to Wichita County and helped clean up the boomtowns there (by far the worst). Lawlessness was rampant and local officials often had to call in state and sometimes federal help. In 1921 he went to work full-time as a prohibition agent of the Treasury Department. Back to the Rangers in 1927 and on to booming Kilgore in 1930. His reputation by this time was one of fearlessness as a hands-on law enforcement officer. Most of the undesirable elements would leave town as soon as Lone Wolf appeared. He worked alone for several months at Kilgore with good success. Well outfitted, he would sometimes show up at a disturbance with his bullet proof Chrysler, two 45 caliber automatic pistols and a sub-machine gun. He quit the Rangers in 1933 when the liberal governor, Ma Ferguson was elected, rejoining in 1940 after serving as head of the State Bureau of Investigation. At the time of his retirement in 1951, Lone Wolf had been shot seven times and knifed many more. He admitted to killing twenty-two men although reports of seventy- five were not considered unlikely. This biography contains very good stories of the colorful Lone Wolf, but they apparently do not go beyond what Gonzaullas wanted in his biography. Edition of 650 is limited, signed and numbered. The hardbound limited edition is scarce.

**B267. (SWEDE STERNBERG) Sternberg, C.W., PIPELINING....WAY BACK WHEN, U.S, Privately Published, (1972), 63pp.**
Sternberg gives very little biographical information on himself, but begins his story at the time his pipelining saga began in 1915. He was an itinerant pipeliner for hire until retiring from the business in 1950. He provides a continuous stream of stories as he moved from company to company, job to job and town to town, where the work was. The history is invaluable as to the descriptions of the primitive equipment and pipeline construction techniques used to connect oil and gas sources to refineries and from the refineries to markets. He worked in many of the important oil and gas fields across the country. Very good history. The hardbound first edition is becoming scarce.

**B268. (ESTHA STOWE) Stowe, Estha Briscoe, OIL FIELD CHILD, Fort Worth, TX, Texas Christian University Press, (1989), 181pp.**
Stowe wrote this biography in 1989 primarily to relieve frustration built up over the years by the negative image oil field people were given in books and movies. She decided to present the story of her own, quite normal family. She was born in 1916 and shuffled from one boomtown to another as her father followed the work, just like hundreds of other families involved in Oklahoma and Texas drilling activity. Stowe provides vivid descriptions of the living conditions in the camps and small towns starting in Ardmore, Oklahoma where her Dad worked for Amerada as a pipeline gang pusher. They proceeded on to Big Lake, Texas where he worked on the Santa Rita discovery well there. He also worked at Wortham, McCamey, Crane, Wink and Odessa. Her father quit the oil fields in 1928 and opened a grocery store at Graford, Texas. Estha Stowe writes with a sensitivity and compassion of the poverty, terrible living conditions and constant accidents which plagued the early oil people. The softbound first edition is common.

**B269. (ALBERT FALL) Stratton, David H. (editor), THE MEMOIRS OF ALBERT B. FALL, El Paso, TX, Texas Western Press, 1966, (Southwestern Studies Vol IV, #3). 63pp.**
Fall's autobiography is presented in the first person as dictated by him in the summer of 1925. He was born in 1861 in Kentucky. He studied to be a teacher like his father, then studied law. He moved to Texas and roamed through Indian Territory and New Mexico doing whatever came to mind.

He met Ed Doheny while prospecting for minerals in West Central New Mexico and practiced law while living at his ranch near Tularosa. He specialized in Mexican law and handled extraditions and land title problems as well as criminal cases. In 1890 he was elected to the territorial legislature, later serving in the Spanish-American war. He was elected to the U.S. Senate in 1912 and Harding appointed him to the cabinet post of Secretary of the Interior in 1921. Fall was well respected by his peers and the general public until the "Teapot Dome" scandal erupted. He left the Department of Interior in 1923 and was plagued by Senate investigations over the next several years as they revealed his suspicious handling of public lands. He made two oil and gas lease arrangements of known government oil reserves to Harry Sinclair (Teapot Dome, Wyoming) and to Ed Doheny (Elk Hills, California). In return, the government was traded refined products at Atlantic, Caribbean and Pacific ports for the Navy. Fall's reasoning was fairly sound in that the reserves of oil were at the time being drained from the government lands by offset operators and the oil that would be produced by Doheny and Sinclair was to be reserved for the Navy and stored in tanks. Unfortunately, Fall "borrowed" $404,000 from Doheny and Sinclair about the time of the lease arrangement. Fall went to prison for a year in 1931 and died at El Paso in 1944, destitute. Quoting from the introduction: "Albert B. Fall's life was an American success story with an unhappy ending". This is one of many good accounts of Mr. Fall's prominent career. The softbound first edition is becoming scarce.

**B270. (JOHNSON CAMDEN) Summers, Festus P., JOHNSON NEWLON CAMDEN: A STUDY IN INDIVIDUALISM, NY, G. P. Putnam's Sons, 1937, xi, 605pp.**

Born in the backwoods of West Virginia in 1818, Camden was able to attend the U. S. Military Academy and subsequently became a lawyer, surveyor, merchant and land speculator. He was an oil pioneer and producer in Pennsylvania shortly after the Titusville discovery. Selling out to Rockefeller in 1875, he later became president of three Standard Oil subsidiaries. Definitely a low key oil history character, Camden was very successful in any venture he touched. After the career with Standard, he went on to build several important railroads. Lots of history and politics in this well written book. Cited in Swanson on page 176. The hardbound first edition is scarce.

**B271. (WIRT FRANKLIN) Sweet, George Elliott, GENTLEMAN IN OIL, Los Angeles, CA, Science Press, (1966), 170pp.**

Wirt Franklin was born in Kansas in the early 1880's. His father was an attorney-politician and moved his family to Washington DC in 1888. Wirt studied law at Georgetown and while still a student went to work for the Dawes Commission and subsequently had much to do with assignment of Indian headrights in Oklahoma Territory. Without a degree, he passed the bar and in 1906 formed the law firm of Apple and Franklin in Ardmore. By 1913, Franklin had formed Crystal Oil Company along with Apple, Edmund Galt and Roy Johnson. In that same year they discovered the giant Healdton Oil Field near Ardmore. This discovery essentially started the southern Oklahoma drilling activity. A geologist's nightmare....for a lawyer to discover one of the largest oil fields in the U.S. Franklin had been aware of significant oil shows in a 1888 dry well located a half mile north of his drillsite and based his rank-wildcat prospect on that. Franklin later operated as an independent and incorporated in 1927. He was a credible speech maker and spokesman for the oil and gas industry. Very active in politics, he promoted the oil industry at every chance. Good reading about a good man. The hardbound first edition is becoming scarce.

**B272. (IDA TARBELL) Tarbell, Ida M., ALL IN THE DAY'S WORK: THE AUTOBIOGRAPHY OF THE FOREMOST MUCKRAKER OF HER TIME, NY, Macmillan, 1939, 412pp.**

Born in Pennsylvania in 1857, Tarbell's father was a teacher, carpenter and farmer and her mother was a teacher. Evidently her father was also an opportunist, for as he was preparing to move the family to their new home and farm in Iowa, he heard about the 1859 Drake discovery well at Titusville and decided to check it out. There were already several wells drilling and he knew that the oil would require storage and that he was an expert on building leak-proof wooden tanks. A perfect situation existed so he started his business on the spot and prospered there, expanding to Pithole when it came gushing in. Wood tanks soon went out of favor so he began building steel tanks and expanded into oil operations. Rockefeller gained control of the price of crude and refined products through special rebates from the railroads, then the primary form of oil transportation. He crushed most of the small producers and refiners in the oil regions, including Ida Tarbell's Dad. Although only eight years old at the time, she did not forget. She became a woman's rights advocate, a reporter, journalist, editor and writer of note. Although well respected for her biography of Abraham Lincoln, her claim to fame lies in her so called muckraking journalism and relentless pursuit of monopolist John D. Rockefeller and his Standard Oil. She published "The History of the Standard Oil Company" in 1904 and gladly made speaking engagements where she related the unsavory facts of Rockefeller's monopoly. Rockefeller was finally defeated by the government and his trust broken into individual companies in 1911. It is believed that Tarbell's efforts were the single most important factor in gaining public and government anti- Standard attention. Her book has been reprinted numerous times but remains difficult to come by. It is an extremely fair company history when compared to the slanted work of today's investigative journalists. In the early twentieth century she was probably the most respected woman in the U.S. She wrote this autobiography at age eighty-two, five years before her death and allowed it to be published only because she needed the money. Great reading about a great woman. Not cited in Swanson. Reprinted in softbound by G.K. Hall Co., 1989. xiii, 412pp. The hardbound first edition is scarce.

**B273. (WILLIAM TAVOULAREAS) Tavoulareas, William P., FIGHTING BACK, NY, Simon and Schuster, (1985), 253pp.**
His father came over from Greece and moved to Brooklyn where Tavoulareas was born, around 1915. William worked his way through business and law school. He started work for Mobil as a junior accountant in 1947 and was in on the ground floor of Mobil's developing Middle East operations. By 1961 he was head of Mobil's Middle East Planning Division and was promoted to president of the company in 1969. A negotiator and planner of exceptional skill, Tavoulareas set up a crude oil shipping deal with Saudi Arabia which became controversial when Washington Post reporters wrote an article about the deal in 1979. They attacked Tavoulareas personally, as well as his son who was working for the shipping entity. The overly ambitious investigators of the powerful Washington Post got carried away with their story and Tavoulareas sued them for libel. At the time of this book he had won several levels of court verdicts, but also lost a few. He was anticipating that the Post would carry the suit to the Supreme Court. This is a rambling story of power gone awry. Tavoulareas thinks he was the little guy in this litigation, although he was president of one of the largest U.S. corporations. He is shocked by

the power of the press. The press was shocked by his power. The hardbound first edition is common.

**B274. (ARTHUR THOMPSON) Thompson, Arthur Beeby-, BLACK GOLD: THE STORY OF AN OIL PIONEER, NY, Doubleday & Co., 1961, 544pp.**

The author published his autobiography and history of oil at age eighty-eight. It is written in the first person and details the highlights of one of the world's most important petroleum engineers during the oil growth period encompassing the first half of the twentieth century. His father was an important scientist in England and head of a science school which specialized in geology and physics. As a youngster he was nationally exposed to his father's skilled teaching. After a rapid scientific schooling and apprenticeship he spent two years as water consultant in the Gold Coast of Africa where he almost died of a fever. In 1898 he went to the Baku oil fields of Russia where he consulted for an oil company. He worked with the great engineers, geologists and heads of state around the world and soon became known as a first rank oil man and trouble shooter. He published numerous definitive reference books between 1904 and 1952 dealing with oil exploration and development. This autobiography is very well illustrated with photos which have not been published elsewhere. It is a bit British in style, but the very readable story of a top oil man. British edition by Sidgwick & Jackson is titled OIL PIONEER. The hardbound first edition is scarce.

**B275. (EVERETTE DeGOLYER) Tinkle, Lon, MR. De: A BIOGRAPHY OF EVERETTE LEE DeGOLYER, Boston, Little, Brown and Company, (1970), xv, 393pp.**

DeGolyer was born in 1886 in Kansas, raised in Oklahoma and died at his home in Dallas in 1956. He was known as "The Father of American Geophysics" or "The World's Greatest Geologist", depending on your preference. Based on what he accomplished, there is not much argument about either title. In 1906, Charles N. Gould (Covered Wagon Geologist) who was in charge of the University of Oklahoma Geology School recommended DeGolyer for a government job as cook on a geological field party. He quickly picked up geology and within two years was assistant to Dr. Hayes at the U.S.G.S. In 1909 he and Hayes took a leave of absence to consult for Lord Cowdray's Mexican Eagle Oil Co., in Tampico. After two years DeGolyer recommended drilling the #4 Potrero del Llano wildcat. The well came in as one of the largest producers in the world, producing 60,000 barrels per day, thus insuring DeGolyer's immortality as discoverer of the giant Golden Lane Field. He humbly returned to O.U. to complete the requirements for a geology degree. With the help of Lord Cowdray, he formed Amerada Petroleum in 1919. Amerada's extremely successful wildcatting record was made possible by DeGolyer's use of the latest geophysical methods. In 1930 he formed Geophysical Services, Inc., a premier seismic company even today. Later he founded DeGolyer and McNaughton, a top ranked petroleum consulting firm in the U.S. His personal book collection which included many different subjects reportedly exceeded 89,000 volumes, most of which were donated to large universities after his death. Exciting reading. The hardbound first edition is becoming scarce.

**B276. (HOWARD HUGHES) Tinnin, David B., JUST ABOUT EVERYBODY VS. HOWARD HUGHES, Garden City, NY, Doubleday & Co., Inc., 1973, xi, 462pp.**

The author reviewed thousands of pages of court testimony from some of Hughes' most prominent court cases. His concept for the book was to reconstruct Hughes' life during the 1960's from court records. Always a fascinating subject, the super rich oil heir whose reclusive and bizarre lifestyle were becoming public knowledge. The more the public knew, the more they wanted to know about him. A well constructed biography, but only as reliable as the court records. The hardbound first edition is common.

**B277. (GUSTAF TOERNUDD) Toernudd, Gustaf, THE LAND OF OIL AND WINDS: LETTERS FROM BAKU, NY, Vantage Press, (1975), 79pp.**

The author was born in Finland in 1832. He became interested in engineering and traveled to England where he flourished in that field. His reputation was such that he was offered the job of manager of the Nobel family's oil complex in Baku. He learned fast and became an innovator, redesigning the refinery facilities and procedures. Elanor Virgil translates the Swedish letters written home from the massive, yet primitive oil operation. The time period is 1881 to 1886 and the descriptions are vivid as Tornud speaks of this famous oil area and its development. A very important part of oil history. The hardbound first edition is becoming scarce.

**B278. (TOL BARRET) Tolbert, Frank X., THE STORY OF LYNE TALIFERRO BARRET: WHO DRILLED TEXAS' FIRST OIL WELL, TX, Texas Mid- Continent Oil & Gas Association, 1966, 16pp.**

This work was well researched as the author had access to Barret's private letters and contracts which were provided by his granddaughters. Nicknamed Tol, Barret was born in Virginia in 1832 and moved to Texas when he was ten years old. According to this commemorative pamphlet, he had a lease for oil and gas and was prepared to drill on the Skillern property early in 1859. Only the lack of machinery kept him from drilling North America's first producing oil well. Located south of Nacogdoches, Texas, amid numerous oil seeps, the well was finally completed in 1866 (the first in Texas) at a depth of 106 feet, producing ten barrels of oil per day. Barret tried to raise money to develop the property, however there was virtually no demand for oil in the area then. Also, the financial climate in Texas during reconstruction was rather poor. He could not get support even from Pennsylvania oil people as they were involved in a problems of their own, namely market-glut. Barret eventually lost his lease, his investment and his savings. He went on to run a mercantile store at Melrose, Texas and spent much of his time overseeing his wife's farms. He died in 1912. Very good history. The softbound first edition is scarce.

**B279. (NOBEL FAMILY) Tolf, Robert W., THE RUSSIAN ROCKEFELLERS: THE SAGA OF THE NOBEL FAMILY AND THE RUSSIAN OIL INDUSTRY, Stanford, CA, Hoover Institution Press, (1976), xv, 269pp.**

Tolf had his work cut out for him as he unraveled the Nobel family history and assembled biographical information. The Nobel family invented and developed the Russian oil industry. Unfortunately, in 1917 the Bolsheviks destroyed the records of all Nobel family accomplishments in Russia. The family had not only been the leader in oil industry development, but was very prominent in several other phases of industrialization that had helped bring Russia to world power. All Swedish expatriates, Immanuel, born 1801 was the patriarch. He and his three sons participated in petroleum ventures as well as explosives development, manufacturing, banking, marketing, and refining. Ludwig, born 1831 was responsible for most of the oil industry ventures. Alfred, born 1833 was the explosives man, and Robert, born 1829 handled refining. Alfred was a sickly, aloof pacifist with an incredible work drive. Robert was neurotic. Ludwig was a good-old-boy genius who raised eight children and ran the 50,000 employee business. This book is a major work clearly demonstrating that the Nobels stood on equal footing with the Rockefellers, the Rothschilds, and Royal Dutch-Shell on the world oil scene prior to 1917. The hardbound first edition is becoming scarce.

**B280. (ROBERT HEFNER SR.) Trafzer, Clifford Earl, THE JUDGE: THE LIFE OF ROBERT A. HEFNER, OK, University of Oklahoma Press, (1975), xviii, 298pp.**

The first of the Oklahoma Heritage Association's "Trackmaker Series", Mr. Hefner was a great asset to Oklahoma. He was born to a poor farming family in Texas in 1874. He worked his way through to a law degree at University of Texas, graduating at the top of his class. He settled there as an attorney and businessman in 1902. He soon left for Beaumont to take advantage of the Spindletop oil discovery activity. He formed a law partnership with Oswald Parker and within a few years was one of the top area lawyers and a knowledgeable oil man. He moved to Ardmore, Oklahoma in 1908 with the idea of helping Indians gain their headrights and also to get in on the rapidly growing oil and gas business there. By 1927 he had acquired 33,000 acres of mineral rights in prime oil country. The oil property value was up and down, but eventually generated tremendous royalty income, all free of overhead and drilling costs. In 1912 he prepared a legal instrument which became the first mineral deed, severing those rights from the fee simple. His form later became the industry standard. He was elected mayor of Ardmore and then to the state supreme court in 1926, for which he moved to Oklahoma City to serve the six year term. In 1939 he became a rare two-term mayor of Oklahoma City, serving until 1947 with distinction. His many charitable and civic works are well known to Oklahomans. Good oil and Oklahoma history. The hardbound first edition is common.

**B281. (BEN HOGAN) Trainer, George Francis, THE LIFE AND ADVENTURES OF BEN HOGAN, THE WICKEDEST MAN IN THE WORLD, NY, No publisher, 1878, 276pp.**

This book was not available for review by Neff, but it is regarded as important. Quoting from Swanson: 'Hogan, in collaboration with Trainer, outlined his early career in Pennsylvania oil regions, and described his conversion". Cited in Swanson on page 176. The hardbound first edition is very scarce.

**B282. (DAVE TRUE) True, H.A. "Dave", THE TRUE STORY, Philadelphia, The Newcomen Society, 1991, 24pp.**
True gave this address to the Newcomen Society in August of 1991 at Casper, Wyoming. He worked in the oil fields for Texaco from 1937 to 1948. He was offered a job at Reserve Drilling Company which was then operating one contract drilling rig. True was allowed to buy into the company. By 1951 the company had five rigs running and the owners wanted out. True and a new partner bought their interests and formed True and Brown Drilling. Constantly expanding and diversifying into other fields, they began developing oil production with moderate success. True bought out his partner in 1954. Then running eight rigs and deeply in debt, he continued exploratory drilling. True's wildcat ventures paid off as he discovered six new oilfields in the Rockies during the early 1960's. With this success he continued diversification until he owned oil field supply houses and pipeline companies in addition to his producing, exploring and drilling companies. The softbound first edition is common.

**B283. (H. L. HUNT) Tucille, Jerome, KINGDOM: THE STORY OF THE HUNT FAMILY OF TEXAS, Ottawa, IL, Jameson Books, (1984), xii, 384pp.**
In 1983 the Hunt family was still riding high, even after a gigantic loss in silver speculation. Their thirteen billion dollar 1980 fortune was still largely intact. From the dust jacket: "The founding father, Haroldson Lafayette Hunt, grew up in Illinois and then drifted west. He became one of the world's best poker players, a womanizer and bigamist, hustler, high- rolling gambler, right-wing eccentric, religious fundamentalist, author of two strange novels, and a health-food freak who believed that Communists were poisoning the nation's food". Tucille refers to Hunt as Arizona Slim or just Slim throughout the book. That was a nickname he gained in his poker playing exploits during the early 1920's and was used only by his close fiends. The author indicates that H.L. drove his first son, Hassie, crazy. Hassie was probably a genius like his dad, plus he had a remarkable ability to find oil. He was regarded as smart in the drilling business as well, but H.L. overrode Hassie's decisions, consistently wrecking his self esteem and confidence. Much on his other sons Bunker and Herbert forcing the price of silver from $3.00 per ounce in the early

1970's to $50.00 per ounce in the early 1980's. A good biography, but it is difficult to capture the real Arizona Slim and his family in so few pages. An excellent bibliography of oil book titles is included. The hardbound first edition is common.

**B284. (C. C. JULIAN) Tygiel, Jules, THE GREAT LOS ANGELES SWINDLE: OIL, STOCKS, AND SCANDAL DURING THE ROARING TWENTIES, NY, Oxford University Press, 1994, xviii, 398pp.**

The author makes a scholarly attempt to unravel the life and circumstances surrounding the oil promotion career of Courtney Chauncey Julian. This is at least the fourth serious work on Julian, being by far the most comprehensive and fair. Julian was a sign of the times, The Roaring 20's. He was an innovative promoter by nature and knew almost nothing about oil when he came to Los Angeles in the early 1920's. He had some mining promotion background. Four years later he and his associates were embroiled in a $150,000,000 swindle which included pyramid schemes, stock watering and almost every other illegal securities violation imaginable. The Julian Petroleum stock issue was oversubscribed, among other things. Business and politics are deeply interwoven into the tragic story. It appears that he missed becoming an important legitimate oil producing company only because of greed and deviant thinking. Excellent reading about a fabulous oil character. The hardbound first edition is common.

**B285. (ROBERT S. KERR) United States Congress, ROBERT SAMUEL KERR: LATE A SENATOR FROM OKLAHOMA, MEMORIAL ADDRESS, Washington, Government Printing Office, 1963, 190pp.**

The memorial contains addresses or eulogies by all of the U.S. Senators and many Representatives. Also included are obituaries and editorials from important newspapers. Much more. Standard memorial publication for a Congressman. The hardbound first edition is common.

**B286. (MORDELO VINCENT) Vincent, Mordelo L., Jr., A MAN REMEMBERS, NY, Vantage Press, (1970), 246pp.**

Vincent was born in Tampico, Mexico in 1903 to parents who had migrated there from the U.S. His Dad was a farmer and lease broker. Vincent spent his childhood in the turmoil of revolutionary activity with his first exposure to oil being a raging blowout and fire in Tampico in 1908. His Dad hit a well in Golden Lane Field in 1917. He and his partners sold out to Gulf Oil for over a million dollars plus overriding interests. Vincent then went to school in the U.S., first at Richmond Academy and eventually Cornell for a Civil Engineering degree. His father had a huge oil discovery in Louisiana and sold out to the majors for several million more. Vincent and a partner founded Vincent and Welch which drilled wildcat and development wells mostly in Louisiana with very good success. The book is difficult to follow as the author jumps around and gives seemingly conflicting dates and areas, which coupled with flashbacks and no chapter headings is unusual. However, a continuous stream of colorful stories of oil and family and WW II are told in a very interesting way. The hardbound first edition is becoming scarce.

**B287. (WALTER TEAGLE) Wall, Bennett H., and George S. Gibb, TEAGLE OF JERSEY STANDARD, New Orleans, LA, Tulane University Press, (1974), xxii, 386pp.**

Known through most of his life as "The boy picked to fill John D.'s shoes", Teagle was born in 1878 in Cleveland. At that time his father had spent several years fighting Rockefeller for a niche in the oil refinery and market boom there. He had negotiated a profit and facilities sharing deal with Rockefeller, but continued to vigorously expand beyond that. In court, he won against a Rockefeller attempt to restrain him. Walter was a large and good humored youngster and was active in campus politics when he studied chemical engineering at Cornell. He graduated in 1899 and began work for his father's company, Scofield, Shurmer & Teagle, at the bottom. The company was successful, but internal strife between the partners was distracting from the main fight with Rockefeller. In 1901 Teagle Sr. was in bad health and accepted an offer where Standard would absorb his company. The renamed Republic Oil became a secret subsidiary of Standard. They hired young Teagle to run the company at an exorbitant salary, thus beginning his steady rise to power within the Rockefeller empire. Surviving the 1911 dissolution of the Standard Trust, Walter became president of Standard (New Jersey) in 1917 and served through 1937. He died in 1962 after suffering several years of ill health. He was one of the most brilliant leaders on the world oil scene during twenty of the most exciting years in oil industry exploration and growth. Excellent, well researched book. The hardbound first edition is scarce.

**B288. (FRANK PHILLIPS) Wallis, Michael, OIL MAN: THE STORY OF FRANK PHILLIPS AND THE BIRTH OF PHILLIPS PETROLEUM, NY, Doubleday, 1988, xxii, 480pp.**

Born the first male of ten children to a farm family in Iowa, Frank left the farm at fourteen to become a barber. Known as one of the best in his field, he opened a high class barber shop in Creston. He married very well, into a banking family. Brother L.E. taught school, sold insurance and later became a traveling salesman. At age thirty, Frank heard about the Osage Oil Boom near Bartlesville, Oklahoma and went there, where he formed Anchor Oil & Gas with his father-in-law and brother L.E. By 1905, they had acquired some oil and gas leases and drilled three dry holes on them. Fortunately, the fourth well, #1 Anna Anderson came in flowing. Amazing success followed. Frank and L.E. soon organized a bank in Bartlesville to help finance the growing oil industry. Kid brother Waite joined them working in the field on some of their oil leases. Phillips was one of many who exploited the Osage oil area including Sinclair, Skelly, Cosden, Marland, Barnsdall and Getty. The book goes on to discuss Phillips 66 and its extensive marketing network along with its part in developing the Oklahoma City field. The early Osage lease auctions, the role of Boots Adams in the company, Philmont Scout Ranch, and the famous Woolaroc ranch retreat are also covered. Frank died in 1950 with Phillips Petroleum in excellent condition and in good hands. He gave most of his personal wealth to charities and foundations. Very good biography. The hardbound first edition is common.

**B289. (WINTHROP ROCKEFELLER) Ward, John L., THE ARKANSAS ROCKEFELLER, Baton Rouge, LA, Louisiana State University Press, (1978), xv, 218pp.**

The author is a newspaper editor in Arkansas and was Winthrop's 1968 gubernatorial re-election campaign manager. Fourth son of John D., Jr., Winthrop's career is followed from 1953 to 1970. After developing a worldwide reputation as a playboy, the Yale dropout, divorced Bobo Sears and moved to a hilltop in Arkansas. He remarried in 1956 and was already being drawn into politics in this 85% Democrat state. Even though Winthrop was a Republican, the powerful governor, Faubus appointed him chairman of the Arkansas Industrial Development Commission, where he served well for several years. Winthrop lost his run for governor against Faubus in 1964, but defeated him in 1966 and was re-elected in 1968. He finally lost the job back to Faubus in 1970.

During his tenure, Winthrop managed to have more new legislation enacted than anyone believed possible and that went a long way towards establishing the two party system in the state. This is a very positive success story for a man who had talent and money, simply needing a place to apply them. The hardbound first edition is common.

**B290. (THOR WARNER) Warner, Thor, THE SIEGE OF LOPENO, San Antonio, TX, The Naylor Company, (1954), xvi, 220pp.**

Warner was born in 1885 and became a field geologist. He explored Canada and South America and much of California. In 1932 he decided to drop everything and pursue a drilling prospect in Zavala County, Texas, near the town of Lopeno. In the book, he changed the names of the people and companies involved in the story, probably to avoid libel suits. He spent twenty years of very difficult land and legal work trying to get someone to drill his prospect. Unimaginable difficulties were encountered, not the least of which was the flooding of Lopeno when the massive Falcon Reservoir was formed. In 1952 multiple gas reservoirs were tapped from the Wilcox formation on the Lopeno anticline and Warner emerged a wealthy hero. Once more, persistence pays! The hardbound first edition is becoming scarce.

**B291. (ROBERT WATCHORN) Watchorn, Robert. THE AUTOBIOGRAPHY OF ROBERT WATCHORN., Oklahoma City, OK, Robert Watchorn Charities, Ltd., 1958, xix, 218pp.**

Watchorn was born in England in 1859. He had little schooling and was working in the coal mines at age eleven. A tough kid, he came to America at age twenty-two and worked in the coal mines around Pittsburgh. He became involved in the miners union where he fought against child labor and gained high office in the national organization. He began working in the public sector, and in 1905 became commissioner of emigration at Ellis Island. After four years he went to work as an assistant to Lyman Stewart at the fledgling Union Oil of California. He was treasurer of Union Cal in 1909 but left them in 1913 (age 54) to begin his own business in the oil fields of Oklahoma. He formed Watchorn Oil and Gas, Watchorn Petroleum and Alma Petroleum, all successful oil companies. Alma was famous for rank-wildcatting around the Mid-Continent and Permian Basin areas. He was involved in the early development of Healdton field of Oklahoma, East Texas field and the South Cowden field in West Texas. Alma sold out in 1942 to Phillips and Sunray. He died in 1944, but the Watchorn companies survive. He was a considerable philanthropist and that work continues today in the form of his charities. Cited in Swanson on page 176. The hardbound first edition is becoming scarce.

**B292. (LEWIS WEEKS) Weeks, Lewis G., "...A LIFELONG LOVE AFFAIR: THE MEMOIRS OF LEWIS G. WEEKS, GEOLOGIST, Westport, CT, (1978), xii, 138pp.**

Born in 1893, Weeks attended college in Wisconsin and later Harvard. For the next several years he worked for various oil companies, becoming known as an expert in foreign oil exploration. Many years with Standard Oil (Esso) allowed him to exchange ideas with the best exploration minds in the world. He was responsible for locating significant oil and gas reserves. He also directed much innovative research and the resulting publications are required reading for petroleum explorationists. The very important "Habitat of Oil", published in 1958 by the A.A.P.G. is a massive symposium which he edited and authored in part. He was president of that organization from 1959-1960. After retiring from oil company life he consulted for Broken Hill Proprietary of Australia where he recommended they enact an oil exploratory play in the Bass Straits in 1965. His old company Esso (now Exxon) joined BHP as a partner in the exploratory work and leasing and drilling of the rank wildcat prospect. It turned out to be a major oil find consisting of several giant fields, and another testimonial to the exploration skill of Weeks. As a finders fee, he was assigned an overriding royalty interest in the project. The override later became worth hundreds of millions of dollars. Weeks died in 1977. The slipcased hardbound first edition is scarce.

**B293. (ARMAND HAMMER) Weinberg, Steve, ARMAND HAMMER: THE UNTOLD STORY, Boston, Little, Brown and Company, (1989), x, 501pp.**

This is a very objective biography of Dr. Hammer. The author received no cooperation from him, but conducted seven hundred interviews with Hammer's friends and enemies. He reviewed hundreds of thousands of pages of documents as shown in his large notes and bibliography sections. Hammer filed a libel suit against Weinberg for writing the unauthorized biography. Good photos are included which were not in earlier biographies. The book opens with the 1976 conviction of Dr. Hammer for illegal campaign contributions to Nixon and ends with George Bush pardoning him in 1989. The judge issued an unbelievably light sentence of one year probation and $3,000 fine for the serious campaign law violation. Quoting from Weinberg: "Looking terminably ill, Hammer was wheeled out of the courtroom, upset at this blot on his reputation but, later, he was running the annual meeting of Occidental Petroleum Corporation in his usual velvet-fisted manner. Shortly thereafter he flew in his personal Occidental luxury airplane to Moscow, where he met with Soviet leader Leonid Brezhnev. Louis Nizer, Hammer's lawyer, called his clients recovery a miracle." Two themes run through the book that are not obvious in Hammer's other biographies. First is the Hammer inclination, if not preoccupation to curry favor of every U.S. president from F.D.R. to Bush, and every Soviet leader from Lenin to Gorbachev. He would do almost anything to gain influence, including sending money and gifts. A second theme is Hammer walking the fine line within the law. He had constant run-ins with the F.T.C., I.R.S., S.E.C., etc. Litigating was his career and he usually won. This is one of the top oilman biographies. The hardbound first edition is common.

**B294. (BILL WELLS) Wells, W.B., WILD BILL THE DRILLER, Oklahoma, Privately Published, (1977), ix, 139pp.**
Born in Ohio in 1895 to a farming family, Wells migrated to Cleveland, Indian Territory, about the time Osage oil was booming (1903). His father worked as a carpenter and Wells grew up around oil operations, where as a youngster, he peddled groceries in the Glenn Pool area. He started work in the oil fields at seventeen, first for Barnsdall then Gypsy in Oklahoma. He bounced from job to job and company to company for several years, finally becoming a cable-tool driller. He relates his personal experiences in the Oklahoma boomtowns of the 1920's and 1930's. He and a partner acquired their own drilling rig and began taking interests in the wells they drilled as partial payment for the contract. Slow but steady success over the several years is described with interesting detail. Wells is something of an oil historian and in the last part of the book presents his version of the important discoveries and events in the industry. A persistent prospector usually prevails. The hardbound first edition is becoming scarce.

**B295. (BET A MILLION GATES) Wendt, Lloyd and Herman Kogan, BET A MILLION: THE STORY OF JOHN W. GATES, NY, The Bobbs-Merrill Co., (1948), 357pp.**

Born into an Illinois farming family in 1855, he had some education and was playing poker for fun and profit by age seventeen. Two unsuccessful businesses preceded his venture into barbed wire, as a traveling salesman. He worked for the inventor-manufacturer of this new product in the Texas area. In 1877, with brains and bravado, he conquered early loud objections of the Texas ranchers and sold barbed wire like crazy. He quit and went into competition with his monopolistic employer, using every means available, and making a fortune at it. He moved headquarters to Chicago and diversified and expanded. His one oil exposure mentioned in the book is truly exceptional, earning him untold millions. Lucas had run out of money drilling at Spindletop and had Joseph Cullinan trying to raise more money for him. Cullinan talked Gates into taking 46% of the deal as his partner in what is now Texaco, Inc. Gates' reputation as being a brilliant gambler was well founded. He got the nickname "Bet a Million" from a one-day episode in 1900 at a London racetrack where he and partner John Drake won at least that much. Gates died in 1907 after an exhausting legal battle with steel monopolist Carnegie. Great story of a compulsive gambler who was also a genius. Not in Swanson. The hardbound first edition is becoming scarce.

**B296. (STEPHEN PECKHAM) White, Gerald T., BAPTISM IN OIL: STEPHEN F. PECKHAM IN SOUTHERN CALIFORNIA 1865-1866, San Francisco, CA, The Book Club of California, 1984, 138pp.**

Born in 1839, Peckham attended Brown University and studied chemistry. In 1861 he started an oil refining business in Providence with strong financial backing, but had no great success. He served as a pharmacist-chemist for three years in the Civil War, after which he spent a year, 1865-66, in California to fulfill a one year contract as refiner-chemist for a new oil company, California Petroleum Company. The state was in the middle of an oil boom and exploitation of the Ojai area was frantic. The Ojai hills were virtually covered with oil seeps and the famous chemistry professor Benjamin Silliman, Jr. had highly touted the area for development. The first part of the book is Peckham's journal after leaving the Union Army, through his year in California where he tested and evaluated oil for refining purposes. He was basically looking for additional products which could be saved from crude distillation. The second part of the book consists of correspondence between Peckham and Thomas Bard, who was struggling as he would for years in the Ojai area. Peckham's additional correspondence with professor Silliman, Brewer and others is interesting as he begins to unravel the makeup of the California oils. In the last section, Peckham's later career is described as teaching chemistry for several universities where he was well respected and published many articles. He left teaching in 1880 and spent four years completing a major work for the government, titled "Report on the Production, Technology, and Uses of Petroleum and its Products". He went back to California as a chemist for his friend Tom Bard at Unocal in 1894 looking again for commercial crude by-products...with no luck. He eventually became an asphalt expert and published another major work titled "Solid Bitumens" in 1909. Good biography of an early oil figure. The hardbound first edition is scarce.

**B297. (BEN WHITE) White, Ben P. III, FROM RICHES TO RAT RACES, Lubbock, TX, Robert Nicholson, (1975), viii, 88pp.**
Born in 1895, the seventh child of a poor farming-ranching family, White tells of his ancestry and of growing up on a sleepy ranch in the south central Texas hill country. He became a tool and dye maker before WW I, but as oil demand increased, he and three of his brothers acquired a rig and began to drill oil wells as contractors. They had good success and prospered in the Burkburnett area. He later took a job with Standard Oil. By 1928, he and a geologist leased some land near Big Lake and bought a cable tool rig where they made a good deal of money. The Depression loomed and caused him to take his profits and family to the hills of New Mexico where they homesteaded and resided before retiring to Lubbock, years later. The softbound first edition is becoming scarce.

**B298. (CHARLES CANFIELD) Whitney, Caspar, CHARLES ADELBERT CANFIELD, Boston, The Merrymount Press, 1930, viii, 217pp.**
This book was not available for review by Neff, but is regarded as important. Quoting from Swanson: "Partner of famous California firm Canfield & Chanslor; important pioneer in California oil production, 1890-1915; best remembered for his leading role in opening up Mexican oil fields". Cited in Swanson on page 177. The hardbound first edition is very scarce.

**B299. (POTTER WICKWARE) Wickware, Potter, CRAZY MONEY: NINE MONTHS ON THE TRANS-ALASKA PIPELINE, NY, Random House, (1979), 228pp.**

The eight hundred mile trans-Alaska pipeline was constructed in record time after the political problems were solved. Prudhoe Bay to Valdez, it flows two million barrels a day. It has been compared to building of the Suez and Panama canals as an extraordinary feat. Wickware is a welder on the pipeline, making more money than he ever expected to. His stories are interesting and generally humorous, covering the seven month period when he was working there. It is a story of people in a boom town atmosphere doing all kinds of jobs for high pay. Pretty exciting. He also educates the reader to his main interest in life, welding. He is also knowledgeable about pipelines and their construction, passing this along to the reader as well. Prior to his career as steam fitter and welder, he attended college in Switzerland, also Tufts and U.C. Berkeley. The hardbound first edition is common.

**B300. (HUEY LONG) Williams, Harry T., HUEY LONG, NY, Alfred A. Knopf, 1970, xxxvi, 884pp.**

Williams is a highly regarded reporter, much published historian and professor at L.S.U. He did considerable research and conducted 300 personal interviews of family and Long organization members who had known him intimately. This is the definitive work on Long. Huey Pierce Long, Jr. was one of the most fascinating political figures in U.S. history. He was born near Baton Rouge in 1893 to an educated and intellectual farming and landholding family. Huey was very bright and although he did not finish high school, attended Tulane Law School for enough course work and study to pass the bar at age twenty-one. He successfully practiced law for several years, defending the little guy against the big guy. This protective attitude would later become his stated mission in life. In 1918 he was determined to enter politics and won his first elected office, Railroad Commissioner. It was through this office that he began a lifelong series of attacks on Standard Oil of Louisiana...never letting up. After being elected governor in 1923, the attacks became more public at every juncture, primarily on oil leasing and taxes. Although Standard remained in the background, they were apparently involved in the aborted impeachment proceedings against Long in 1929. Long had a brief and fiery term as U.S. Senator, leaving a liberal mark on Roosevelt's agenda. Long's untimely death in 1935 at the age of forty-two came at the hands of a young doctor who had become obsessed with opposition to Long's politics. The hardbound first edition is becoming scarce.

**B301. (CHILTON WILLIAMSON) Williamson, Chilton, Jr., ROUGHNECKING IT, NY, Simon and Schuster, (1982), 288pp.** Williamson is an editor, journalist and writer. His main thrust in this book is to try and make a case of similarity between gold rush and oil boom. He writes this book as a journal, following the lives of oil drilling roughnecks in the Wyoming overthrust oil play, at the height of activity there. The oil development and his story are centered at Kemmerer, Wyoming. He follows the workers through their daily routine and gets to know what their problems and pleasures are. Williamson even worked on the rigs and hunted with the workers. This particular drilling boom became so hot that it received national media attention and all kinds of problems resulted from Dan Rather's reporting of it. Williamson's oil boom/gold rush analog is valid and would probably be so for most any boom scenario. People seem to get a little crazy when big money is everywhere. The story is largely presented in dialogue and vernacular. The hardbound first edition is common.

**B302. (WILLIAM BURTON ET AL) Wilson, Robert E., PIONEERS IN OIL CRACKING, Chicago, The Newcomen Society, 1946, 27pp.** This Newcomen pamphlet includes the address of Robert Wilson, Chairman of the Board of Standard Oil (Indiana), given at Chicago in 1946. A discussion of the importance of oil cracking which enabled the refiners to increase gasoline yield from eighteen percent to near fifty percent. Three men in the Standard organization were responsible for work at the Whiting refinery, William Burton, Robert Humphreys and F.M. Rogers. Wilson discusses their many contributions to the field, not the least of which was the Burton Still which was an important early refining technology breakthrough. Cited in Swanson on page 109. The softbound first edition is becoming scarce.

**B303. (JOHN D. ROCKEFELLER) Winkleman, B.F., JOHN D. ROCKEFELLER: THE AUTHENTIC AND DRAMATIC STORY OF THE WORLD'S GREATEST MONEY MAKER AND MONEY GIVER, Philadelphia, The John C. Winston Co., (1937), viii, 310pp.** Winkleman's book was published upon the death of John D. Rockefeller, May 23, 1937. The popular book had no new revelations at the time and is consistently favorable to him. The country was slowly changing its attitude towards the ninety-six year old Rockefeller, largely due to the philanthropic efforts of his son, John D., Jr. Born in 1839 to William and Eliza Rockefeller, John D."s brief family history precedes the telling of his business career. The spectacular career goes from errand boy at Hewitt and Tuttle who were commission merchants in 1855, to his retirement as head of Standard Oil around 1900. The last half of the book is low key and covers his golf and landscaping hobbies and the Rockefeller Foundation. There is a short discussion on the dissolution of the Standard Oil Trust. Several excellent photos and political cartoons are also included. The author was a published Philadelphia attorney and references Tarbell, Josephson and Flynn in his footnotes. Cited in Swanson on page 168. The hardbound first edition is becoming scarce.

**B304. (JOHN D. ROCKEFELLER) Winkler, John K., JOHN D.: A PORTRAIT IN OILS, NY, The Vanguard Press, 1929, 256pp.**
This book was the first complete Rockefeller biography to reach a large public audience since Ida Tarbell's attack on him in her "The History of the Standard Oil Company". Winkler's book was originally serialized in Cosmopolitan, not exactly muckraking journalism, it does not favor Rockefeller. Winkler evidently had no access to Rockefeller or his associates, but pieced the book together from previously published material and news accounts and guesswork. John D. was born in 1839 to William A. and Eliza Rockefeller. William was a rover, a gambler and a quack doctor, and always had money. They moved a lot. John learned business and sharp trading from his Dad's example. He went to business college for a short while, but at age sixteen took a job with Hewitt and Tuttle, Cleveland merchants, as a general flunky. In two and a half years he learned more about business while making good money and keeping it invested. In 1858 he formed a partnership with Clark and Gardner, as commission merchants on the Cleveland docks. Under John's financial guidance, the firm prospered. He borrowed and invested and became obsessed with making money. He recognized the oil industry potential immediately after the 1859 Drake discovery well came in at Titusville and went there to decide how he could exploit this new industry. He and Clark bought a refinery and it became his passion. He expanded rapidly and bought out his partner. Henry Flagler, another refiner who was just as aggressive as John became his next partner. Through every straight and crooked method imaginable they gained control of refining and thus the

marketing of oil. They dealt illegally with the railroads, receiving rebates which eventually forced their competition out and gave them control over oil transportation. The book briefly covers the Standard Oil rise and breakup and rise again, up to just before the Depression. Winkler does not present John D. as bad or evil, only as brilliant, manipulative and above the law. In today's investigative journalism era, this sensational biography hardly rates as an attack. Easy reading. A good old book. Cited in Swanson on page 168; Hardwicke page 45. The hardbound first edition is common.

**B305. (CLINT MURCHISON) Wolfe, Jane, THE MURCHISONS: THE RISE AND FALL OF A TEXAS DYNASTY, NY, St. Martin's Press, (1989), 505pp.**

Clint Murchison (pronounced "murky sun"), founder of the family dynasty, was born in 1895 at Athens, Texas (near Tyler) to a business oriented family. His father and grandfather were bankers. As a child, Clint was off-beat and extremely industrious. He and local running-mate, Sid Richardson learned business by dealing and trading the hard way as they grew up. Clint attended college but disliked it and went to work at his Dad's bank. He astounded the bank personnel by working complex mathematical problems in his head in seconds which would normally have taken hours. He served in the army in WW I and then followed his friend Sid to the oil fields at Wichita Falls, Texas. Clint and partner Ernest Fain found oil and gas all over Texas and by 1925 Clint was a millionaire, retiring to San Antonio to raise cattle. That did not last and he expanded his oil and gas explorations very successfully. He formed Southern Union Gas Company to build and manage a very successful gas transmission and marketing effort in the San Juan basin of New Mexico. Six months after the East Texas field discovery, Clint made a giant speculative lease play and eventually drilled several good producers in the field. He ran hot (illegal) oil after proration was established. He bitterly fought oil conservation as did most independents. Clint died in 1969 and his two sons took over the business and made even more money than Dad, although borrowing more than they should have. The oil business decline in early 1981 along with a falling real estate market ruined the over-extended Murchisons. Clint Junior sold his beloved Dallas Cowboys in the early 1980's for eighty million dollars which likely did not begin to put a dent in his debt. Much family infighting followed with lawsuits and counter suits preceding the bankruptcy in 1985. The book title tells all. Excellent reading. The hardbound first edition is common.

**B306. (JOHN WOOD) Wood, Captain John M., TEXAS RANGER IN THE OIL PATCH, Austin, TX, Woodburner Press, 1994, 123pp.** Wood tells his stories of running down oil patch criminals and thieves. Several moonlight ventures trying to trap oil or equipment thieves in the Permian Basin are told in Wood's own vernacular with an unusual sense of humor. Truth again is stranger than fiction. Wood served the Texas DPS in West Texas from 1942 to 1949 at which time he joined the Texas Rangers, serving in the field until 1962 when he was promoted and transferred to Austin. This is a fun and interesting book. The softbound first edition is becoming scarce.

**B307. (W.E. YOULE) Youle, W.E., SIXTY THREE YEARS IN THE OIL FIELDS, Taft, CA, Midway Driller Publishing Co., 1926, 122pp.**
This book was not available for review by Neff, but it is regarded as important. Quoting from Swanson: "Author's work and experiences in Pennsylvania oil fields, 1863-76; move to California, where for 50 years he was one of the state's most famous drillers and discoverer of several of California's earliest oil fields. California oil history is that prior to 1900". Cited in Swanson on page 177. The hardbound first edition is rare".

**B308. (HUEY LONG) Zinman, David, THE DAY HUEY LONG WAS SHOT: SEPTEMBER 8, 1935, NY, Ivan Obolensky, Inc., (1963), xiii, 289pp.**
Always a popular subject in Louisiana and in politics in general is Huey Long. The powerful Long was one of the all time most liked and disliked public officials. As governor and U.S. senator, his liberal politics and anti big business philosophy kept him in the public eye constantly. His unfortunate murder in 1935 put an end to a spectacular career. In this book, Zinman makes a long drawn-out case for believing that a cover up of facts occurred in the assassination of Long. His thinking is that although the perpetrator, young Dr. Weiss, was carrying a handgun, their fist fight and argument resulted in an accidental shooting. The case is weak, but conflicting testimony along with Long's bloody lip leave the door open for speculation. Dr. Weiss certainly didn't tell, because Long's

bodyguards and police shot him thirty or more times at the scene. The hardbound first edition is common.

**B309. (SHAH OF IRAN) Zonis, Marvin, MAJESTIC FAILURE: THE FALL OF THE SHAH, Chicago, University of Chicago Press, (1991), x, 350pp.**
The author was a personal friend of the Shah and is a leading authority on Iranian history,. He tries to analyze the failure of the Shah's thirty-seven year government as it yielded to an Islamic revolution which he could easily have crushed by military force. Zonis makes several psychological points about the situation and the Shah himself which seem reasonable. Evidently Shah lost the confidence of the people who had become tired of his affluent Western ways. At the same time he also lost the support and confidence of the U.S. government. As a final blow, he was found to have cancer which to Iranian thinking, was an indication that Divine Protection was no longer in effect. The combination of these things killed first his self esteem and then him. Zonis presents a great deal of history, biography and politics to validate his thesis. Interesting reading about a historic oil figure. The hardbound first edition is common.

# PART III - COMPANIES

**C1. (GEOLOGICAL SURVEY OF CANADA) Alcock, F.J., A CENTURY IN THE HISTORY OF THE GEOLOGICAL SURVEY OF CANADA, Ottawa, National Museum of Canada, 1947, vii, 94pp.**
This tribute was published on the 105th anniversary of the Geological Survey of Canada. WW II had shut down the planned centennial celebration for which this booklet was originally prepared. Better late than never. In 1841, Canada was growing wildly and mineral wealth was being uncovered across the country. Out of necessity the government decided to fund a survey with the commission to document Canada's resources. It stands as one of the oldest surveys in the world. Sir William Logan was the founder and first director, serving twenty-seven years. Logan was probably the first, in 1848, to comment that gas seeps occurred on surface anticlines. Exploratory mapping of the vast country occupied full attention of the survey employees. As the great mineral wealth became known, exploration accelerated from government and private interests, booming in the 1850's. This is a well illustrated, brief history with early photos. The softbound first edition is becoming scarce.

**C2. (SCHLUMBERGER) Allaud, Louis A. and Maurice H. Martin, SCHLUMBERGER: THE HISTORY OF A TECHNIQUE, NY, John Wiley & Sons, (1977), xxv, 333pp.**
The definitive work on Schlumberger is presented here in three parts. Part one discusses Conrad Schlumberger's original methods of subsurface exploration by electrical methods which was revolutionary in the 1920's. Part two covers the period 1927-1945 and shows how electrical logging of wells became an indispensable tool for the oil industry. Part three discusses modern methods of logging in terms laymen can understand. Reviewed are the nuclear, acoustic, electric and mechanical logging technologies currently in use around the world. The hardbound first edition is becoming scarce.

**C3. (AAPL) American Association of Petroleum Landmen, PATH TO PROFESSIONALISM: THE AAPL STORY, Noplace, The Association, 1985, 82pp.**
This book commemorates thirty years of the association. Most of the research for it was done by John McConal (published author) through

interviews and study of association documents over a two year period. Much on the background of the first president of the organization, George Brown, is given. He started his career as a church janitor in 1925. History of the predecessor organizations such as Mid-Continent Oil Scouts and Land Men's Association is told. All of the regional oil areas had landmen's associations and the need for national coordination was obvious. In 1954, a group of top oil company landmen met in Denver and then proceeded to solicit an initial 600 members from across the country. They set the first meeting in the future headquarters town of Fort Worth, Texas. Lots of history and photos. The hardbound first edition is becoming scarce.

**C4. (ANGLO-IRANIAN) Anglo-Iranian, OUR INDUSTRY: ANGLO-IRANIAN OIL CO., LTD, London, Anglo-Iranian, 1947, 304pp.**

The company was formerly called Anglo-Persian and is currently British Petroleum. Basically this is the first edition of a handbook for employees or prospective employees which covers every phase of the oil industry from its start in Pennsylvania to modern marketing. It is a valuable reference for an employee in one particular phase of Anglo's integrated structure so that he can learn how his job or department fits into the overall scheme. Chapters on production, refining, transportation, distribution and engineering are technical and could be used by an employee of any company for enlightenment as to what the state of the industry was in 1947. Several chapters on the various end products of crude oil. Contains a glossary, maps and good bibliography. Second edition expanded to 368pp in 1949; Third edition by British Petroleum, to 472pp in 1958; Fourth edition to 528pp; Fifth edition in 1970 to 600pp in a larger format, much updated with photos and drawings in color with their part the giant discovery at Prudhoe Bay, Alaska. Cited in Swanson on page 178. The softbound first edition is becoming scarce.

**C5. (ARAMCO) ARAMCO, ARAMCO AND ITS WORLD: ARABIA AND THE MIDDLE EAST, Dhahran, Saudi Arabia, ARAMCO, (1980). xi, 275pp.**

Aramco came into being in the early 1940's and grew very rapidly as the Arabian American Oil Company found and developed giant oil fields in Saudi Arabia. The book cited above is actually the fifth edition of the Aramco Handbook. It is a very well done combination history and handbook with numerous photos and maps of the Saudi Arabia country and oil industry. The four part book is laid out as follows: Part 1, Before Islam; Part 2, Islam and Islamic History; Part 3, Saudi Arabia, The Heartland; Part 4, Oil and Aramco. This volume is designed for English speaking employees and prospective employees as an educational tool. Excellent photos and a significant series of Middle East maps from 2500 BC to the modern oilfield maps and concessions of 1980 are included. It contains a useful bibliography of Middle East reference books. This fifth edition evolved from what were originally handbooks for American employees in Arabia. The first edition was published in 1950 as five spiralbound handbooks covering different phases. Information demand became so great that in 1952, all of the information was concentrated into two, hardbound 8.5x11" books, both of these second editions were edited by Roy Lebkicher, designed by Ray Graham Associates and published by Russell F. Moore Co., of New York. Volume 1 is titled ARAMCO AND WORLD OIL, with viii, 112pp, and contains much oil history and details of the company and its operation. Volume 2 is titled THE ARABIA OF IBN SAUD, with xiii, 179pp, and containing much on Arabian history and culture. The handbook was further consolidated into one hardcover volume in 1960 (the third edition), titled ARMCO HANDBOOK, covering updated material including that in the foregoing. It is also in large format with 343pp, and edited by Lebkicher, et al. What essentially is the fourth edition was issued in 1968 with 279pp, also titled ARAMCO HANDBOOK with no editor. Cited in Swanson on page 59 and 171. The softbound first editions are very scarce.

**C6. (ATLANTIC) Atlantic, THIS IS ATLANTIC, Pa, The Atlantic Refining Company, (Ca 1958), 36pp.**

This is a propaganda pamphlet, styled generally like a corporation annual report without many statistics. Atlantic Refining Company was one of the oldest companies in the oil industry, incorporated in 1870. Atlantic founders were the first to export oil to London in 1860. Atlantic became a big refiner but lacked its own oil. Now it is a fully integrated major company, engaging in all industry functions, exploration, production, manufacturing, transportation, marketing, research and development. Good photos of these operations in 1958. Motto "Keeps Your Car on the Go". The softbound first edition is becoming scarce.

**C7. (SCHLUMBERGER) Auletta, Ken, THE ART OF CORPORATE SUCCESS: THE STORY OF SCHLUMBERGER, NY, G.P. Putnam's Sons, (1984), 184pp.**

Author Auletta is a journalist who writes for the New Yorker. In 1981 he was looking for a company to profile for the magazine. He quickly found Schlumberger, of which he had no prior knowledge. In 1981 Schlumberger had the highest profit margin of any corporation of size. The author wrote a two phase article, first on the dynamic Jean Ribaud, Schlumberger's chief executive and second on the company. Long recognized by company analysts as one of the world's best companies, Schlumberger is low key and the only publicity or advertising they do is directed specifically at potential clients in need of their diverse oil services. A couple of pages are given to the history of Schlumberger with Conrad and Marcel inventing and perfecting electronic exploration tools used first for surface mineral deposit detection then subsurface mineral and oil exploration. The author is concerned with Ribaud and how Schlumberger achieved and maintains their industry dominance and profitability. His conclusions as to why Schlumberger succeeds is that they have dominated their wireline logging industry from day one by their company philosophy that science and research are first priority. He also admits that they also stay ahead of their competitors through excellent corporate leadership. The hardbound first edition is common.

**C8. (PERTAMINA) Bartlett, Anderson G. III. et al, PERTAMINA: INDONESIAN NATIONAL OIL, Djakarta, Amerasian Ltd., (1972), xi, 420pp.**

Bartlett researched this book almost entirely through interviews as little has been written of the events leading up to the formation of Pertamina. He used a team concept including interpreters to record the many hours of information. Indonesian oil was first discovered in 1871, later produced and refined in Sumatra, Java and Kalimantan by Royal Dutch-Shell. Standard began producing in 1912, later merging with Mobil to form Stanvac. Standard of California began producing in 1930 and by 1936 merged with Texaco to form Caltex. The big three. By 1940 Indonesia was the fifth largest producing area in the world. After WW II, Indonesia gained independence and as production fell, the net profits from concessions dropped to near zero as exploratory activity dwindled. Pertamina was formed in 1968 out of several state owned companies which handled separate phases of the business in Indonesia. Pertamina became a state owned vertically integrated major oil company. By 1963 Indonesia had converted all oil operating companies concessions to work contracts by virtue of 1960 law No. 44. As of this writing, Pertamina has taken full control of every phase of their oil business from exploration through drilling, production, refining and marketing. This book is a highly pro- Indonesian, anti major oil company political work designed to bring history and pride to the Indonesian people. It is a well constructed company history with appendices, biographies, glossary and bibliography. The hardbound first edition is becoming scarce.

**C9 (SHELL) Beaton, Kendall. ENTERPRISE IN OIL: A HISTORY OF SHELL IN THE U.S., NY, Appleton-Century-Crofts, Inc., 1957, xiii, 815pp.**

Although Beaton was a Shell employee, he did a fine job of presenting the American Shell Oil Company in this exhaustive work. He opens with Pennsylvania oil history then moves to the Royal Dutch-Shell history. In 1911 Deterding broke into the American markets by importing Sumatran gasoline to the San Francisco area. Long a competitor of Standard Oil, he quickly teamed up with Union Oil of California, another rival of Standard, forming Shell Union Oil. The marketing arm of American Gasoline Company was formed then, and began building market share. By 1913 they had acquired one of California's largest producers which owned big oil reserves at Coalinga and formed California Oil Fields Ltd, a British corporation. They bought refineries and more oil companies, changing the name in 1914 to Shell Oil Company of California. Deterding organized Roxana Petroleum Company to explore for and develop oil reserves in the booming Oklahoma area in 1912. He had been trying to break into Oklahoma since 1909 but was not successful in buying significant oil producing properties until 1912. By 1914 Roxana was a big Oklahoma producer and was exploring on their own. They forced their way into the monster Cushing and Healdton fields with cash and acquired still larger oil reserves. They bought more refineries and by the 1920's California began booming. The Shell #1 Alamitos discovery at Signal Hill was just an indication of what lay ahead in California. Shell- Union and Roxana were both very successful and with the guidance of the Dutch quickly achieved major company status. The Roxana name was dropped in favor of Shell in 1928. At this writing, American Shell was expanding into foreign markets to broaden their exposure and become a world power on their own. Included are sources, chronology and appendices. Cited in Swanson on page 168. The hardbound first edition is common.

**C10. (Tide Water) Buente, Frances M., AUTOBIOGRAPHY OF AN OIL COMPANY, NY, Tide Water Oil Co, 1923, 81pp.**
This book was not available for review by Neff, but it is regarded as important. Quoting from Swanson: "Commemorates 45th anniversary of company which began Nov., 1878 as Tide-Water Pipe Company Ltd.: account of laying first pipe line across Alleghenies; manufacture of barrels, their use in oil transportation". Cited in Swanson on page 169. The first edition is rare.

**C11. (KANSAS GAS) Cadman, Wilson K., KANSAS GAS AND ELECTRIC COMPANY: EIGHTY YEARS OF INNOVATION, Philadelphia, The Newcomen Society, 1989, 28pp.**
This address was delivered to a meeting of the society in Wichita, Kansas in 1989. Cadman joined KG&E in 1951 and by 1979 was elected president. Several electric companies started up in Wichita during the late 1880's and KG&E is the survivor of over 100 companies which served Kansas. They sold the gas system in 1925, although much of the electric power is still generated from natural gas burning. The company was forced into nuclear power as supplementary energy source prices rose during the gas supply crunch of the late 1970's. The softbound first edition is common.

**C12. (CAMERON IRON) Cameron Iron, CAMERON IRON WORKS, INC. TWENTY-FIVE YEARS: 1920-1945, Houston, TX, Privately Published, 1946, 112pp.**
Celebrating twenty five years of impressive manufacturing growth and service to the oil industry was a good reason for this book to be published. However, a more compelling reason was to recount Cameron's part in the WW II war effort. Cameron was drawn into wartime specialty manufacturing by the Navy, prior to Japan's attack on Pearl Harbor. The first project was to make depth charge projectors which would be critical in combating the German submarine threat to shipping. Overtime work and innovation were required to produce the quantities required as the war began and then raged. Throughout the war period, Cameron continued providing their normal product line of oil well blowout preventers and other oil equipment as the industry was trying to produce

more oil. Cameron went on to make gun mounts and rockets along with many other specialty items. Their outstanding service was acknowledged by numerous Navy awards. The last one-third of the book gives the Cameron Iron history from its founding by Harry Cameron and Jim Abercrombie in 1920 along with short biographies of the current officers. Good company. Not mentioned in Swanson. The hardbound first edition is scarce.

**C13. (TEXAS EASTERN) Castaneda, Christopher J. and Joseph A. Pratt, FROM TEXAS TO THE EAST: A STRATEGIC HISTORY OF TEXAS EASTERN CORPORATION, College Station, Texas A&M University Press, (1993), xix, 296pp.**
A fine academic work on one of the major gas transmission companies. The company as it exists today had its start as the original Texas Eastern, which had been organized by Herman and George R. Brown in 1947. Many other Houston companies were beginning to recognize the east coast as a market for natural gas. The same natural gas which was so abundant it was being burned or vented to the atmosphere as an unwanted byproduct of oil production across Texas. Brown & Root, a massive construction firm spearheaded the organization which created the two "Inch" pipeline systems for the U.S. government. These oil lines contributed much to the WW II war effort, as they transported crude and products eastward. Texas Eastern bought the lines and converted them to transport natural gas, and Texas was off and running. The author had access to all company records and spent much time in evaluating the success and failure of company strategy through the acquisition of the smaller Panhandle Eastern in 1989. The epilog is most interesting as it details the hostile takeover attempt by Coastal Corporation and subsequent "White Knight" merger with Panhandle. Although Panhandle had fewer reserves, it had twice the pipeline network. The high cost of the merger forced the immediate sale of most of Texas Eastern's non pipeline assets, such as their North Sea oil production and other refining and producing properties. The hardbound first edition is common.

**C14. (SOUTHERN UNION) Chesnutt, N.P., SOUTHERN UNION, El Paso, TX, Mangan Books, 1979, 216pp.**
Chesnutt knows his company well, starting with them as a field and meter man in the 1930's. He retired as president of the company in 1976. Clint Murchison started the company almost by accident. He was a very

successful oilman in 1927 and actively developing an 80-acre oil lease near the instant boomtown of Wink in West Texas. He started building a pipeline to move his oil to the railroad south of Wink. Refiners ran a competing line and left Murchison sitting there with his own pipeline to town. He and partner Roy Moyston decided to form a natural gas distribution system for the town so the line was changed to handle that product, and Wink Gas Company was thus formed. The customers had to lay their own lines to the main and pay $5.00 per year for unmetered gas. Finding this a big success, Murchison expanded the gas utility business to Oklahoma and South Texas. Southern Union was created in January 1929 by absorbing the stock of all of separate Murchison utility companies. During the same year, Murchison and his brother Frank began acquiring property and utility rights in the San Juan Basin of New Mexico, around Farmington. A large gas supply was developing there and Southern Union became a primary transporter of that gas around the country. They struggled through the stock market crash and Depression to emerge as one of Americas leading energy companies. The hardbound first edition is becoming scarce.

**C15. (STANDARD OIL OF CALIFORNIA) Chevron, ONE HUNDRED YEARS OF HELPING TO CREATE THE FUTURE 1879-1979, CA, The Company, 1979, 170pp.**

This is an excellent history and picture book commemorating the 100th anniversary of the company. A brief history of the company is told as Frederick Taylor became the first one to commercially produce oil from Pico Canyon. He had no marketing experience so a fortuitous merger with Mr. Scofield and his group was made to form Pacific Coast Oil in 1879. They astutely built a refinery which could handle their 500 barrel per day production. At about the same time Standard (Ohio) was taking a look at the California market for a way to dominate. They found Pacific Coast and merged with it in 1900. The operations consisted largely of pipelining and refining by 1909, but they soon they began wildcatting and pursued that successfully. Orphaned in the 1911 Standard Oil Trust breakup, Standard of California moved aggressively on to find oil all over California and the West, and eventually became a major player overseas. The picture section of this book is in black and white and very well done in twenty year segments beginning with 1879. Excellent tribute. The softbound first edition is becoming scarce.

**C16. (WESTERN COMPANY) Chiles, H.E., (Eddie), THE WESTERN COMPANY OF NORTH AMERICA: 44 YEARS OF PACESETTING IN THE OIL BUSINESS, NY, The Newcomen Society, 1984, 28pp.**

The controversial chairman of the board, Chiles gives a typical fired-up address to the Newcomen Society. He started Western Company in 1939 in the Permian Basin of West Texas. He had recognized the need for good acidizing and cementing services in that area and did pioneering work for the industry there until the mid 1940's. Western holds many depth and pressure records for fracturing and cementing. Their large expansion was to offshore service in 1967. Good brief history of the company with analysis of its present operations. The softbound first edition is common.

**C17. (CITIES SERVICE) Cities Service, SERVING 3,000 COMMUNITIES: CITIES SERVICE COMPANY, Noplace, The Company, (1927), 43pp.**

Cities Service was established in 1910 by Henry L. Doherty, as a holding company for a hundred or so public utility enterprises he had developed. Cities concentrated in Texas and Oklahoma, although its influence was felt across the country. They ran the gamut from oil exploration to public utility service. Such companies as Dominion, Empire and I.T.I.O. were held and later merged into Cities. This first class publication was probably designed for potential investors and it is impressive. All eight of the major company divisions are given a page writeup of its history and current doings, along with a photo montage. Many of the more important facilities are shown in the historic photos. Not mentioned in Swanson. The softbound first edition is very scarce.

**C18. (CITIES SERVICE) Cities Service, FROM ARROWS TO ATOMS, Noplace, The Company, (1957), 9pp.**

This small brochure is reprinted from the company magazine, SERVICE, July 1957 edition for the Oklahoma Semi-Centennial celebration. The text is authored by James A. Clark, famous oil historian, and follows Oklahoma's energy development from before statehood to the present, with much emphasis on the important part played by Cities Service. Cities predecessor, I.T.I.O., had brought in the wildest well in the Oklahoma City field (Wild Mary Sudik) in March of 1926, flowing some 50,000 barrels per day. It blew oil all over the city and was out of control for eleven days. As Cities part of reconstructing an early oil boomtown for the Centennial, they selected the Sudik tool house as their contribution. This is a fine Oklahoma promoting brochure. The softbound first edition is common.

**C19. (P.G.& E.) Coleman, Charles M., P. G. AND E. OF CALIFORNIA, NY, McGraw-Hill Book Company (1952), x, 385pp.**

At the time this book was published, PG&E was the result of merger, acquisition, etc. of 520 separate predecessor companies over a one hundred year period. Undoubtedly a few more by now. San Francisco in 1852 had a diverse population of about 36,000 souls, largely gold seekers of one kind or another. Growth was accelerating all over the new state but particularly there. San Francisco Gas Company was established in 1852 by Peter Donahue, a forty-niner, bringing the first gas light to San Francisco. Initially the manufacture of synthetic gas from coal imported from Australia was sold for $15.00 per thousand cubic feet. The original street lamps functioned well for eighty years. With the success there, other areas followed, creating hundreds of different companies. Hydroelectric power generated competition and later supplied electric light. By 1904, Martin and de Salba had built and merged and consolidated others into California Gas & Electric, a loosely knit utility serving everywhere but San Francisco. Martin saw the future and knew that customers could not get efficient and economical power until a consolidation took place. The complex merger went through in 1905. The company survived the San Francisco earthquake and fire along with much government intervention to become the prime server for Northern California. Cited in Swanson on page 169. The hardbound first edition is common.

**C20. (TEXACO) Coll, Steve, THE TAKING OF GETTY OIL: THE FULL STORY OF THE MOST SPECTACULAR & CATASTROPHIC TAKEOVER OF ALL TIME, NY, Atheneum, 1987, xiii, 528pp.**

The author is a staff writer for the Washington Post and lives in New York City. He gives a good deal of family history on the Getty side and leaves the impression that they were ripe for a takeover shortly after the death of J. Paul Getty in 1976 and that several companies had investigated that possibility. During 1983 Gordon was co-trustee of the Getty Trust, 40% owner of Getty Oil. He was an opera buff and music composer, having serious problems working with Petersen the head of

Getty Oil. Gordon decided to sell the company to Pennzoil for a tidy sum. Petersen decided to sell to Texaco. Texaco won Getty, but lost the subsequent lawsuit pressed by Pennzoil. The jury was in a punitive mood and lost control of reason during the deliberation. Their eleven billion dollar award to Pennzoil evidently was meant to be a warning to business that hostile takeover mania should cease and that this method of acquiring oil reserves would not be tolerated. The author tells a good story and has impressive notes and bibliography. The hardbound first edition is common.

**C21. (P.S.E.& G.) Conniff, James C.G. and Richard Conniff, THE HISTORY OF P.S.E.& G.: THE ENERGY PEOPLE, Newark, NJ, The Company, (1978), 392pp.**
This company is the third largest combined utility in the nation. In its seventy-fifth year, it is serving nearly eight in every ten of the more than seven million residents of New Jersey. Formed in 1903 by Thomas N. McCarter with the idea of consolidating the many diverse and duplicating small lighting companies, the earliest of which had been formed in 1825. This is a thorough academic work on the company. The hardbound first edition is common.

**C22. (CONOCO) Continental Oil Company, CONOCO 74, Stamford, CT, The Company, 1974, 35pp.**
This is an exceptional issue of the company magazine, published as Conoco approached 100 years in the oil business. It contains historic articles on the 101 Ranch which had been the Marland and Conoco mainstay. Also there are various articles concerning Conoco's history and the part they play in the Athabasca oil sands, methanol from coal and worldwide oil and gas exploration. There is an interesting section which explains how Conoco employees should answer questions about the oil industry from non oil company people. The softbound first edition is common.

**C23. (CONOCO) Continental Oil Company, CONOCO: THE FIRST ONE HUNDRED YEARS, OK, Conoco, (1975), 238pp.**
In 1875 Isaac Blake formed Continental Oil Company at Ogden, Utah with the mission of producing, refining, transporting, selling, delivering and dealing in petroleum and in all of its products. He explored in Colorado and in 1877 formed Continental Trading to handle

transportation of crude and products in California. By the early 1880's Continental built up their marketing and transportation in the western U.S. and challenged Standard there and in many other areas as well. In 1884 Continental affiliated with Standard as their subsidiary, settling for Colorado, New Mexico, Wyoming, Montana and Utah. Continental was cut loose from the Standard in the 1911 Supreme Court ordered breakup. In 1909 E.W. Marland began drilling for oil in Oklahoma and soon had success on some Ponca Indian lands. Marland and his geologist had incredible success in Oklahoma and Kansas by the early 1920's and was involved in refining and marketing. Marland over-expanded and in 1923 J. P. Morgan bankers started a friendly takeover with $90,000,000 in stock purchases. Marland used the latest technology and geology to find immense quantities of oil, but continued to spend himself into insolvency. The Morgan people forced a 1929 merger between him and the conservative Continental Oil. Conoco was the survivor although they retained Marland's red triangle logo. With Marland's incredible oil reserves in hand, and the great oil finder out of the way, Conoco went on growing to become a world oil leader. The hardbound first edition is becoming scarce.

**C24. (CONOCO) Continental Oil Company, WE'RE BREAKING GROUND FOR ANOTHER 75 YEARS, Houston, TX, The Company, (1950), 48pp.**
This is a well prepared, seventy-fifth anniversary company propaganda publication. It was issued on the occasion of starting construction on a new research laboratory at the headquarters in Ponca City. The text is a story about a hypothetical Conoco employee as he tells his grandson about the company. Much history is presented as well as current doings as grandpa tells everything you always wanted to know about Conoco. Cited in Swanson on page 169. The softbound first edition is scarce.

**C25. (ARCO) Cook, Lodwrick M., ARCO AT 125 YEARS: CELEBRATING THE PAST, ANTICIPATING THE FUTURE, Philadelphia, The Newcomen Society, 1991, 22pp.**
This Newcomen pamphlet contains an address dealing with the history of Arco which was delivered at a society meeting in Los Angles. Mr. Cook joined Arco in 1956 as an engineer. He was involved in the North Slope exploration and in the pipeline, then elected to CEO in 1986. Charles Lockhart founded Atlantic Petroleum Storage Company in 1866 and

shortly merged into Standard Oil. The modern Arco was brought together by Robert O. Anderson as he merged with Richfield a West Coast refiner in 1966, then with Sinclair, a fully integrated major in 1968. Richfield had been exploring in Alaska for several years with moderate success. They held a huge acreage position which Anderson recognized as very valuable. Almost immediately after the merger, they discovered Prudhoe Bay, the ten billion barrel Alaskan oil field which now produces twenty-five percent of the domestic U.S. oil requirement. The softbound first edition is common.

**C26. (SHELL) Cox, Ian (editor), THE SCALLOP: STUDIES OF A SHELL AND ITS INFLUENCES ON HUMANKIND, London, Shell Trading & Transport, 1957, 135pp.**
Eight authors were involved in preparing this fascinating coffee table book which presents all aspects of the Shell Oil Company logo, the symbolic mollusk, Pecten. Pecten is commonly called a scallop. On Shell's sixtieth anniversary, the book was published and presented to key employees and stockholders. Everything about the scallop is shown in glorious color photos, from its anatomy and sex life to its appearance in artwork throughout history. Also covered is the scallop at the table, how it is prepared and served around the world. Beautiful book. The hardbound first edition is scarce.

**C27. (CREOLE) Creole Petroleum, WORKING WITH CREOLE IN VENEZUELA, NY, Creole Petroleum Corporation, (1956), 77pp.**
In the mid-1950's Creole was recruiting Americans to work in Venezuela. The book is a bit of company propaganda giving everything from a short history of the company to what to do with your pets when you move to Maricaibo. In 1956 Creole (subsidiary of Standard Oil of New Jersey) was producing over a million barrels of oil per day from Venezuela through the work of their 15,000 employees. It was operating as a major company although only a part of Standard. Creole Petroleum Corporation was formed in 1920 and in 1943 a merger between Standard of Venezuela and Lago Petroleum completed the company as it currently operates. The company headquartered in Caracas and eighty percent of their oil reserves were developed in Lake Maricaibo. A good employee's handbook. Not mentioned in Swanson. The hardbound first edition is common.

**C28. (TECO ENERGY) Culbreath, H.L., THE TECO ENERGY STORY: FUELING GROWTH FOR 90 YEARS, Philadelphia, The Newcomen Society, 1988, 23pp.**

This pamphlet contains an address dealing with the history of TECO Energy. It was delivered at a society meeting in Tampa. Culbreath formed the company in 1957 and remained CEO until 1972. Tampa Electric was formed in 1899 to consolidate the trolley system. The project and company was largely funded by Boston investors. Consumers Electric owned the electric service from 1887 and had a hydroelectric plant which was expanded in 1897. They used fuel oil for electricity generation for many years but changed over to coal when high prices forced the change. Gatliff Coal Company is now a subsidiary supplying most of the Tampa needs. The softbound first edition is common.

**C29. (STANDARD OIL OF INDIANA) Dedmon, Emmett, CHALLENGE AND RESPONSE: A MODERN HISTORY OF STANDARD OIL COMPANY (INDIANA), Chicago, The Mobium Press, 1984, xiv, 324pp.**

This work by Dedmon, an important American writer, was published shortly after his death. The first one third of the book is a rehash of Giddens' earlier massive work, although Dedmon tried to concentrate more on people, personalities and motivations. The company began as the mid-western refiner for Standard oil in 1889. Standard (Indiana) grew rapidly and was fully integrated by 1950, challenging Standard Oil of New Jersey for the number one spot. The challenges mentioned in the title are several but mainly they include competition, foreign exploration, government intervention and alternate energy sources. Of course Standard (Indiana), known widely as Amoco by the early 1970's, met and exceeded their challenges with appropriate responses, although dropping in the polls from number two to number seven or eight in U.S. oil company size. Mainly company hype, the subsequent chapters stress the company's role in technological changes and foreign exploration. They join the wave of the day by pursuing merger and acquisition as the way to increase reserves and solve all problems. The book is dry, but a part of industry history which must be understood before one can successfully navigate to the future. No source notes but Appendix 4 gives the 1911 Standard companies and their 1982 names....some surprising. The hardbound first edition is common.

**C30. (TEXACO) Dennison, Charles Stewart, AROUND THE WORLD WITH TEXACO, Houston, TX, The Company, (1925). viii, 165pp.**

Beginning in Japan and ending in New York City, Dennison covers 56,000 miles in his travelog. Quoting from the text: "The success which has attended The Texas Company's expansion in the foreign field is a fitting tribute to the wisdom, skill, courage, and vision of the men who direct its affairs...who have made a reality of the fact that the sun never sets on the Texaco Star". The book is a compilation of articles printed in the company magazine "The Texaco Star", entitled Around the World With Texaco. The series was so popular the public demanded more. Dennison wrote of all of the places he visited while studying the results of Texaco's marketing and developments in foreign lands. With minor Texaco plugs, he describes the geography, people, culture, customs etc. of: Japan, Korea, Manchuria, China, the Philippines, Siam, the Malay States, Java, Australia, New Zealand, South Africa, East Africa, India, Ceylon, Egypt and a short look at Europe via Vienna and Paris. Cited in Swanson on page 169. The hardbound first edition is scarce.

**C31. (STANDARD) Dodd, S.C.T., TRUSTS: AN ADDRESS DELIVERED BY S.C.T. DODD BEFORE THE MERCHANTS' ASSOCIATION OF BOSTON, JANUARY 8, 1889, Noplace, Standard Oil, 1889, 17pp.**

Dodd reviews the history of law regarding trusts, combinations, associations, partnerships, etc. from early English law to the modern trust. He was the Standard attorney who drew up the original trust agreement of 1882, which consolidated and protected the original 1872 Standard combination. Dodd is generally credited with inventing the holding company. He defended the trust by showing that oil production had greatly increased and prices had dropped....at that time. Quoting from Swanson: "Addresses by Mr. Dodd; includes history of Standard Oil Company, ten years of Standard Oil Trust". Cited in Swanson on page 161. The hardbound first edition is scarce.

**C32. (HOUSTON PETROLEUM CLUB) Donahue, Jack, THE FINEST IN THE LAND: THE STORY OF THE PETROLEUM CLUB OF HOUSTON, Houston, Gulf Publishing Company, (1984), ix, 336pp.**

Author Donahue whose writing credits include the biography of Halbouty, "Wildcatter" and the novel "Grady Barr", here does a nice job of relating the Houston Petroleum Club history. An easy reading history of the oil business after WW II to the oil boom peak year of 1981 is interwoven with the history of the club. Good stories of the club's most prominent members hold interest. In 1945 Wilbur Ginther and Howard Warren decided to form a club for oilmen in the city which was clearly the oil capital of the world. Beginning in 1946 the more affluent members began a six year search for satisfactory quarters. One false-start after another finally resulted in leasing the top of the Rice Hotel and construction of a fabulous facility there. Heralded by the press and membership as "The Finest in the Land", the club overcame snubbing by many independents and all of the majors, and by completion date they were all begging for membership. Much oil history and biographical information on Houston's oil wheelers and dealers. The hardbound first edition is common.

**C33. (OHIO-MARATHON) Donnell, J.C., II, THE EVOLUTION OF AN INTERNATIONAL OIL CORPORATION: MARATHON OIL COMPANY, NY, The Newcomen Society, 1962, 28pp.**
An address reviewing company history, delivered by a member of the Donnell family dynasty which ruled Ohio-Marathon for generations. Given at Detroit before the Society to celebrate the company's seventy-fifth anniversary. Also the name was officially changed from Ohio Oil Company to Marathon Oil Company at this time. Good brief history from 1887 and the old Standard Oil Trust. The company's main claims to fame are the discovery of the billion barrel Yates field of West Texas in the 1920's and major success in Libya during the 1950's. The softbound first edition is becoming scarce.

**C34. (CITIES SERVICE) Ellis, William Donohue, ON THE OIL LANDS WITH CITIES SERVICE, Tulsa, OK, Cities Service Oil and Gas Corporation, (1983), 483pp.**
This is a very comprehensive company history and excellent story of the beginnings of Oklahoma oil. Most likely the book was stimulated by the 1982 takeover of Cities by Occidental Petroleum Company which put an end to Cities' self determination. The story begins with Henry Foster's 1896 blanket lease of the entire Osage Indian reservation of 1,470,559 acres. The lease was to expire in 1906 and bore a ten percent royalty and required payment of fifty dollars per year per well. Foster died in mid-negotiation, but his brother signed for the heirs and formed Phoenix

Oil Company as their operating company. Three dry holes were drilled near the numerous surface oil seeps along the Kansas border in 1897. Phoenix finally found some shallow oil and merged with Osage Oil Company, later becoming ITIO. Logistical problems were enormous and the lease was going to expire. Henry's son, H.V. saved the lease by subleasing first to Guffey & Galey and then as the play caught on, to others anxious to exploit the Osage such as Barnsdall, Getty and Phillips. By 1904 the Foster's lease had produced about five million barrels and they had made 175 subleases. By 1912, Doherty Company of New York, an extensive gas and electric utility had acquired control of ITIO as well as Barnsdall. Doherty shortly brought all of his Oklahoma and related companies under one name, Cities Service Oil Company. By 1982, Cities was one of the top major oil companies with worldwide operations. Cities main claim to fame was the ITIO discovery of the Oklahoma City Field in December 1928. Included is a chronology of merger activity from 1980 with suitors Conoco, Texaco, Mesa, Gulf and finally Oxy. The Oxy-Cities fit was probably the best of the bunch since Oxy's primary oil operations were overseas and Cities largely had their operations in the U.S., including distribution and marketing facilities. Good historic photos and history. The hardbound first edition is becoming scarce.

**C35. (MAORT) European Gas & Electric Co., STANDARD OIL COMPANY (NEW JERSEY) AND OIL PRODUCTION IN HUNGARY BY MAORT 1931-1948, NY, The Company., (1949), xv, 65pp.**

This book was prepared by Jersey in response to the expropriation of two of its subsidiaries which were MAORT Gas Trading Company and Hungarian-American Oil Company in September of 1948. Evidently the Hungarian Communists were becoming established in government and took the trouble to fabricate a case of industrial sabotage against the Standard employees. They extracted phony confessions from them to justify the seizure. This is a well documented case against expropriation, although it went to little effect. Cited in Swanson on page 80. The hardbound first edition is common.

**C36. (EXXON) Exxon, CENTURY OF DISCOVERY: AN EXXON ALBUM, NY, Exxon Corporation, (1982), 112pp.**

Great picture and history book from the early days of Standard Oil (New Jersey) with Henry M. Flagler the first president. The Standard Oil Trust was formed in 1882 and is considered the beginning of Exxon, thus the commemorative first century book. Early company functions were limited to manufacture and marketing of kerosene as a household illuminant. Currently this vertically integrated company performs every activity from exploration to marketing a diverse product line on a worldwide basis, being the largest U.S. oil company. Many historic photos in this well composed company synopsis. The hardbound first edition is becoming scarce.

**C37. (KERR-McGEE) Ezell, John Samuel, INNOVATIONS IN ENERGY: THE STORY OF KERR-McGEE, Norman, University of Oklahoma Press, (1979), xiv, 542pp.**

This is a major academic work which tries to bring the human element into the history of a company. Robert S. Kerr and his brother-in-law, James L. Anderson formed Anderson and Kerr Drilling Company. They acquired some drilling rigs and suffered their way through the Depression as they expanded their interests into exploring and producing. Anderson left in 1937 and Kerr brought in a partner named Robert Lynn as executive vice president and another, Dean McGee as chief geologist. These men were carefully picked since Kerr was intent on going into politics. Kerr-McGee grew to be worldwide in scope and became fully integrated. Kerr went to the U.S. Senate and McGee built the modern day company. Kerr-McGee pioneered offshore drilling and was a major uranium player in the 1960's. The Karen Silkwood case which remains a blight on the company, is covered at great length and fairly objectively. The hardbound first edition is common.

**C38. (A.P.I.) Fanning, Leonard M. (editor), THE STORY OF THE AMERICAN PETROLEUM INSTITUTE: A STUDY AND REPORT (WITH PERSONAL REMINISCENCES), NY, World Petroleum Policies, (1959), 168pp.**

The A.P.I. was incorporated March 20, 1919 and in 1959 celebrated its fortieth birthday. It was the oil industry's first national trade association. It had consolidated numerous regional and local oil associations. Alfred Bedford and Robert Welch are given credit for doing most of the early work. The industry had been deeply divided on all subjects between the independents and Standard Oil, each going their separate and usually opposite ways, thus precluding a national oil organization. Doherty with Cities Service and Josh Cosden were early to speak out on the need to bring the industry together. When it was finally done, the industry for the first time had statistics which were meaningful, such as supply and demand numbers along with the reports on crude stocks, refinery runs, product stocks and production. The A.P.I. would probably not have been created until years later except for the WW I mobilization of the industry which brought Standard in line with the independents and with the government as they all worked together to supply the allies. Fanning is one of the most knowledgeable oil industry spokesmen and historians with numerous books to his credit. He knew the A.P.I. people well and served as its first director of information in the 1920's. It is doubtful that anyone else would have done a better job on this history as he relates much personal dialogue with the participants and of the events as it came together. Cited in Swanson on page 169. The softbound first edition is becoming scarce.

**C39. (ALABAMA POWER) Farley, Joseph M., ALABAMA POWER COMPANY: DEVELOPED FOR THE SERVICE OF ALABAMA, Philadelphia, The Newcomen Society, 1988, 24pp.**
This Newcomen pamphlet contains an address dealing with the history of Alabama Power. It was delivered at a society meeting in Birmingham. Mr. Farley is an attorney who represented the company for thirteen years before joining them as executive vice president in 1965. He became CEO in 1969. Alabama Power was formed as a hydroelectric plant in 1906, largely funded by English investors. By the 1920's it had become the largest industrial development company in the U.S. By 1935 they covered Mississippi, Alabama, Georgia and Florida. They used coal and oil fired steam generated electric power in the 1950's and moved into nuclear in the 1970's. Governor George Wallace almost sunk the company in the 1970's by anti utility rate legislation. The softbound first edition is common.

**C40. (HOUSTON NATURAL GAS) Fellows, Kenneth, HOUSTON NATURAL GAS CORPORATION: ITS FIRST FIFTY YEARS 1925-1975, Houston, The Company, 1975, viii, 163pp.**
Fellows was a HNG corporate officer and undertook this massive project shortly after his retirement in 1973. The fifty year commemorative book, published in 1975 finds HNG to be a company expanded far beyond its Texas roots, into Canada, South America, Europe, the Middle East and Asia with diversified products and services. In 1925 Houston Oil

discovered a shallow natural gas field while looking for oil on the Twin Oaks Ranch in Live Oak County of South Texas. Houston Pipe Line Company was formed to build a 200 mile, eighteen inch line to service the growing city of Houston. There was much dissent within company ranks for this successful oil company to venture into the new and unknown gas business. Houston Pipe Line Company was a great success and Houston Natural Gas Company was formed to deliver gas to cities along the main route. By 1927 they served all of the Houston area. By WW II they had 55,000 customers, by 1953, 200,000. In 1948 HNG began to acquire leases and drill for oil and gas reserves. By the early 1970's the oil and gas exploration and development company had become a huge success and was on their way to becoming a major producer and transmission company, particularly in the Permian Basin. Largely this was accomplished by merger and acquisition and by smart management. A 32 page version of this fine book was provided to key employees and stockholders. Both slipcased hardbound first editions are scarce.

**C41. (JULIAN PETROLEUM) Finney, Guy W., THE GREAT LOS ANGELES BUBBLE, ETC, Los Angeles, Milton Forbes Co., 1929, 203pp.**
This book was not available for review by Neff, but it is regarded as important. Quoting from Swanson: "Financial speculation in area 1920-1927; mainly of organization and promotion of Julian Petroleum Corporation". Cited in Swanson on page 151.

**C42. (BECHTEL) Finnie, Richard, BECHTEL IN ARAB LANDS, San Francisco, CA, The Company, 1958, 149pp.**
This is a fifteenth year review of the company's engineering and construction projects. Bechtel has historically been the largest oilfield construction firm operating in the Middle East. Largely Saudi Arabia and Kuwait, the company presents a photographic history of their accomplishments there. Pipelines, storage facilities, refineries and port facilities all over the Middle East are very impressive. If it was a gigantic job, Bechtel got it, not only because of their performance, but also their ability to get along with the various diverse governments and oil company partners in these ventures. Bechtel company policy favors employing the locals whenever possible. Not mentioned in Swanson. The hardbound first edition is scarce.

**C43. (TEXASGULF) Fogarty, Charles F., THE STORY OF TEXASGULF, NY, The Newcomen Society of N.A., 1976, 40pp.**
Charles Fogarty in 1976 was chairman and CEO of Texasgulf. He presented a rosy picture of his company which is, one of the four large U.S. sulphur companies. Texasgulf was formed in 1909 upon the expiration of the Frasch sulphur extraction process patent. Originally called Gulf Sulphur Company, it was incorporated by two St. Louis businessmen, Einstein and Harrison, who were interested in mining a dome called Big Hill in Texas. The company was under capitalized at $250,000 and by 1916 Bernard Baruch and J.P. Morgan became interested in the venture after noting the success of Freeport Sulphur as they developed a gulf coast salt dome. By 1918 the company was firmly capitalized and changed their name to Texas Gulf Sulphur. In 1919 they began production from the twelve million ton Big Hill dome. In 1927 they began developing the largest deposit in the world, Boling Dome. By 1953 Texasgulf was one of the big four sulphur companies and began diversification into other mining extraction projects and oil. They started as a small producer and developer gaining access to the industry through extraction of sulphur from sour crude and high hydrogen sulfide natural gas fields. A major mineral deposit find at Kidd Creek, Ontario, Canada was the subject of an SEC court battle over insider stock trading. It created a landmark case called the "Perpetual Jeopardy Decision" where some employees and speculators were convicted on appeal of the original favorable verdict. This is a good history of a small company which grew to world stature. The softbound first edition is becoming scarce.

**C44. (ROYAL DUTCH) Forbes, R.J. and D.R. O'Beirne, THE TECHNICAL DEVELOPMENT OF THE ROYAL DUTCH-SHELL 1890-1940, The Hague, E.J. Brill, 1957, 670pp.**
According to the preface, this book presents a survey of the technical progress made within the companies belonging to the Royal Dutch-Shell Group from 1890 to 1940. Gerretson's history of the company and Beaton's history of Shell were also published this same year and deal with the company's organization and cooperation between themselves. All of these commemorate the fifty year anniversary of the Royal Dutch-Shell merger. Since Royal Dutch was one of the top three oil companies in the world, they participated in virtually all new technology and technique as it was developed, either by themselves, Standard, British Petroleum or even the independents. Royal Dutch's part is documented and profiles of

many of its major contributors are also given. The story begins with a history of the industry and quickly turns to oil and gas prospecting methods, explaining how they have progressed from drilling near oil seeps, to the use of high resolution seismograph surveys. Next, the drilling and production techniques are followed from spring pole drilling to cable tools to modern deep drilling rotary rigs and directional drilling. Much is also given on ancient and modern reservoir engineering theory and practice. Refining is covered from the simple early stills developed to recover kerosene only, to modern catalytic and chemical cracking facilities which yield a variety of hydrocarbon products. Various oil and hydrocarbon chemical products are covered from early times, but concentration is on the modern petrochemical industry which flourished from the late 1920's on. The transportation and storage story from early barrels through pipelines and modern oil tankers is well covered. All of the chapters have useful bibliographic references and appendices. The book is well illustrated with photos and drawings including many of the early drilling and producing equipment from original patents. Cited in Swanson on page 169. The hardbound first edition is scarce.

**C45. (BOW VALLEY) Foster, Peter, FROM RIGS TO RICHES: THE STORY OF BOW VALLEY INDUSTRIES LTD, Calgary, Canada, The Company, (1985), 203pp.**

This is the fourth book on the Canadian oil business by this best selling author-journalist. The other three were written for his own account and were rather critical of the industry. This effort at documenting Bow Valley's history from its beginning in 1949 to a billion dollar worldwide corporation by 1984 was commissioned by the company. Foster delves deep into company files and is able to create a readable history on this vibrant company. The history is largely that of the company founder, Doc Seaman, who built the diverse giant. Daryl Seaman was raised in Saskatchewan, the son of a dirt contractor. He served as a pilot in the Canadian Air Force during WW II, later finished college and gained his nickname. About the time the Leduc oil gusher came in (1947), he and his brother, B.J. were hired by Western Geophysical. Seismic crews were exploring everywhere for another Leduc. Less than two years later, he and a partner bought a brand new truck mounted, shot hole rig and their business boomed. Within another year they were running three rigs and began to operate a complete seismic service company. By the mid-1950's he was diversified into road construction, water well and oil well drilling.

That led to drilling his own wells in 1957, which resulted in phenomenal success. Bow Valley became heavy in oil exploration starting in 1962 and with success there, undertook more diversification. The hardbound first edition is common.

**C46. (DOME) Foster, Peter, OTHER PEOPLE'S MONEY: THE BANKS, THE GOVERNMENT AND DOME, Toronto, Canada, Collins, (1983), 286pp.**
This book was released at the same time as Lyon's "Dome". It is a more comprehensive and less friendly view of Jack Gallagher and his brilliant rise to fame with Dome. The book mainly deals with Dome after 1982 when OPEC lost its stranglehold on the oil industry and the crude price began its descent. At that time Dome was drilling in the Beaufort Sea and with its acquisition of Hudson's Bay Company, was seven billion dollars in debt. Foster looks at the various government people and political problems involved with the crashing of Dome. He also looks at the banks and various other entities which held their notes. Canada's largest banks held about half, but the remainder was held by many of the major banks of the world. The resolution was still a long way from settlement at the writing of this interesting book. The hardbound first edition is common.

**C47. (GALENA OIL WORKS) Galena Oil Works, A BRIEF HISTORY OF PETROLEUM, PA The Company, 1876, 16pp.**
This brief pamphlet was probably authored by John J. McLaurin, who wrote Sketches in Crude a few years later. The little pocket booklet was issued as an advertising gimmick for the Philadelphia Centennial Exposition and contains only the briefest of oil history. Another edition was issued the same year with 20pp and again in 1883 with 35pp. All softbound editions are rare.

**C48. (PENNZOIL) Gentry, Mickey, and Kimberly Patrick, PENNZOIL: THE FIRST 100 YEARS, Houston, TX, Pennzoil Company, (1989), 88pp.**
Oil industry and company history combined into a slick centennial brochure. South Penn Oil came out of the Standard Oil breakup and later acquired Pennzoil which had considerable refining and marketing. Many others merged into the modern entity. Good biographical material on leader Hugh Liedtke who directed the company from the 1960's. The softbound first edition is becoming scarce.

**C49. (ROYAL DUTCH) Gerretson, F.C., HISTORY OF THE ROYAL DUTCH, The Netherlands, E.J. Brill, 1953 Vol. I, 287pp, 1955 Vol II, 355pp, 1957 Vol III & IV, 307 & 362pp.**

This massive four-volume work was researched and compiled by Dr. Gerretson. He reviews oil history and explains the right of the Netherlands to mine their colonial conquest's mineral wealth. Supervisor of a Dutch tobacco company there, he became aware of the oil seepages there and gained a lease to explore for oil from the Sultan of Sumatra. On his second try, Zijlker and his Sumatra Petroleum Company drilled a poor producer, #1 Telager Toenggal. This was in 1885 in the Netherlands East Indies Sumatra. Sumatra was already importing Rockefeller kerosene for lighting. Zijlker was very persistent and drilled several holes and while most were dry, the poor well began producing better and better until it reached 164 barrels per day. The project had potential but was not doing well. In 1890, he returned to Holland and formed Royal Dutch with partner VanBerg to raise money and pursue the venture. J. A. de Gelder was elected president and Zijlker died within six months. Royal Dutch bought producing equipment and a rig from Oil Well Supply Company out of Pennsylvania. They also hired men from there to operate their new fields. August Kessler took over operations as assistant president and began building a refinery which by 1892 was producing 300 barrels per day of kerosene. He began to produce "Crown Oil" in competition with American and Russian Oil. Over the next few years problems became severe including a native uprising, manufacturing breakdowns, falling prices, falling exchange rates, a sunk ship and customs delays. By 1894, he had the company running smoothly. Standard takeover attempts were constant as they tried to dominate the Eastern markets. In 1896, Kessler hired Henri Deterding, a young aggressive government banking official to be his key marketing man. Deterding changed everything and was a hands-on innovator, rapidly expanding facilities, markets and products in a wider and wider area. Kessler died suddenly in 1900 and Deterding, moving his headquarters to The Hague, took over. In 1902, Shell, The Rothschilds (Bnito) and Royal Dutch formed Asiatic Petroleum Company to handle Far East marketing for the three, thus allowing them to concentrate on competing with Standard, rather than each other. In 1907, Shell and Royal Dutch merged. Deterding and Royal Dutch became the 60% owner. The Hague became the technical center of operations and

London, the financial. Definitive. Cited in Swanson on page 169. All hardbound editions are scarce.

**C50. (GETTY OIL) Getty, J. Paul, HISTORY OF THE OIL BUSINESS OF GEORGE F. AND J. PAUL GETTY, 1903-1939, Noplace, The Author, 1941, 501pp.**

This book was not available for review by Neff, but is regarded as important. Quoting from Swanson: "Company's start in Mid- Continent, growth, investments in Tidewater-Associated Oil Company, work of George Franklin Getty; Paul Getty diaries. Cited in Swanson on page 170. The hardbound first edition is very scarce.

**C51. (STANDARD OIL OF NEW JERSEY) Gibb, George Sweet and Evelyn H. Knowlton, THE RESURGENT YEARS: HISTORY OF STANDARD OIL COMPANY (NEW JERSEY) 1911-1927, NY, Harper & Brothers, (1956), xxix, 754pp.**

The second of four Standard books takes the newly orphaned Standard Oil (New Jersey) which had been the holding company for all Standard affiliates, and follows the incredible growth as it copes with the traumatic change. The dissolution decree of 1911 was thought by most observers to be the end of Standard dominance, however, Jersey Standard was ready for the challenge of rapidly increasing worldwide demand for fuel and products. John Archbold was president from 1911 to 1916, Cotton Bedford until 1917 and Walter Teagle served from 1917 to 1937. Teagle was a dominating world oil figure. One of the main problems Jersey Standard had to overcome was the persistent resentment held by average Americans towards Standard. Also a problem of some magnitude was the government's restrictions as to how and where they could operate. This era sees government control in the oil industry come into its own as federal regulating concepts evolved. Standard participated heavily in fueling WW I. The search for new oil reserves worldwide, and major labor problems at home are well covered in this massive volume. Much credit is given to Standard's exploration company, Carter Oil Company, which was very aggressive and had considerable success in the mid-continent field, finding and developing millions of barrels of oil. Also mentioned is the rise of their Texas producing, refining and marketing company, Humble Oil & Refining. Many photos, drawings and tables are presented. Cited in Swanson on page 170. The hardbound first edition is becoming scarce.

**C52. (STANDARD OIL OF INDIANA) Giddens, Paul H., STANDARD OIL COMPANY (INDIANA): OIL PIONEER OF THE MIDDLE WEST., NY, Appleton-Century-Crofts, Inc, (1955), xviii, 741pp.**

Giddens was the most respected oil historian in the U.S. A Standard grant to the Allegheny College where he was professor of history and political science compelled him to undertake this definitive work on the company. Giddens was given free access to company personnel and files during his two and a half year leave of absence from the college to prepare the book. This abstract is largely taken from his comprehensive preface. Organized in 1889 by the Standard Trust, Standard of Indiana became a big part of the industrial and agricultural development of the middle west. They built the largest and most sophisticated refinery in the west at Whiting, Indiana to handle crude from Ohio and Indiana. They supplied oil products to the Standard marketing companies. In 1892 they began acquiring marketing companies to establish a web of sales all over the middle west. By the time the auto was catching on, they had service stations scattered about their territory which helped stimulate that new industry. After the 1911 dissolution of Standard Oil, Standard (Indiana) found itself with refining and marketing assets only, thus being subject to shortages of crude supply. Prairie Oil & Gas was the primary supplier to them. Standard's (Indiana) research people pioneered crude cracking and doubled the refinery net gasoline output per barrel. By 1917 they began acquiring producing properties and companies such as Midwest Refining. They also bought Sinclair Pipe Line and Sinclair Crude Oil Purchasing as well as the giant Pan American Petroleum. By 1932 Stanolind, the exploration subsidiary was having phenomenal success across the country. At this writing, Standard (Indiana) was the number two U.S. oil company and fully integrated (now Amoco). Many illustrations, photos and charts are included along with an important bibliographical note. Impressive work. Cited in Swanson on page 170. The hardbound first edition is common.

**C53. (STANDARD OIL) Grad, Morton, HOW ROCKEFELLER BUILT THE STANDARD OIL COMPANY, Girard, KS, Haldeman-Julius Company, (1931), 36pp.**

Grad's pamphlet is an attempt to put the Standard Oil story and Rockefeller in perspective as they relate to the progress of society and industry. Short chapters cover the beginnings of the industry, foundation of Standard Oil, the oil wars of 1872, monopoly, the Standard Oil Trust and dissolution of the company are well written and objective. Grad gets to his point in the last chapter where he states that Standard's dissolution changed nothing except it increased U.S. vulnerability to economic and political manipulation by foreign government-controlled oil interests such as Royal Dutch-Shell. He believes it likely that with these forces in action, the Standard companies will be reunited "perhaps under the auspices of the government of the United States - in the form of a vast super-corporation in order to counteract the policies of the British government manifested in the aggressive activities of the Royal Dutch-Shell group..." The last thirty pages of this pamphlet contains advertisements for Haldeman-Julius books for sale including 1,500 different titles of their "Little Blue Books" series at five cents each and covering every subject imaginable. Not mentioned in Swanson. The softbound first edition is scarce.

**C54. (PACIFIC PETROLEUMS) Gray, Earle, WILDCATTERS: THE STORY OF PACIFIC PETROLEUMS AND WESTCOAST TRANSMISSION, Toronto, Canada, McClelland and Stewart, (1982), 306pp.**

This is the fourth major book on the Canadian oil industry by Gray. He is a knowledgeable oil journalist and writer. Frank McMahon was born the son of a poor miner-prospector and grew up believing he would become a millionaire. He dropped out of college to become a hardrock driller for Standard Oil of California. By 1927, He had gained experience in oil ventures and core drilling, so he decided to start his own core drilling company. Unfortunately the Depression hit and he and his several operating rigs were shut down. He founded Columbia Oil to promote wildcat drilling. Lots of dry holes preceded his successful drilling at Turner Valley. He got in on that boom by drilling a risky, expiring lease. He was fully extended with loans beyond his credit, however tremendous success was had in 1938. He then formed Pacific Petroleums with the idea of pipelining natural gas from the oil fields of Alberta to the west coast of British Columbia. In 1949 West Coast Transmission was formed to pipe natural gas to Vancouver for export to the U.S. The political and logistical saga that took place in completing these two pipeline projects makes interesting reading as well as providing a classic lesson in persistence. The hardbound first edition is common.

**C55. (GSI) Geophysical Service, Inc., GEOPHYSICAL SERVICE, INC., Dallas, The Company, 1955.**

This is a reprint of an excellent article in the company magazine, "The Grapevine", Issued September 1955 as Volume 11, No. 8. It reviews the first twenty-five years of DeGolyer's seismic company, GSI. It contains a brief history of the company plus some field crew reports from 1930. Good photos and history of the pioneering geophysical exploration company. The softbound first edition is scarce.

**C56. (PRADO) Haas, Paul R., et al, ENTREPRENEURSHIP IN OIL: MR. PAUL R. HAAS AND THE PRADO OIL AND GAS COMPANY, Austin, University of Texas, 1971, 199pp.**

This work was sponsored by a Moody Foundation grant and the person conducting the four interviews was Larry Secrest. Quoting from the introduction: "This oral history is the first of a series that will focus on entrepreneurial activity in the Southwest, the formation and management of successful new business ventures". The Paul Haas story is short and sweet, being a real oil company success story. The book follows Prado's entire existence from its formation in 1960 to its eventual sale to Sohio, in 1964. Prado was a highly leveraged company which was formed with one purpose in mind, to buy an oil and gas lease and exploit the Prado field which was located in Jim Hogg County, Texas. Interesting situations took place as development work found the field to be much larger than anyone expected. The text is typewritten and delivered in the original question and answer format. The hardbound first edition is scarce.

**C57. (TEXAS UNIVERSITY LANDS) Haigh, Berte R., LAND, OIL AND EDUCATION, El Paso, TX, Texas Western Press, 1986, vi, 351pp.**

The Permanent University Fund began when Texas was still a Republic and the publishing of this book coincides with the Republic's sesquicentennial. In 1839 President Lamar enacted the legislation giving fifty leagues (221,400 acres) as an endowment for higher education. 2,000,000 more acres of West Texas were later added (half in 1876 and half in 1883), prior to the discovery of oil on the acreage at Big Lake in 1923. The first fifty leagues were located largely in Fannin, Cooke and Grayson counties of north Texas, near the Red River. These were sold in 160 acre parcels at around $3.00 per acre to generate funds for the University System. The subsequent 2,000,000 acres lie within the

Permian Basin oil province and on its barren western flank. In 1917 Texas allowed exploration of these lands through permit. Rupert Pickrell filed on 431,360 acres in Crockett, Irion, Reagan, and Upton counties at ten cents per acre. He turned the acreage to Krupp and Pickrell who with New York funding and Texon Oil and Land Company drilled #1 Santa Rita, discovery of the Big Lake field. In 1929 the University Board of Regents commenced their sealed bid oil and gas lease sale system with five year term leases on 160 acre tracts, much of it in prime oil country. By 1936 the University began their oil and gas lease auction sales which continue today. The vast University of Texas system currently has a two billion dollar reserve and funds the state colleges & universities. The hardbound first edition is common.

**C58. (SHAMROCK) Haley, J. Evetts, STORY OF THE SHAMROCK, Amarillo, TX, The Company, (1954), 75pp.**
Haley is a widely acclaimed southwestern writer, rancher and oilman. This is his only attempt at an oil company history. Company histories are usually dry, statistical propaganda works. This one is different. Haley uses his expertise in the first several pages as he describes the temper of the time as it related to the 1918 Panhandle field discovery which was followed by wide spread drilling and development. John J. Sheerin was of Irish descent and in the 1920's gave up law practice to become an oil man. He was the motivator, idea-man force which caused Shamrock Oil Company to be formed in 1925 after having considerable success on his own. Next he formed The Shamrock Petroleum Company in 1926 which was very successful in exploration and sold out to Prairie. Then the Shamrock Company was formed in 1927. The company suffered ups and downs until a merger with Mr. Fownes and Mr. Orr who were established brokers and financiers. The new, merged company was chartered in 1929, called the Shamrock Oil and Gas Company and consisted of Shamrock, Cockrell-McIlroy, Plains Drilling, Gardner Brooks & Collins, U.S. Oil, and other interests centered around the Texas Panhandle. They became the major marketer of natural gasoline in the panhandle. The product had previously been wasted, burned or vented to the atmosphere prior to their reclaiming it. With the stock market crash, Shamrock almost folded and was desperate by 1933. They expanded into carbon black production along with various other petroleum products which eventually pulled them out. Marketing and service territory was expanded to include the contiguous states. "Shamrock, A Mighty Good Brand" was the company

banner for years. Cited in Swanson on page 170. The slipcased softbound first edition is scarce.

**C59. (STANDARD OIL OF NEW JERSEY) Hidy, Ralph W. and Muriel E. Hidy, PIONEERING IN BIG BUSINESS: HISTORY OF STANDARD OIL COMPANY (NEW JERSEY) 1882-1911, NY, Harper & Brothers, (1955), xxx, 839pp.**
This book is the biography of the Standard Oil combination, not just Standard Oil Company, New Jersey alone. It is the first of four comprehensive and massive volumes which study the policies and practices of the combination during its incredible growth. Rockefeller and his partners brought order out of chaos in the new oil industry. They were dynamic men who brought people and companies together by any and all means available and then systematized their administration. They centralized the administration under the Standard Oil Trust and later Standard Oil Company, New Jersey. Modern business practices and procedures developed in the 1870's generally are still being followed today. The book covers the entire range of Standard Oil operations from exploration to marketing. A full color map showing the 1911 marketing territories of Standard Oil units in the U.S. is impressive and a little awe inspiring that one company dominated marketing of oil products over the entire country. The Supreme Court thought it was too much and in its May 15, 1911 opinion served notice that such combination was unreasonably restraining trade. Ninety pages of notes along with tables, drawing and photos are included in this major work. Cited in Swanson on page 170. The hardbound first edition is becoming scarce.

**C60. (STANDARD OIL) House of Representatives, HOUSE REPORTS 1887-88, VOL 9: REPORT ON INVESTIGATION OF TRUSTS, Washington, Government Printing Office (1888), 956pp.**
This official House of Representatives volume also contains 211 pages of testimony on the sugar trusts as a prelude to delving into Standard Oil. Testimony of 57 oil people for and against Standard included John D. Rockefeller, John Archbold, Henry Flagler and John Teagle. Over 900 pages of testimony and exhibits were presented to the Congressional Committee on Manufactures and was reported in July 1888 as Report No. 3112. Great dialogue here was essentially the start of twenty-two years of constant Standard badgering which ultimately ended in breakup by the Supreme Court. Cited in Swanson on page 167. The Leatherbound first edition is very scarce.

**C61. (STANDARD OIL) Howland, Harold J., STANDARD OIL, NY, The Outlook, 1907, 20pp.**

This is a special 1907 reprint of a comprehensive favorable article describing Standard's operations taking place in that year, from the magazine "The Outlook". Quoting from Swanson: "Author's visits to various properties and plants of the company, views of employees interviewed". Howland is very knowledgeable about Standard's operations and is probably on their staff. His story begins at the site of an Illinois development well which has reached the pay zone and is being prepared for shooting. He describes the scene as 200 quarts of nitroglycerin are detonated beneath the earth. Oil and mud, a hundred feet in the air. That is the most exciting, although he provides interesting reading about the tank farms, refineries, tanker ships, etc. A very thorough propaganda piece. Cited in Swanson on page 162; Hardwicke page 34. The softbound first edition is becoming scarce.

**C62. (STANDARD OIL) Hubbard, Elbert. THE STANDARD OIL COMPANY., East Aurora, NY, The Roycrofters, (1910), 24pp.**

Hubbard's booklet was published just before the Supreme Court ordered dissolution of Standard Oil. The first part of his story explains all of the good Standard Oil has done and is doing for the American people and the world. The middle part of his story berates Ida Tarbell. He calls her a great and talented woman, but claims that she misuses her abilities while getting her facts confused in the McClure's articles and her 1904 book, The History of the Standard Oil Company. The last part of his book discusses his belief that people and government are coming to their senses and will most likely not punish the successful, benevolent company for a few past indiscretions, but will allow Standard to continue to serve and benefit the nation. He guessed wrong. Interesting, historical booklet. Cited in Swanson on page 162; Hardwicke page 35. The hardbound first edition is scarce.

**C63. (SKELLY) Ironside, Roberta Louise, AN ADVENTURE CALLED SKELLY: A HISTORY OF SKELLY OIL COMPANY THROUGH FIFTY YEARS 1919-1969, NY, Appleton-Century-Crofts, (1970), x, 147pp.**

The author was a Skelly employee and graduate journalist on their Public Relations staff when assigned to this commemorative publication.

Although this is a company history, it is more the story of one man, W. G. "Bill" Skelly, the founder. He was born near Titusville, Pennsylvania in 1878 to hardy English-Irish parents. He left home at sixteen to seek his fortune in the oil fields near Oil City, Pennsylvania. By twenty-six he had an interest in two drilling rigs and had established some oil production in Illinois. He was an early player in areas such as Electra, Desdemona, Burkburnett and Ranger...then Tulsa in 1912. He participated in drilling deals and moved in the same crowd with Josh Cosden, J. Paul Getty, Frank and L. E. Phillips, Harry Sinclair and E. W. Marland. He had a big hit at Healdton, Oklahoma in 1915 assuring him success in his further drilling and by 1917 he was refining at El Dorado, Kansas. Skelly Oil Company was born there in 1917. He built the organization around the varied abilities of his co-founders, Herndon, Pielsticker and German. Skelly headquartered in Tulsa and became a major factor in refining, producing and marketing within the Kansas, Texas and Oklahoma areas. Erratic oil prices and heavy debt just about sunk Skelly during the depression. In a complex move, he borrowed money from Standard Oil and offered controlling interest in Skelly as collateral. He defaulted and Standard got the controlling stock. To avoid anti-trust activity, Standard placed the Skelly shares in Mission Resources, a California holding company which also owned significant amounts of Tidewater Oil stock. J. Paul Getty bought control of Mission in his quest to acquire Tidewater and got Skelly as a bonus. Bill Skelly retained the presidency for thirty-seven years and operated the company as a separate Getty entity. The hardbound first edition is common.

**C64. (LUFKIN) Jackson, Elaine, LUFKIN: FROM SAWDUST TO OIL, Houston, TX Gulf Publishing, (1982), xiv, 241pp.**
Well written and illustrated company history celebrating eighty years. The company was started by Kurth, Henderson, Wiener, and the Kavanaughs in 1902 to repair sawmill and locomotive equipment. They moved into the pumping unit business in the 1920's and soon came to dominate the field. The hardbound first edition is becoming scarce.

**C65. (TEXACO) James, Marquis, THE TEXACO STORY: THE FIRST FIFTY YEARS 1902-1952, US, Texaco, (1953), 115pp.**
Texaco went all-out for their golden anniversary company history. They hired one of the best authors in the country to put it together and make the generally boring growth of a large multinational integrated oil company

seem interesting. James does a good job with the aid of historic black and white photos. Shortly after the Spindletop discovery, Joseph Cullinan and his Texaco had a plan. They developed large storage facilities and refineries in order to cushion the radical market fluctuation which could quickly devastate an independent operator. Further oil discoveries and development in other areas assured Texaco of major company status early on. The rest is history. Texaco has always stood apart from the other majors and large independents. Usually they operate in a non-cooperative, go-it-alone mode and must have been uncomfortable in their roll as one of the Seven Sisters. This company history book was given to all officers, employees and stockholders of Texaco, the hardbound version for the high ranking and softbound to others. Cited in Swanson on page 70. The hardbound first edition is common.

**C66. (MAPCO) Jarman, Rufus, THE ENERGY MERCHANT, NY, Richards Rosen Press, (1977), 279pp.**
Jarman relates the history of the relatively new company and tries to avoid making it a dry statistical academic work. He partially succeeds. Robert E. Thomas essentially built MAPCO through smart moves and hard work. Although it was coming together through the late 1950's, MAPCO was not born until 1960, when its first 2,000 mile LPG pipeline was completed and began functioning. It carried propane, butane, ethane and natural gasoline under pressure from West Texas to the Middle West. It was a first. Because of the touchy nature of transporting highly volatile gasses at high pressure through interstate commerce, approvals had to be gained from countless sources. With rapid economic success in this venture, Thomas directed the company into other fields such as oil and gas exploration...all very successfully. The hardbound first edition is common.

**C67. (SUN) Johnson, Arthur M., THE CHALLENGE OF CHANGE: THE SUN COMPANY, 1945-1977, Columbus, OH, Ohio State University Press, (1983), xvii, 481pp.**
Johnson is a Harvard professor and has written several books on the oil industry. His study of Sun looks at the problems and changes from after WW II into the modern oil boom of the late 1970's and early 1980's. The Sun Company was founded in 1901 by Joseph Newton Pew to exploit the Spindletop oil discovery at Beaumont, Texas. He had a diversified background in oil, and upon his nephew Henry's recommendation,

directed the family assets and efforts there. Pew's second son John Howard was an engineer from MIT and joined the company that same year. He was a major influence over the years as he directed Sun into becoming an integrated major company. In 1966 Sun was sixteenth in the American petroleum industry. A competitor, Sunray DX ranked eighteenth. Both companies were having a hard time developing sufficient crude supply and wanted to expand. A merger was negotiated and approved in 1968, moving Sun into the top ten American oil companies. The book is somewhat dry, but shows what an integrated oil company faces in dealing with majors and how they solve the constant problems. The hardbound first edition is common.

**C68. (TEXAS POWER & LIGHT) Johnson, Robert L., TEXAS POWER & LIGHT COMPANY 1912-1972, TX, Texas Power & Light Company, (1973), ix, 209pp.**
Texas Power & Light was primarily formed by three men, Strickland, Templeton and Goodwin in 1911 to consolidate a ragged industry in Texas. In the late 1800's, electrification of Texas was a slow and economically hazardous process on a city by city basis. TP&L began with Waco, Waxahachie, Cleburn, Bonham, Strawn initially. They also controlled several natural gas producing properties which were later disposed of. Within a year of formation, they had acquired other city electric companies and were nearing monopoly in the state. Their large GE generators were first diesel, then lignite, and eventually natural gas fired. TP&L was a major factor in oilfield electrification across the state. At this writing, TP&L serves 500,000 customers, largely in rural Texas. The hardbound first edition is common.

**C69. (RICHFIELD) Jones, Charles S., FROM THE RIO GRANDE TO THE ARCTIC: THE STORY OF RICHFIELD OIL COMPANY, OK, University of Oklahoma Press (1972), 364pp.**
Keeping track of the name changes and mergers within Richfield is quite a problem and it is resolved in this book. Six original owners of Kelly Oil merged with Los Angeles Oil in 1911 to form Richfield. They were acquired by Atlantic Refining Company in 1966. The book covers all of the excitement in between. The author became a director of Rio Grande Oil in 1918 and when it merged with Richfield, he became president of the surviving new Richfield. The company found oil and gas in California and refined it and their marketing became famous up and down the west

coast. Harry Sinclair and Anson Jones of Cities Service were both deeply involved in the management of Richfield from the early 1930's. The main history left by Richfield was in their aggressive exploratory efforts on the North Slope of Alaska, over many years, which eventually led to the giant Prudhoe Bay oil discovery of 1968, two years after they were absorbed by Atlantic. This one oil field accounts for twenty-five percent of the U.S. domestic oil production. The hardbound first edition is common.

**C70. (COOPER INDUSTRIES) Keller, David Neal, COOPER INDUSTRIES: 1833-1983, Athens, OH, Ohio University Press, (1983), xii, 387pp.**

Brothers Charles and Elias Cooper formed the company in 1833 as an iron foundry. They expanded the foundry and by 1842 were selling farm machinery. By the early 1850's they were manufacturing steam powered compressors, then steam engines and turbine generators for oil well drilling and the railroads were a main product line. These gave way to gas engines so Cooper and Bessemer vied for the gasoline extraction compressor business in 1910. In 1929 they merged to dominate the market. Cooper is heavily involved in all phases of petroleum from exploration to distribution. They make Gardner-Denver drilling rigs, Demco mud pumps and Martin-Decker drilling instrumentation. Their compressors are used in production, transportation, storage and distribution through their subsidiary companies. Much documentation in appendices. The hardbound first edition is common.

**C71. (EAST OHIO GAS) Kelso, J. Richard, THE SPIRIT OF PROGRESS: THE STORY OF THE EAST OHIO GAS COMPANY AND THE PEOPLE WHO MADE IT, Philadelphia, The Newcomen Society, 1988, 24pp.**

This Newcomen pamphlet contains an address by Richard Kelso, president of the company, relating the history of East Ohio Gas. It was delivered before the society at Cleveland in 1988. The company is ninety years old and serves over a million customers. It is one of the gas marketing companies previously owned by Standard Oil, and serves Northeastern Ohio. Rockefeller's pipeline company, National Transit was put in charge of developing a natural gas business in 1883, and began acquiring producing properties and building pipelines for gas transportation. In 1898 Standard organized their effort and formed five companies to produce and market gas. East Ohio was a marketer and built their first line 100 miles to Akron, then Cleveland and most of Ohio. Standard was ordered to divest itself of gas utilities in 1935 and by 1943 formed Consolidated Natural Gas (CNG) as a holding company for the five entities and spun it off. East Ohio stayed on the marketing fore front and was one of the first to have their auto fleet powered by natural gas. The softbound first edition is common.

**C72. (OG&E) Kennedy, Donald S., PIONEERS IN PUBLIC SERVICE; THE STORY OF OKLAHOMA GAS AND ELECTRIC COMPANY, NY, The Newcomen Society, 1972, 28pp.**
Address given at Oklahoma City before the Society by Chairman of the Board. Brief history of the company from its incorporation in 1902 with coverage of its many mergers which effected the consolidated present day stage. Also included are several interesting stories of energy generating ideas that were tried and failed throughout Oklahoma. Chronology of major accomplishments and innovations. The softbound first edition is common.

**C73. (TEXACO) King, Kerryn, A SHORT HISTORY OF TEXACO INC. (1902-1977), NY, The Company, 1977, xxxpp.**
This short history, containing a cover letter showing intercompany distribution to department heads, indicates this to be the authorized version of the company history and should be used as a guide for answering general requests for information about Texaco. This is also a very good, although brief history of the entire oil business since Spindletop. The critical roll of Joseph Cullinan, Arnold Schalaet, John Gates, the Lapham interests and the Mitchell family played in the funding and guidance of early Texaco are recapped. While probably ninety-eight percent of the companies born at Spindletop have long since expired, Texaco survives and thrives. The primary stated reasons are that they quickly built storage to reduce the effect of price fluctuations and they also moved into exploring other salt domes with good success which compensated for the rapid depletion at Spindletop. The history is presented in sections: 1. Frontier Years (1902- 1913); 2. Gasoline Era (1914-1933); 3. Expansion (1933-1940); 4. War Years (1941-1945); 5. Dynamic Development (1945-1970); 6. Texaco in the Seventies. Much discussion of exploration, development, products, and marketing of this fully integrated world oil leader. The softbound first edition is becoming scarce.

**C74. (STANDARD OIL) Klein, Henry H., STANDARD OIL OR THE PEOPLE?, NY, Henry H. Klein, (1914), 134pp.**

Klein was the publisher of a weekly Socialist newspaper in New York City and continually attacked the rich. A flyer touting his book states "It tells why there can be no further general prosperity without a limitation of private fortunes". The flyer goes on: 'It shows how Standard Oil controls the industrial life of the nation". A newspaper ad for the book states "The greed of Standard Oil is destroying economic freedom in America. If you want to know how and why, read the book". Klein uses most of the book to discuss his version of how Standard got what it has and who controls what. He takes several of the prominent Standard Oil owners and officers and dissects their financial holdings in detail. He then makes a case that Standard controls all business and consequently the basis of American culture. The Only solution, he believes, is redistribution of wealth by restructure of private fortunes to an arbitrary $10,000,000 limit, the balance to go to the government to pay off city, state and national debt. Much good statistical information in the appendices. The wealthiest stockholders are listed along with much financial information on the original Standard companies in the back. Cited in Swanson on page 163; Hardwicke page 35. The hardbound first edition is scarce.

**C75. (BROOKLYN UNION) Larson, Elwin S., BROOKLYN UNION GAS: FUELING GROWTH AND CHANGE IN NEW YORK CITY, Philadelphia, The Newcomen Society, 1987, 23pp.**

This Newcomen pamphlet consists of an address dealing with the history of Brooklyn Union. It was delivered at a meeting of the society at New York City. Larson began with Brooklyn Union in 1947 at a time when gas had been heating and lighting their city for almost a hundred years. Brooklyn Gas Light Company was founded in 1925 to light the city. Citizens Gas Light Company was formed in 1873 and innovated an effective heating system. The two and three other competitors combined in 1895 to form Brooklyn Union. Coal gasification was used until the 1940's when natural gas became the standard. In the 1970's and 1980's the company became interested in oil and gas exploration through its subsidiary Fuel Resources, Inc. The company also serves Queens and Staten Island. The softbound first edition is common.

**C76. (HUMBLE) Larson, Henrietta M. and Kenneth Wiggins Porter, HISTORY OF HUMBLE OIL & REFINING COMPANY: A STUDY IN INDUSTRIAL GROWTH, NY, Harper & Brothers Publishers, (1959), 769pp.**

From the dust jacket: "This is the history of a small Texas company, organized in 1917, which became the largest oil producer in the United States, a leading pipeline carrier, and an important refiner". Prior to the legal founding of Humble, the independent Texas producers were having a difficult time marketing their crude and crude products. Major companies Magnolia, Texaco and Gulf controlled 70% of the refining and most of the transportation, thus the markets. The independents decided to band together and pool their production, refineries and facilities to compete with the majors. This very successful group of unknowns included Paraffine Oil Company, Reliance Oil Company, Bafler & Farsh, Schulz Oil Company, and Ardmore Oil Company. Shortly Southern Pipe Line Company and Globe Refining Company were added and the corporation capitalized at $4,000,000, about half of its real value at the time. Their areas of operation were in the major oil pools of Oklahoma, Wichita Falls, Corsicana, Houston and Beaumont. Humble immediately became very successful and competitive in exploration but could not balance income and expenses. The company had to consider borrowing money to keep up their lease acquisition and development drilling program, particularly at Ranger. Seeking money in the East, Humble met with Standard Oil and after a few months of negotiation, sold them half of the company stock for $17,000,000. The acquisition-merger was a perfect fit. Humble had too much crude and Standard had too much refining and marketing capacity. The book covers Humble history to 1948 by which time their presence was felt to varying degrees in many large oil and gas field developments in the U.S. This is one of the best company histories. Cited in Swanson on page 171. The hardbound first edition is becoming scarce.

**C77. (STANDARD OIL OF NEW JERSEY) Larson, Henrietta M., Evelyn H. Knowlton, and Charles S. Popple, NEW HORIZONS: HISTORY OF STANDARD OIL COMPANY (NEW JERSEY) 1927-1950, NY, Harper & Row, Publishers, (1971), xxv, 945pp.**

This is the third of the four large books covering the company history of Jersey Standard and includes the years from 1927 to 1950. The company greatly expanded its resources and facilities to meet the ever increasing demand for energy. During WW II Standard was able to make remarkable adjustments to production and refining, greatly increasing production for the effort. In 1927 they had 6.5 percent of the world crude oil production and by 1950 they had 13.7 percent. During the period, research and technology moved to the forefront in all phases of their company operations. A major Jersey Standard discovery during this time was the technique of fluid catalytic cracking which revolutionized the refining industry. Jersey merged with Mobil in the Far East (Stanvac), found significant Middle East oil, developed a major reserve in Venezuela, and through Humble found large oil deposits throughout Texas. Expropriation of Standard interests in Mexico and Bolivia were severe setbacks. Lots of photos, tables, maps and charts along with one hundred pages of appendices and notes. The hardbound first edition is common.

**C78. (ENRON) Lay, Kenneth L., THE ENRON STORY, Philadelphia, The Newcomen Society, 1990, 24pp.**
This Newcomen pamphlet contains an address dealing with the history of Enron Corporation. It was delivered at a meeting of the society at Houston. Lay was elected CEO of Enron upon the merger of InterNorth & Houston Natural Gas in 1985. Northern Natural Gas was formed in Omaha in 1930, from which they rapidly expanded their pipeline system and service area through the 1950's. During the 1960's they served most of the Mid West. Their name was changed to InterNorth in 1980 and they absorbed Belco Petroleum in 1983, who had been deeply involved in exploratory and development drilling. Houston Natural began its life in 1925, serving gas to Houston. They expanded and developed a Southern States and West Coast pipeline network. After the merger, Enron became the largest natural gas major (oil) company. The softbound first edition is common.

**C79. (AMERICAN METER COMPANY) Lief, Alfred, METERING FOR AMERICA: 125 YEARS OF THE GAS INDUSTRY AND AMERICAN METER COMPANY, NY, Appleton-Century-Crofts, Inc., (1961). 154pp.**
This is not just a company history, it contains the history of the late blooming natural gas industry to 1960. Early efforts to illuminate with natural gas were thwarted by technical, financial and public fear problems. Lighting was certainly possible using gas, but the regulation and method of charging for it were slowly resolved. They even tried

charging an annual fee for unregulated use. For efficient and affordable gas delivery a meter had to be developed and early "wet meters" were imported from England. They were subject to evaporation and freezing of the water which controlled throughput. Baltimore and Philadelphia were going to gas and looking for American manufactured meters. They found a frustrated meter repairman who began constructing his own meters. Theopolis Code began manufacturing in 1936 and quickly expanded his operation. He partnered with Samuel Down in New York and began producing a full line of essential gas apparatus and the modern dry meter by 1853. These men and several others combined in 1863 to form American Meter Company. Coal gas was used almost exclusively until the 1870's when water gas, a cleaner product also derived from coal began taking over the market. In 1891 the market was using natural gas. American Meter Company stayed at the forefront of technology, developing meters to handle tiny volumes as well as gigantic high-pressure gas streams for pipelines. The excellent photos of early day industry and equipment are credited at the end. Much history. The hardbound first edition is common.

**C80. (STANDARD OIL) Lloyd, Henry Demarest, WEALTH AGAINST COMMONWEALTH, NY, Harper & Brothers, (1894), xiv, 563pp.**

Here Lloyd wrote the earliest, widely circulated exposure of Rockefeller and his Standard Oil Company. He had been after Rockefeller and monopolies in general for several years in his articles published at the Chicago Tribune, with which he was associated. A previous book of his titled "A Strike of Millionaires Against Miners" was a huge success as it attacked the coal mine owners for locking out some Illinois workers in 1889, "Starving them to lower wages". In Wealth Against Commonwealth, Lloyd claims to be attacking monopoly, however he spends about one percent of the book discussing the whiskey, beef and coal monopolies. He goes straight for Standard, as an example of how all monopolies operate for the benefit of capitalists and for subjugation of the laboring class. From the dust jacket: "Mr. Lloyd has traced the growth of this institution from its origin, in the sagacity of one or a few business men of Cleveland, to its vast proportions now, when it controls the oil markets and trade of a continent, and is perhaps the most influential and most dreaded financial power in America". Most of his story is taken from serious study of the public record, primarily the Hepburn

Commission Report of 1879 as well as numerous state and federal lawsuits. Railroad rebates and South Improvement Company (1872) are the key subjects of interest to Lloyd as he considers them the most conspiratorial of Rockefeller's methods used to corner transportation and refining of oil. Much dialogue and some sensationalism. Cited in Swanson on page 163; Hardwicke page 36 The hardbound first edition is very scarce.

**C81. (BRITISH PETROLEUM) Longhurst, Henry, ADVENTURE IN OIL: THE STORY OF BRITISH PETROLEUM, London, Sedgwick and Jackson, (1959), 286pp.**

Written to commemorate fifty years in the oil industry, the first half of the book contains a comprehensive company history with dialogue and details of critical meetings integrated into developing world situations. The last half is company propaganda telling what they are doing, and where. Company name history is Anglo-Persian from 1909 to 1935, Anglo-Iranian from 1935 to 1954 and subsequently British Petroleum. In 1900 Knox D'Arcy, a retired Queensland gold millionaire was convinced by several promoters to seek a drilling concession in Persia. He was granted a sixty year, 480,000 square mile lease arrangement and formed a company with the Persians owning 16%. By 1904 they had a big oil discovery. D'Arcy funded the company individually, but within six months the well gave out and he was in need of financial help. Burmah Oil needed crude oil desperately to fulfill it's English navy contract and agreed to help fund Anglo-Persian. Dry holes were drilled until May 1908 when they hit the mother lode at 1,180 feet. After WW I, Anglo-Persian along with Royal Dutch- Shell, a French company and a U.S. company joined with Mr. 5% Gulbenkian to form Iraq Petroleum Company. In 1927 they also hit it big. Much more oil success, wars, politics, etc. Great History. Cited in Swanson on page 171. The hardbound first edition is becoming scarce.

**C82. (INTERSTATE) Loos, John L., OIL ON STREAM: A HISTORY OF INTERSTATE OIL PIPE LINE COMPANY, 1909-1959, Baton Rouge, LA, Louisiana State University Press, (1959), xvi, 411pp.**

Interstate Oil Pipe Line Company was actually formed in 1944 out of a merger of Oklahoma Pipe Line Company and Standard of Louisiana's pipe line department. This book is the history of these separate entities, their coming together and subsequent operations. It covers construction and operations of the companies in the nine states where they operate. From the dust jacket: "...the construction of the original trunk line from Tulsa, Oklahoma to Baton Rouge, Louisiana; early living and working conditions; the growth of the system to 4,500 miles..." Lots of maps and photos. Much on Oklahoma and Louisiana oil booms as production is gathered into their lines and moved to their Baton Rouge refinery. Fine Company. Bibliography and notes along with maps and photos. Cited in Swanson on page 171. The hardbound first edition is common.

**C83. (LUMMUS) Lummus Company, NEW HORIZONS, NY, The Company, (1954), 106pp.**

Largely operating in the U.S. and Europe, the company does whatever kind of petroleum processing the clients require. It was formed in the late 1930's to operate in this highly specialized field. They design and manufacture and operate an incredible complexity of refining techniques such as: distillation, cracking, coking, polymerizing, blending and extracting hundreds of products through their systems. They are heavily involved in petrochemical production such as the chemicals that produce plastics, detergents, fibers and synthetic rubber. The first one third of the book describes the company and its mission. The last part describes the various major facility types they create and how they do it. Simplified explanations of the various processes with drawings and photos. Most clients are major oil companies. Cited in Swanson on page 106. The hardbound first edition is becoming scarce.

**C84. (DOME) Lyon, Jim, DOME PETROLEUM: THE INSIDE STORY OF ITS RISE AND FALL, NY, Beaufort Books, (1983), 225pp.**

Lyon wrote this book while Dome was in a precipitous fall from grace. The company history was unauthorized, but he did receive a few interviews and was allowed to view some of their important Arctic facilities. Jack Gallagher was born about 1917 and received a geology degree from the University of Manitoba. He worked summers for the Canada Geological Survey and later hired on with Shell Oil in California although he had never seen an oil well. Between Shell and subsequent work at Standard Oil, he traveled all over the world, gaining tremendous exposure to oil finding. He joined Imperial in Canada shortly after the 1947 Leduc discovery as assistant production manager. In 1950 he was convinced by investors to join Dome Exploration, a new oil company which he would run. Dome Exploration was owned by Dome Mining, a 1910 gold producing company. He opened the office in Calgary and proceed to put together wildcat drilling prospects. The first two hit; one, an oil producing field at Drumheller and the other a gas discovery at Provost. They also became involved in development drilling at Redwater field. Much more success was had and Dome became known as a very shrewd exploration company, known for Gallagher's sharp trading. Gallagher was obsessed with drilling a prospect in the Beaufort Sea where the logistical, political and environmental problems were incredible. These problems and this venture turned out to be debilitating to Dome. Worse, in 1981, Dome borrowed 1.68 billion dollars to buy a controlling interest in Hudson Bay Oil & Gas Company and committed another 1.8 billion for the remaining outstanding interest. Short term notes were called and during 1982 as the industry itself was rapidly shrinking. Late 1982 saw a punitive bank and government bailout of Dome where they essentially took over the company. Gallagher was out of the picture in 1983 and Dome was in the hands of businessmen and politicians. Not much chance for success under those conditions. The hardbound first edition is common.

**C85. (EL PASO NATURAL) Mangan, Frank, THE PIPELINERS: THE STORY OF EL PASO NATURAL GAS, El Paso, TX, Guynes Press, (1977), 1st ed. 354pp.**

Mangan was in charge of advertising and public relations for El Paso for many years and his book is well conceived and compiled. He uses interviews and historical events to take El Paso Natural's progress from a one man show to a major U.S. corporation. Paul Kayser was a bright young Houston lawyer who saw the potential of natural gas in 1928. The Permian Basin was expanding with new giant oil fields and during the oil production the accompanying natural gas was being wasted. Born about 1888 near Tyler, he practiced law for several years prior to meeting H. G. Frost a successful oil wildcatter and head of Salt Dome Oil Company. The two bought royalties in the Permian Basin and sought New York financing to form a royalty company. About that time they came up with a better idea. El Paso, Texas was the only city of significant size which had not been supplied with natural gas. After some work they acquired a gas source from wells in southeast New Mexico and a franchise from the city of El Paso 204 miles away. Kayser hired A. L. Forbes to ramrod the pipeline job. Smith Brothers of Dallas was the contractor to lay the sixteen inch line. There were many hardships in this remote construction project but the first gas deliveries were made in June of 1929. Surviving the great depression, El Paso Natural went on to build major lines from the Permian Basin to California in the 1940's. El Paso began exploratory drilling and was very successful in their source area. They also tapped big gas reserves in New Mexico's San Juan Basin and piped that gas to northern and southern California. Oversupply of natural gas in the 1960's gave way to shortages beginning in the mid-1970's as this book was being completed. A 1984 Update is included with later printings. Good history. The hardbound first edition is common.

**C86. (CANADIAN HUNTER) Masters, John A., THE HUNTERS: SEARCHING FOR OIL AND GAS IN WESTERN CANADA, Canada, The author, (1980), 106pp.**

Canadian Hunter in 1980 was only six years old, having been started by two geologists, the author and Jim Gray. Both had been successful exploration geologists with Kerr-McGee in Canada. They left to form their own company and convinced Noranda to finance their exploratory ventures. Within a year they had drilled several prospects and three of them were oil discoveries. Much drilling activity and success followed in 1974. In 1975 the Elmworth play was initiated and farmouts from Texaco and Arco were negotiate by Hunter. Elmworth is a gas basin within which oil and gas production had been scattered over the years. It was nothing special and was being ignored by majors and independents alike. Hunter's first well was a re-entry of an abandoned well which showed pressure and led to eventual establishment of up to ten trillion cubic feet of gas reserves to Hunter's interest. Amazing history for any oil and gas company in modern times. The hardbound first edition is common.

**C87. (HOME-STAKE) McClintick, David, STEALING FROM THE RICH: THE HOME-STAKE OIL SWINDLE, NY, M. Evans and Company, (1977), 335pp.**

Author McClintick followed the Home-Stake story and reported on it while working for the Wall Street Journal during 1971 and he later finished research for this book. This is the story of Robert Simons Trippet and his Home-Stake Production Company. From the dust jacket: "...the most spectacular swindle in U.S. history, a scam aimed exclusively at the rich and financially astute people, employing the most irresistible of lures - the legal tax dodge." Bob Trippet was born in Enid, Oklahoma in 1916. His Dad was a prominent banker there and later at Bartlesville. Trippet began a law career, writing oil contracts for Cities Service in Tulsa shortly after his graduation from the University of Oklahoma and Columbia law school as a Phi Beta Kappa. A real overachiever, he began a private practice in 1948 and became obsessed with making money. He came in contact with some crooked oil promoters who inspired him. Primary targets for these promoters were "effete eastern snobs or kinky Hollywood types". He started Home-Stake Production Company in 1955 using a similar name to two other well respected Tulsa-based oil companies. It is hard to put this book down as the saga unfolds, slowly at first with some genuine leasing and drilling success. Subsequently, greed and a Ponzi-like pyramid scheme prevailed throughout the 1960's and into the early 1970's when it was finally discovered that most of the bought-and-paid-for oil wells were never even drilled. Eight pages of moguls and movie star investors are listed in the back section along with source notes. Fascinating reading. Trippet actually spent one day in jail for his dastardly deeds. The hardbound first edition is common.

**C88. (KERR-McGEE) McGee, Dean A., EVOLUTION INTO TOTAL ENERGY: THE STORY OF KERR-McGEE CORPORATION, NY, The Newcomen Society, 1971, 24pp.**

The Newcomen pamphlet contains an address given by Dean McGee, Chairman of the Board of the company at a meeting of the society in Oklahoma City during 1970. Robert S. Kerr and his brother in law, James Anderson started Anderson & Kerr Drilling Company in 1929. By 1930 they owned four drilling rigs and the Oklahoma City field was rapidly developing. They wisely began taking producing interests in wells as payment for contract drilling services. They went public in 1935 and changed the name to Kerlyn Oil Company as they expanded into oil and gas exploration and production. They drilled the discovery well for Magnolia field in Arkansas in 1938 and were involved in the 1943 discovery at West Edmond field. Kerr decided on politics, and leaving the guidance to McGee, served as Oklahoma governor from 1943-1947 and U.S. Senator from 1948 until his death in 1963. In 1945 they owned twenty-two drilling rigs and bought a refinery. The company name was changed to Kerr McGee in 1946. In the 1950's they moved into refining and marketing. In the late 1950's and early 1960's they did much pioneer work in offshore drilling. They later became a major uranium producer after uncovering a major deposit at Ambrosia Lake, New Mexico. The softbound first edition is becoming scarce.

**C89. (VARIOUS COMPANIES) McLean, John G. and Robert Wm. Haigh, THE GROWTH OF INTEGRATED OIL COMPANIES, Boston, Harvard Graduate School of Business Administration, 1954, xxiv, 728pp.**

This is a massive scholarly study of oil company integration before the Standard breakup of 1911 and afterward. How various companies formed and integrated from exploration to marketing and how others reverse integrated from marketing to exploration. Harvard tries to answer: 1) Why the large vertical oil companies dominate rather than small non-integrated units; 2) What are the fundamental differences in these companies; 3) What changes have affected the small refinery business. Over half of the book relates to the last question. Cited in Swanson on page 146. The hardbound first edition is becoming scarce.

**C90. (McKEE & COMPANY) Mills, Charles E., FIFTY YEARS OF ENGINEERING, Cleveland, OH, Arthur G. McKee & Co., (1955), 96pp.**

A fifty year commemorative book for the company. History of steel industry and oil industry followed by the company. McKee was joined by Baker and Herr in setting up corporate offices in the Rockefeller building in Cleveland. They built blast furnaces and crude stills all over the world. Refining and distillation plants from the 1920's to the most modern cracking units of the mid- 1950's. The hardbound first edition is becoming scarce.

**C91. (STANDARD OIL) Montague, Gilbert Holland, THE RISE AND PROGRESS OF THE STANDARD OIL COMPANY, NY, Harper & Brothers, Publishers, 1903, v, 142pp.**

This is the oldest oil company history to receive wide circulation in book form and was written by Harvard economics scholar Montague. It was first presented in the Quarterly Journal of Economics for Harvard in two major articles during 1902 and 1903. His sources were the official investigating committee reports, three in particular: The New York "Hepburn" report on railroad abuses of 1879, the congressional testimony of 1888 and the president's Industrial Commission of 1898 and report of

1900. Rockefeller's entry into refining in 1865 at Cleveland, Ohio came at a time when every phase of the new industry was in chaos from drilling to marketing. He began buying out, or merging with competitors there and by 1870 formed Standard Oil of Ohio with his partner Henry Flagler. Standard efficiency in every phase of refining achieved the lowest cost per barrel of product. To make Standard even more profitable, transportation costs from the oil fields had to be reduced. South Improvement Company was formed with the Cleveland, Pittsburgh and Philadelphia refiners to coerce the railroads into giving rebates and drawbacks to them as high volume customers. They netted fifty percent less than the going rate for crude and seventy-five percent less on refined kerosene compared to the posted price which had to be paid by refiners and producers outside of their company. A precedent had been set for these rebates in coal and cattle shipments, but monopoly was not involved in those instances. By 1879 Standard controlled the refining in the oil regions as well as the major cities....also the transportation facilities and agencies. Independent Tidewater built a 300 mile pipe line from the oil regions to the east coast refineries in 1879 after overcoming Standard's countless obstacles. A big success, Standard quickly constructed a pipeline system of their own...bigger and better. The Standard Oil Trust was formed in 1881 to manage the vast empire and its thirty-nine separate companies. The nine trustees included John D. and William Rockefeller, Henry Flagler and John Archbold. During 1887, eighty-five percent of the 14,000 independent oil producers in the Oil Regions made an agreement with Standard to try and stabilize the industry by stifling production and drilling in hopes of forcing the price up. It worked temporarily for both parties. Within a few years Standard was controlling prices by buying only what it wanted. The Trust was dissolved in 1892 but Standard Oil Company survived intact as of this 1903 writing. The Pure Oil Company, a strong independent is mentioned as virtually the only competition, however, Sun, Texaco and Gulf were emerging from the Spindletop field in Texas and were well on their way to reducing the pervasive Standard influence. Montague concentrates on the railroad rebates throughout the book and as an economist sees this as the main factor which allowed the monopoly to exist. He believes that as general policy, large shippers should receive freight breaks even from transporters who are classified as common-carriers. Cited in Swanson on page 163: Hardwicke page 36. The hardbound first edition is common.

**C92. (A.A.P.G.) Morley, Harold T., A HISTORY OF THE AMERICAN ASSOCIATION OF PETROLEUM GEOLOGISTS: FIRST FIFTY YEARS., Tulsa, OK, AAPG, 1966, AAPG Bulletin, Vol 50 #4 (Specially bound), p668-820.**

This is a compilation of AAPG history to 1965. AAPG is the largest geological society in the world. From humble beginnings, the association was formally organized at a Tulsa, Oklahoma meeting of February 1916. No lists of those attending the meeting survived, however papers were presented and later published in the Bulletin. The 125 founders are listed and contain many of America's great oil finders. Petroleum geologists were already a tight-knit group and had met several times before, but needed the organization for information sharing. The history in five parts: I) Founding and Foundation (1915-1926); II) Years of Growth (1927-1941); III) War Period (1942-1946); IV) Post War Expansion (1947-1958); V) Recent Years (1959-1966). Reminiscences of founders Raymond C. Moore, Wallace Pratt and Elmer Thomas as well as details of the early meetings are enlightening as to the foresight of the men in the relatively new field of petroleum geology. An appendix listing previous association officers and honorary members is included. The hardbound first edition is common.

**C93. (OIL WORKERS UNION) O'Connor, Harvey O., HISTORY OF OIL WORKERS INTERNATIONAL UNION (CIO), Denver, CO, Oil Workers International Union (CIO), 1950, xiv, 442pp.**

O'Connor was an accomplished writer with several books to his credit. His 1933 Mellon's Millions was widely circulated as intensely anti-business and anti-mogul propaganda. From the forward: "The oil industry always has been wealthy, powerful and feudalistic. It has bitterly battled unionization of its workers. Organizing efforts have been carried on by oil workers for more than half a century and all of this time the road has been stony". The first unions appeared early in Pennsylvania but Standard oil "snuffed them out". Ohio suffered the same fate as well as California in 1899. Guffey Oil tried to cut wages at Beaumont from \$3.00 to \$2.50 per hour. The workers struck and it spread to Baston, Saratoga, Sour Lake and to Humble....the first major oil workers strike. The strikers prevailed. Unionism spread to refineries. A big upsurge in union membership throughout the oil fields occurred in 1917. The unions asked for a conference with the operators to discuss better wages, hours and conditions. They got no response. By spring of 1917, the Texas and

Louisiana unions fought, but were ineffective. However, California operators gave in to the union and the eight-hour day was established. By 1918 the International Union was formed with locals in the various oil cities. The first convention was held at El Paso, Texas where a constitution was created. The resolution called for nationalization of railroads, communications, oil, coal, water, electric power and shipping. Union participation and effectiveness waxed and waned over the years along with the economy and state of the oil industry. Much oil company bashing. Cited in Swanson on page 171.

**C94. (LITTON) O'Green, Fred W., PUTTING TECHNOLOGY TO WORK: THE STORY OF LITTON INDUSTRIES, NY, The Newcomen Society of the U.S., 1988, 24pp.**
In 1988, O'Green was retired president, CEO and Chairman of the Board of Litton Industries, Inc. He presents the company history and future prospects for the Newcomen Society. Litton today is a multi-industry, multi-national manufacturing company. Chuck Thornton started Litton in California in 1953. He had primarily been a high achiever with Ford and later Hughes Tool Company. Electronic technology was to be his concentration. First components, then products, then systems was his plan. Radar and navigation technology led him to acquire other firms in the 1960's. From cash register companies to shipbuilding, by the late 1970's Litton had over 100,000 employees. Several entities in the oil and gas and energy field were acquired and now comprise one of their three core businesses. Manufacturing items from well logging instruments to seismograph equipment has made Litton a big factor in the oil industry. The softbound first edition is common.

**C95. (PURE OIL) Oil City Derrick, PURE OIL TRUST vs. STANDARD OIL COMPANY: BEING THE REPORT OF AN INVESTIGATION BY THE U.S. INDUSTRIAL COMMISSION, Oil City, Derrick Publishing Co., 1901, lxxvi, 606pp.**
The fifty-fifth Congress formed the United States Industrial Commission to examine and recommend legislation to meet problems presented by labor, agriculture and capital. Five senators, five congressmen and nine members of industry comprised the committee. The authorizing bill had been pushed for several years by Congressman Phillips who was in favor of fair rights and treatment for common laborers, or so he said in 1898. He became vice chairman of the committee and ran it almost single

handedly. As the committee began its work the first order of business became the investigation of trusts and industrial combinations. Of course it was aimed strictly at the Standard Trust. As many government investigations, this one was a classic farce. It seems that Phillips was also an independent oil man in competition with Standard and was out to wreak whatever havoc he could upon them. He called an impressive list of witnesses, most related to the Pure Oil Trust, which was Standard's only competition, and weak at that. On the Standard side, John D. Archbold and H. H. Rogers were the primary witnesses. As an expert statistician and knowledgeable oil man, Patric Boyle (editor-owner of Oil City Derrick) testified to the tune of 100 pages. Boyle was so incensed at the unfair attack on Standard that he published and distributed this volume of the proceedings. Of course, he edited it some and is pro-Standard, believing they were responsible for keeping the price of oil and its products down. Although much of the testimony is hearsay, it does provide a good picture of the pervasiveness of Standard on the world scene. Cited in Swanson on page 164; Hardwicke page 27. The hardbound first edition is scarce.

**C96. (OKLAHOMA NATURAL) Oklahoma Natural Gas Company, SERVING NATURAL GAS, Oklahoma City, The Company, 1943, 145pp.**

This is an employees handbook which includes the training program adopted by the company. It served as a textbook and source of reference and covers everything to do with natural gas, from the idea to the customer. There is a fine section on the company history which began near the same time as oil began in Oklahoma. In 1903, the famous explorationists Guffey and Galey, laid a two-inch gas line from a well just north of Tulsa to a brick plant in that city. By 1905 they were serving 400 customers and organized Arkansas Valley Oil and Gas Company to operate the system. Only three small oil pools existed in Oklahoma at the time, all of them around Tulsa. In 1905, T.N. Barnsdall and G.T. Braden, also Pennsylvania oil men purchased the system, along with which came a 155,000 acre lease on the Osage reservation. They formed Osage and Oklahoma Company and funded it through stock offerings in Pittsburgh. In the next few years oil boomed in central and northeastern Oklahoma. Oklahoma Natural Gas was formed to bring gas from the Cleveland field to Oklahoma City. As oil was produced, gas was burned or vented and there was a scramble to try and conserve it as major pipelines from the fields to the cities were constructed. Around 1917 ONG became the aggressor and bought up the small distribution companies they had been serving. From that time and through the 1930's they came to monopolize the Oklahoma markets. Not mentioned in Swanson. The softbound first edition is becoming scarce.

**C97. (PAN AMERICAN) Pan American Petroleum & Transport, MEXICAN PETROLEUM, NY, The Company, 1922, xiv, 300pp.** This book is a tribute to Doheny and Canfield and their incredible oil finding success in Mexico. The book opens with an introduction by editor, W.J. Archer, where he provides a brief history, along with some historic photos. Chapter two is an address by Edward Doheny which was delivered at the second American API meeting in Chicago in 1921 entitled History and Future of the Oil Industry in Mexico. Doheny was the president of Pan American and had been almost entirely responsible for their entry into Mexico. He talks of his prospecting trip with C. A. Canfield in 1900 to the jungles thirty-five miles from Tampico where they observed active oil springs. No question of oil in the area, but logistically it was a bad situation, being miles from civilization. In fact, at that time the nearest oil well supply station was in Pittsburgh. Betting on the future they bought one large ranch after another within the area of major oil seeps and springs, acquiring by 1901 some 600,000 acres. In may 1901 the first well hit oil at 545 feet, producing large quantities. Over the next twenty years they had phenomenal success with gigantic producers. They overcame all kinds of problems to lay pipelines to Tampico for shipment out. He describes the exploration method used to locate the large pools by simply following the lineation of the oil seeps and projecting that to the subsurface. Chapter four tells the story of four gushers. First, the Drake 1859 well at Titusville, producing about twenty-five barrels per day; Second, the Russian Baku 1883 gusher which produced oil at the rate of 40,000 barrels per day; Third, the Lake View California 1910 gusher at 60,000 barrels per day; Fourth, Mexican Petroleum's #4 Cerro Azul well at 260,000 barrels per day in 1916. Chapters follow on transportation, refining and distribution. Much supplementary data is included along with index. Cited in Swanson on page 51; Hardwicke page 38. The hardbound first edition is common.

**C98. (PARKER DRILLING) Parker Drilling Company, RIGGED UP: PARKER DRILLING COMPANY'S FIRST FORTY-FIVE YEARS, Tulsa, OK, 1979, 30pp.**

At the height of the late 1970's oil boom, Parker was the largest land drilling contractor in the world. In 1914, Gifford C. Parker was seventeen years old and left his parents Illinois farm to take a roughnecking job. He made his way to the booming Tulsa area and was established there by 1919. He formed a small drilling partnership in 1922 with which he operated four cable tool rigs. By 1925, rotary drilling had taken over because it was cheaper and faster...he was out of business. Parker was in and out of the contract drilling business over the next few years, even trying other ventures. Barely surviving the Depression, he had acquired two rotary rigs by 1934 and still owned the four cable tool rigs, so he formed Parker Drilling Company. The rest is history, and it is told here with good stories and illustrations. The softbound first edition is scarce.

**C99. (TEXASGULF) Patrick, Kenneth G., PERPETUAL JEOPARDY: THE TEXAS GULF SULFUR AFFAIR, A CHRONICLE OF ACHIEVEMENT AND MISADVENTURE, NY, The Macmillan Company, (1972), xvii, 363pp.**

Texasgulf was formed in 1909 and today is a diversified natural resource extraction company operating largely in the U.S. The text of this book is the precedent setting lawsuit brought by the SEC against Texasgulf and several of its employees. Texasgulf had discovered a massive mineral deposit at Kidd Creek, Ontario, Canada. The employees leaked confidential company information and then bought Texasgulf stock prior to the news reaching the public. The question of who in such a situation can do what, when, with and to whom are partially answered in the loss on appeal and conviction of some of the Texasgulf people. Interesting. The hardbound first edition is common.

**C100. (DRESSER INDUSTRIES) Payne, Darwin, INITIATIVE IN ENERGY: THE STORY OF DRESSER INDUSTRIES, NY, Simon and Schuster, (1979), 415pp.**

Payne did hard research to complete this major history of Dresser. In 1976 he was fortunate enough to have a live interview with 101 year old William Graham, a Dresser employee who had worked for the company in the 1800's. Dresser by the late 1970's was a three billion dollar company, a conglomerate of oil and mining equipment and service companies. Corporate mergers beginning in the 1930's and on going, added diversification to their well-run, but not flashy company. Solomon Dresser was born in 1842 to a farming family and was able to attend college in Michigan for a short while. In 1861 he began working in a dry goods store in Indiana. In 1864 he dabbled in Pennsylvania oil and by 1870 he had established some oil production. He moved to Bradford Pennsylvania to take advantage of the boom there in 1877. In 1880 he developed and patented an effective oil production packer to seal crude from unintentionally leaking to the surface. A hard rubber product, it worked simply by expanding from weight when applied. He gained control of that market and went into oilfield explosives which were used to fracture the oil reservoirs by 1884. By 1887 he had developed and patented a pipe coupling which revolutionized pipeline laying by eliminating screw type joints and that remained the company's major product to the 1930's. By the late 1970's, Dresser had acquired such service companies as Boviard Oil Tools, Lane Wells Logging, Guiberson Oil Tools, Security Bit company, PGAC Logging, Kobe Pumps and Waukesha Motors. They are a competitive major oil field service company. The hardbound first edition is common.

**C101. (VARIOUS) Peebles, William K. (compiler), DIRECTORY OF OIL COMPANY NAME CHANGES, Houston, TX, Houston Geological Society., 1993, i, 49pp.**
This fascinating compilation of oil company name changes is indispensable to anyone interested in oil history. The book is revised from time to time as the need arises. It is a comprehensive alphabetical listing of company names from the past and present. It is cross-referenced for easy use and searching either direction. The current softbound edition is common.

**C102. (GENERAL AMERICAN) Perryman, W. L. Jr., GENERAL AMERICAN OIL COMPANY OF TEXAS: INTEREST BY THE BARREL, NY, Newcomen Society of N.A., 1961, 28pp.**
In 1961, Perryman was president of General American. He was presented to the Newcomens by Al Meadows, one of the company founders. Meadows' father was a doctor in Georgia and gave young Al first hand lessons in trading and finance which he later used to good advantage. By the 1920's he had held several jobs but finally found a good one with Standard Oil. By 1928 he was manager of materials in Shreveport,

Louisiana. He got a law degree, passed the bar, and with Ralph Trippett formed a finance company and called it General American Finance System. They bought production payments on the East Texas field at twenty-five cents on the dollar. Very active, they acquired great assets but great debt as well, eventually forcing a merger with J.W. Gilliland, who had solid assets in refineries and gasoline plants. Several difficult years followed as the price of crude plunged to ten cents per barrel. With proration came stability in the market and General American under Meadows guidance began expanding outside of East Texas with excellent results in West Texas. At this writing, General American had merged with several smaller companies and was expanding with production in Canada and the Mid- continent areas. The softbound first edition is becoming scarce.

**C103. (GEOSOURCE) Petty, O. Scott, SEISMIC REFLECTIONS: RECOLLECTIONS OF THE FORMATIVE YEARS OF THE GEOPHYSICAL EXPLORATION INDUSTRY, Houston, TX, Geosource, Inc., (1976), xi, 81pp.**

The seismograph was invented in 1841 and was used for earthquake detection. Germans used the seismograph to triangulate incoming gunfire sources during WW I, effectively. After the war, Dr. Mintrop began using his seismograph to interpret geology through variations in arrival times of seismic energy from dynamite charges. He called his company Seismos. In 1924 Gulf Oil began using one of Dr. Mintrop's crews to hunt for shallow gulf coast salt domes. Scott Petty was a structural engineer living in Dallas. His brother Dabney was a geologist working for the state in Austin. His other brother Van was an attorney and independent oil operator. The three pooled their resources and talents to develop their own seismic method, instrument and company. Scott did the research, Van the financial and Dabney was the key field man. The rest of their fairly wealthy family subscribed to shares in the Petty Geophysical Engineering Corporation. Within a few months their experiments developed new instrumentation and technique. Practicing on known salt domes, they began reading reflection waves as well as the common refraction waves, a real innovation. Their crews worked on the Texas and Louisiana gulf coasts for major and independent clients with good success over the years. Their continuing research kept them at the forefront of technology in this rapidly changing industry. In 1973 Petty merged into Geosource who already controlled Ray Geophysical to form

Petty-Ray Geophysical. The last part of the book has historic photos and describes exhibits in the Petty-Ray museum at Houston. The hardbound first edition is becoming scarce.

**C104. (TEXACO) Petzinger, Thomas, Jr., OIL & HONOR: THE TEXACO PENNZOIL WARS, NY, G.P. Putnam's Sons, (1987), 495pp.**

Author Petzinger was the Wall Street Journal's deputy bureau chief in Houston during the Texaco-Pennzoil-Getty trial. His coverage of the trial earned him a 1985 Pulitzer Prize nomination. His book was the first of three major works to hit the market about two years after the verdict was announced. The book is in three parts: 1. The men, where the lives of the main characters are reviewed and their diverse and varied backgrounds are explored... J. Paul Getty, Gordon Getty, Pennzoil Chairman Hugh Liedtke, Texaco chairman, John McKinley, Getty Chief Sidney Petersen are studied. 2. The deals, which are several verbal and contract negotiations first between Gordon Getty and Pennzoil, then between Petersen and Texaco, then between Gordon Getty and Texaco. 3. The case, where Pennzoil sued Texaco for illegally squelching their done-deal with Gordon Getty for control of the company. At the writing of this book the jury had awarded Pennzoil an 11.3 billion dollar verdict. The author presents an exciting view of the participants and the events as they unfold. This is very good reading and the author goes out of his way to educate as well as stimulate. The jury's award was later reduced by a bankruptcy judge but the punishment was still out of proportion to the crime by any standard applied. Good notes and bibliography. The hardbound first edition is common.

**C105. (PHILLIPS) Phillips Petroleum. PHILLIPS: THE FIRST 66 YEARS., Bartlesville, OK, The Company, (1983), 219pp.**

A fine photo-based company history produced on the 66th anniversary of Phillips Petroleum. This is a well designed chronology of Phillips history from Frank, L. E., and Waite Phillips family photos to post oil boom company goals and philosophy. Historic photos not previously published are presented here to dramatize the growth of Phillips. Persistence by the brothers and their company, Anchor Oil, in the early 1900's made their first success at #1 Anna Anderson wildcat drilled near Dewey in Indian Territory. After three straight dry holes they prevailed. By 1917 Frank and L. E. had become important oil industry bankers and producers.

They consolidated their oil interests into Phillips Petroleum Company in 1917 and proceeded to become a fully integrated major oil company with worldwide interests. This is an excellent informative book. Great photos. The hardbound first edition is common.

**C106. (STANDARD OIL) Popple, Charles Sterling, STANDARD OIL COMPANY (NEW JERSEY) IN WORLD WAR II, NY, The Company, NJ, 1952, xxvii, 340pp.**
This is a comprehensive record of Jersey's contributions to the war effort during WW II. Largely technical and dry, Popple presents an impressive history of gearing up to produce more crude from all of their North American subsidiaries. The primary contribution from Jersey was their large volume increase in production of 100 octane aviation gasoline. Other areas of major help were in butadiene production for manufacture of synthetic rubber. Production of toluene for explosives from their Baytown plant was also critical. Many company tankers were sunk. For a quick look at the impressive record, a brief chronology is included. Cited in Swanson on page 164. The hardbound first edition is scarce.

**C107. (CANADIAN SUPERIOR) Porter, J.W., A BACKWARD GLANCE: A HISTORY OF CANADIAN SUPERIOR OIL, Calgary, The Company, (1986), 62pp.**
Superior Oil Company decided to move into the Canadian oil regions in 1943. Two of their top field geologists, Meyer and Nichols were sent to Calgary to begin the exploratory venture. Rio Bravo Oil Company was chartered by Superior that year. The name comes from an excellent oil field discovered by the company in California. Very slow going and tough field work led initially to several discouraging dry holes. The 1947 Leduc discovery set off a major exploratory play in the area and Rio staffed up. In the early 1950's the company changed their name to Canadian Superior and successfully participated in several discoveries in the Devonian Reef play. In the 1960's, Canadian Superior went world-wide, doing business from the U.S. to Australia with more success. Howard Keck was the driving exploratory force in the big growth years and took the responsibility. In December 1985, Mobil Oil absorbed all of Superior's holdings, foreign and domestic. This fine tribute, was prepared by Jack Porter, who joined the company in 1948 and had been its head of geology for many years. The hardbound first edition is common.

**C108. (DELTA DRILLING) Presley, James, NEVER IN DOUBT: A HISTORY OF DELTA DRILLING COMPANY, Houston, TX, Gulf Publishing, (1981), 543pp.**

This book might also have been titled The Biography of Joe (Guiseppe) Zeppa. Zeppa was one of the five Delta Drilling Company founders, but was the most dynamic and ran the company with little opposition from 1931 until his death in 1975. Born in 1893 at Fubine, Northern Italy, he came to America at age twelve. He studied accounting and finance at NYU at night school while working for a brokerage firm during the day until WW I which he spent in Europe. He returned in 1919 but was not happy back on Wall Street and took an opportunity to join an army buddy at the El Dorado, Arkansas oil boom in 1921. His company, El Dorado Petroleum discovered several gushers but the price of crude was thirty five cents per barrel and the company went bankrupt. Zeppa then joined Natural Gas and Petroleum Company which was quickly absorbed by Cities Service. Cities moved Zeppa to Shreveport in 1928. He then made a big move to become vice president of sales for McMillan Petroleum in California, returning to Texas in 1931 right in the middle of the biggest oil boom in history, the East Texas Field (discovered September 1930). He wanted to form a drilling company to get in the frenzy and convinced Bob Stacy to take one third and Sam Dorfman with his two partners, Sam Sklar and Sam Gold to take one third, funding the purchase of two drilling rigs. Delta was born in November 1931. They used all of the funds to buy two junker drilling rigs and Dorfman put up another $3,000 to run the operation. Zeppa was a hard-nosed accountant and not surprisingly ran Delta with a Rockefeller type touch. At the time of this writing, Delta was under the direction of Joe's son, Keating, and the company had become the largest privately held land based drilling company in the world. They had also established considerable oil and gas production throughout Texas, Louisiana and Arkansas. One of the top oil company histories. The hardbound first edition is common.

**C109. (SOUTH IMPROVEMENT COMPANY) Petroleum Producers' Union, A HISTORY OF THE RISE AND FALL OF THE SOUTH IMPROVEMENT COMPANY, Oil City, PA, The Union, 1873, 126pp.**

This book was not available for review by Neff, but is regarded as important. Quoting from Swanson: "Testimony presented before subcommittee of House of Representatives Committee on Commerce

(1872), to investigate railroad monopolies and combinations. Important document in literature of producers' fight with Standard Oil". Cited in Swanson on page 164. The first edition is very scarce.

**C110. (NEWMONT) Ramsey, Robert H., MEN AND MINES OF NEWMONT: A FIFTY YEAR HISTORY, NY, Octagon Books, 1973, 334pp.**

In 1921 Colonel William Boyce Thompson formed Newmont Corporation in New York as a vehicle for his many promotions. His modus operandi was to locate potentially lucrative mining properties and form companies to develop and manage them. He would then promote the companies to the public as new stock offerings with the resultant income rolled over into greater and greater ventures. It worked. By 1973 Newmont Mining had a billion dollars worth of assets. Acquisitions, mergers, exploration and development made Newmont a major producer of several key metals and minerals, particularly copper. Newmont's oil interest began in 1925 where speculative leases on the Gulf Coast produced and were later traded for stock in Conoco. Newmont subsidiary, Alder Oil Company, participated in oil exploration ventures with Conoco and Amerada in Texas, Louisiana and the Rockies. In 1944 another subsidiary, Newmont Oil Company was formed to speculate in oil leases and royalties. Under the guidance of Phil Kraft the company grew into active oil exploration in the gulf coast, Permian Basin and some foreign ventures. Always the step-child of Newmont Mining, the oil divisions were producing steady profit at this writing. The hardbound first edition is common.

**C111. (P.G. & E.) Redinger, David H., THE STORY OF BIG CREEK, Los Angeles, Los Angelus Press, (1940), xvi, 182pp.**

Originally, this story of the gigantic Big Creek hydroelectric project was prepared as a report for company files. The project was of such magnitude and importance that publication was decided to be in the public interest. Pacific Light & Power group began the project in 1911 with the major design being to create several hydroelectric plants down stream of Huntington Lake and encompassing countless other lakes, dams and power stations. Southern California Edison absorbed PL&E in 1917 and continued the project to its completion in 1929. The 1.5 billion kilowatt capacity facility and complex serves Southern California and the water serves the San Joaquin irrigation systems. The hardbound first edition is becoming scarce.

**C112. (ETHYL) Robert, Joseph C., ETHYL: A HISTORY OF THE CORPORATION AND THE PEOPLE WHO MADE IT, Charlottesville, University Press of Virginia, (1983), 448pp.**
Albemarle Paper Company was formed by five young businessmen at Richmond Virginia in a very competitive market in 1887. It grew and expanded with specialty paper products, exporting a fair percentage to the world. A new regime came in during 1907, headed by the thirty-two year old Watkins Ellerman, who saw to the company's expansion and great success until the late 1950's. Stiff competition and petrochemical products which began replacing their main-stay items forced the company to diversify. Ethyl Gasoline Corporation was chartered in 1924 through the research efforts of Charles Kettering and Thomas Midgley. They developed tetraethyl lead as the practical effective anti-knock gasoline additive. Kettering had been working on anti-knock fuel first for diesel and then attempted it for gasoline engines. The first tetraethyl was actually sold in 1923. Dupont later marketed the tetraethyl from the Ethyl patent which protected them from competition for thirty-five years. Ethyl was dyed red to distinguish it from regular gasoline as seen in the glass pump containers of the day. Ethyl was synonymous with premium fuel until the 1960's. Albemarle and Ethyl began merger talks in 1961 and by 1962 Albemarle had borrowed 200 million dollars and taken over the majority of Ethyl. Albemarle was the surviving company although Ethyl was the surviving name. Ethyl diversified and formed a successful oil and gas division which established significant oil and natural gas reserves in the North Sea, Canada and U.S. gulf coast. Ethyl got out of the paper business in the 1960's and today has a variety of interests from fuels to aluminum. Appendices and notes are included. The hardbound first edition is common.

**C113. (TEXACO) Rodgers, W.S.S., TEXACO'S HERITAGE, also includes THE WAY AHEAD BY Harry T. Klein, NY, (Texaco), 1952, 14pp.**
Rodgers was chairman of the Board of Texaco in the 1950's. Quoting from the forward: "On April 7, 1952, the Texas Company's Golden Anniversary, some 7,000 employees, all of whom had served Texaco for at least 25 of its 50 years, sat down at 29 Old- Timers' dinners throughout

the country". Rodgers speaks of the company from before Spindletop to the present day. He stresses the importance of worldwide operations and Texaco's product "Fire Chief Gasoline" and "Havoline Motor Oil". He cites Texaco as an example of what can be accomplished within the competitive free enterprise system. The last two pages are Mr. Klein's talk where he stresses that the company is a living thing with character (most observers will agree to that). He believes that Texaco's success was due largely to quality products and good service. The softbound first edition is common.

**C114. (ROYAL DUTCH) Royal Dutch-Shell Group, A PETROLEUM HANDBOOK, London, Asiatic Petroleum Company, 1933, 362pp.**

For the employee and prospective employee, the book first gives an overview of the world oil industry and brief history of Royal Dutch-Shell. The history is followed by a detailed technical section on production which includes prospecting as well as drilling and production. Products, transportation and marketing are also treated. Fundamentals presented here are still valid although technology has moved steadily forward. Numerous photos, diagrams and charts are shown along with a useful glossary of terms. The second 1938 edition by Asiatic Petroleum had 473pp. and was much improved and updated as edited by James Kewley. The third 1948 edition by Shell Petroleum is titled The Petroleum Handbook and has 658pp. Fourth edition by Shell is larger format 7x9" with 678pp. Fifth 1966 edition is larger still 8.5x11" format with 318pp. This one is largely in color with useful bibliographies for each section. Cited in Swanson on page 180. The hardbound first edition is scarce.

**C115. (ROYAL DUTCH) Royal Dutch, THE ROYAL DUTCH PETROLEUM COMPANY 1890-1950, The Hague, The Company, 1950, 204pp.**

Produced as a commemorative "Diamond Jubilee Book" this is a less technical version of the 1933 handbook and makes far easier reading. The first chapter is glimpses of the previous sixty years and describes the Royal Dutch origins in 1890 to its major expansion and production in the Netherlands East Indies under Deterding. Deterding took over in 1901 and retired in 1936 after directing the company to world oil leadership. The 1907 alliance with Shell and subsequent expansion to confront and compete with Standard Oil and others is told in good detail. A major part of this book in each chapter is spent pointing out how important this company is to The Netherlands. Half of the company was owned by Dutch interests in 1950. Other chapters are written on Royal Dutch's exploration, production, processing, chemical, transport, refining and marketing. All describe the company's historic position and philosophy in those respective areas. The last chapter is titled Resurgence. In 1950 they had 250,000 employees worldwide and made a case that they are well provided for by the company health and retirement plans. The remote oil camps facilities and conditions are described as well as some large city company clubs. Sounds great. Includes many maps and photos. Cited in Swanson on page 172. The hardbound first edition is scarce.

**C116. (VARIOUS) Sampson, Anthony, THE SEVEN SISTERS: THE GREAT OIL COMPANIES AND THE WORLD THEY MADE, NY, The Viking Press, (1975), xv, 334pp.**
There are lots of big oil companies but none have been more influential in the world of oil than the Seven Sisters. Exxon, Gulf, Texaco, Mobil, Chevron, British Petroleum and Shell (and their predecessors) have dominated every phase of the oil industry since shortly after its beginnings in 1859. An important aspect of this is the fact that they essentially controlled the worldwide price of crude. They were also able to control many political events through this power. The book was written after the 1973 Arab oil embargo and emergence of the power of OPEC. The author assimilated much information to make his story of how the companies were formed, how they developed and how they combined to divide up the world markets. The first half of the book is pre-OPEC history from the Rockefeller influence to the early 1970's. The last half describes OPEC's emergence, nationalization of oil company interests, the embargo and the standoff as this book was published. Much of the material presented here comes from the Congressional Investigation by Senator Church's subcommittee for the Foreign Relations Committee. The hardbound first edition is common.

**C117. (GLADYS CITY) Sanders, Christine Moor, CAPTAIN GEORGE WASHINGTON O'BRIEN AND THE GLADYS CITY COMPANY AT SPINDLETOP, Noplace, The Author, (1992), 44pp.**
The author's short book retells the Spindletop story of Patillo Higgins, Captain Lucas and the many others, however she adds much unpublished information on the part her family played in the saga. Gladys City was

O'Brien's company and they owned the land on which the Gladys City boomtown and the great field was developed. Sanders gives much credit to O'Brien for acquiring the land and forming the company (in 1892) and actively pursuing oil on the dome. He had also acquired land on the Sour Lake dome. The softbound first edition is common.

**C118. (SCHLUMBERGER) Schlumberger, Anne Gruner, THE SCHLUMBERGER ADVENTURE: TWO BROTHERS WHO PIONEERED IN PETROLEUM TECHNOLOGY, NY, Arco Publishing, Inc., (1982), xv, 152pp.**
The author is the daughter of Conrad Schlumberger. Conrad and his brother Marcel invented electrical well logging. They went on to develop diverse techniques with extensive field research, completely dominating the industry they created. Anne Schlumberger defers to the book by Allard and Martin, titled "Schlumberger-The History of a Technique" as the definitive technical book on the company. She presents the human side of the participants as she grew up with her father experimenting and developing a potentiometer which would become a critical petroleum exploration tool. Conrad was born in France in 1887 and later moved to Paris. He died in 1936, just before his company began to make big profits. He was the physicist and idea man. Marcel was also born in France and by 1900 was living in Paris. He died in 1953 after seeing the company dominate well logging on a worldwide scale. He was an engineer, a practical minded innovator and man of action. Together the two made a perfect team. With the help of Henri Doll (Anne Schlumberger's first husband), head of research and engineering departments, the company grew to what is today a multi-billion dollar, multi-national company with 85,000 employees. The "First in the field, first in research" motto has been religiously adhered to and it has paid off. Historic photos from the author's private collection are shown. Her first hand stories and anecdotes of the early days are well told and leave no doubt as to her personal opinions. The hardbound first edition is common.

**C119. (ASHLAND) Scott, Otto J., THE EXCEPTION: THE STORY OF ASHLAND OIL & REFINING COMPANY, NY, McGraw-Hill Book Company, (1968), 450pp.**
The title comes from Ashland's observation that over the years their success came from a different direction than that of most major oil corporations. Ashland was the exception. Most majors found and

developed oil then proceeded to become integrated, vertically. Ashland began as a refiner and still relies primarily on that, although they have slowly moved into transportation, then marketing and lastly oil producing. The book is a comprehensive company history as opposed to BLAZER AND ASHLAND OIL, by Massie, which concentrates almost entirely on his individual efforts and management style. This is a good early history from 1910 and J. Fred Miles oil career as he formed Swiss Oil, bought Quaker Oil and drilled numerous wells in Alabama and Kentucky. Paul Blazer's career is also followed as the two got together, Paul being the refinery expert and Miles the drilling and producing expert. In 1924 both men were having a financially difficult time, but Blazer was convinced to join Miles as general manager with ten percent ownership in the new Ashland Refining Company. Blazer managed the company successfully as they merged and remerged and acquired companies to build what eventually became the giant Ashland. Contemporary world and U.S. history is well incorporated into this company history. The hardbound first edition is common.

**C120. (DOW) Seward, William B., EAST FROM BRAZOSPORT, Midland, Michigan, Dow Chemical, (1974), 191pp.**

The headquarters for Dow Chemical is in Midland, Michigan, however their largest research and production facility is in Texas. Brazosport is an area south of Houston where the Dow chemical complex is surrounded by several small towns. It is a large manufacturing and distributing center for Dow both for foreign and domestic markets. The idea of this book was to take a Dow product from its raw material and follow it through to the ultimate consumer. The concept allows great freedom to the author, who traveled to various Dow facilities around the world and interviewed some of the top management people as well as some of the hourly laborers. Company people with enthusiasm and ideas were sought out and the result is a very upbeat story. Well done company propaganda for employees and outsiders alike with lots of photos. The hardbound first edition is becoming scarce.

**C121. (TEXACO) Shannon, James, TEXACO AND THE $10 BILLION JURY, Englewood Cliffs, NJ, Prentice Hall (1988), xxiii, 545pp.**

Shannon apparently was the driving force on the jury of the Pennzoil vs. Texaco trial. His book was the last of three major works to come out of the trial and he claims that the reason he wrote it was to counter the numerous editorial attacks on him individually as well as on the jury as a whole. They have been described as incompetent, ignorant, unsophisticated and confused. Essentially a memoir, the book takes us from Shannon receiving his jury duty notice, through the five month trial and to the final judgment. Much of the book is actual court testimony along with jury instruction and deliberation. Shannon's wife had lost her job with Conoco after a DuPont takeover and that fact was used to try and discredit Shannon's ability to serve as a juror. Texaco had called it a "jury run amok". This is a well presented account of Mr. Shannon's five month ordeal but does not support a judgment of the magnitude assessed. Further, Gordon Getty, Hays, and Peterson appear to be the real criminals in that they were double-dealers, not using normal business ethics. For those who believe that the end justifies the means, this jury award land-mark case did put an end to the rampant, oil company raiding. Good reading. The hardbound first edition is common.

**C122. (SHELL) Shell Union Oil Corporation, SHELL...SOLDIER AND CIVILIAN, Noplace, The Company, 1945, 78pp unnumbered.**
An excellent company brochure describing the various contributions made by Shell and its allied companies towards the WW II effort. From the foreword: "When a coast-to-coast organization like Shell makes this kind of decision, a tremendous force is unleashed. Refineries, pipelines, trucks and terminals, research and production facilities swing from peacetime operation to single purposed wartime manufacture. When this decision is backed by twenty-six thousand war working employees, the pathway to results is paved with more than good intentions". Striking illustrations are full color drawings of various Shell operations. Cited in Swanson on page 158. The softbound first edition is scarce.

**C123. (TEXASGULF) Shulman, Morton. THE BILLION DOLLAR WINDFALL., NY, William Morrow and Company, Inc., (1970), xiv, 239pp.**
This book is a review of the Texasgulf insider trading scandal of the late 1960's which resulted from their discovery of major new two billion dollar ore deposit in Ontario, Canada (Copper, Zinc, Silver, et al). The author jumped the gun somewhat as there had been several conflicting rulings from various court levels for and against Texasgulf and individuals, but no settlement. Shulman seems fair in his relation of

events and history although some of his dialogue and conversations must have been contrived. Much is from the court records although no other source notes are provided. This is a story of a brilliant discovery and subsequent unethical maneuvers gone out of control. Thirteen Texas Gulf employees were indicted in this landmark case brought by the S.E.C. Shulman's account of the speculative events which followed the discovery make it appear to be more unlawful than just insider trading activity. Motivation for the entire episode appears to be greed. The hardbound first edition is common.

**C124. (AMPOL) Simpson, Colin. SHOW ME A MOUNTAIN: THE RISE OF AN AUSTRALIAN COMPANY, AMPOL, Sydney, Australia, Angus and Robertson, (1961). 253pp.**

Simpson relates the twenty-five year history of Ampol in a personal manner with a close look at the main characters. He uses much dialogue and describes important meetings and significant events with good color. The book is inspiring to one who would want to create a business from nothing and build it into a giant corporation. Founders O'Callaghan and Walkley got together in 1931 when O'Callaghan hired Walkley to work at the Motor Union (Like the American Automobile Association). Walkley was a brilliant, established accountant and O'Callaghan was a retired, successful auto dealer. Walkley's philosophy from his teen years had been "Show me a mountain and I will show you a way to get over it". The major oil companies were abusing Australians and by 1936 the two men formed Ampol to compete with them by importing and marketing oil and allied products. At age 39, Walkley ramrodded the company to overcome one political or technical obstacle after another. They became a major factor in Australia as they diversified into many related fields. In 1952, the #1 Rough Range wildcat well, a joint venture between Ampol and Standard of California known as WAPET, became the 250th well drilled in Australia. All previous efforts had been dry or marginally productive. The wildcat came in flowing 25 barrels per hour at 3,600 feet, establishing the first commercial oil production in Australia and commencing years of moderately successful exploratory drilling in western Australia. By 1961, WAPET had developed a major field there but discouragingly drilled another fifty dry wildcats. Ampol, however continued to grow into an impressive conglomerate. Well written, interesting history. The hardbound first edition is common.

**C125. (PENN SQUARE BANK) Singer, Mark, FUNNY MONEY, NY, Alfred A. Knopf, 1985, xi, 221pp.**

Singer was a writer for the New Yorker and followed the Penn Square Bank failure from the summer of 1982 to 1985. He gives a good rendition of the events leading up to the catastrophic collapse in Oklahoma City. His sometimes humorous description of the perpetrators and victims make them believable and at the same time pathetic. Bill Jennings, the chairman of a small bank and his right-hand man Bill Patterson were for a few short years absolutely out of control. Taking advantage of the late 1970's oil boom hysteria, they first loaned thousands, then hundreds of thousands, then millions of dollars to independent oil operators, promoters and just plain stupid people wanting to get in the oil business....generally without collateral. Supposedly the loans were for low-risk drilling, leasing and well equipment, however, that was usually not the case. Penn Square then passed these loans on to heavier, greedy banks such as Continental Illinois and Seattle First. By the summer of 1982, the steady rise in oil and natural gas prices had turned downward and was being felt as the loans came apart throwing Penn Square and their correspondent banks into what can be described as the usual result of a Ponzi scheme. No source notes in this short history although Singer's personal interviews with the main characters make for very entertaining and colorful reading. The hardbound first edition is common.

**C126. (LUBRIZOL) Smalheer, Calvin V., THE STORY OF LUBRIZOL, Cleveland, OH, The Company, (1972), 88pp.**

This is a success story of an important, low-profile oil industry company. Gasoline and lubricant additives are the primary products of Lubrizol, and they are specialists. Their extensive research has allowed them to spearhead the market on a worldwide basis. Since the company markets only to petroleum and automotive customers, their name was not generally recognized until they went public in 1960. The company was founded by the three sons of Dr. Albert W. Smith. Smith had worked side by side with Herbert Dow for many years. The sons followed in father's footsteps and before long had developed and patented a graphite oil lubricant. They formed Graphite Products Company in 1928. Barely surviving the Depression, they were doing fine by 1933 and changed the name to Lubrizol. The hardbound first edition is becoming scarce.

**C127. (SHELL) Smith, P.G.A., THE SHELL THAT HIT GERMANY HARDEST, London, Shell Marketing Co., 1919, 77pp.**
This book was not available for review by Neff, but it is regarded as important. Quoting from Swanson: "Achievements of Royal Dutch-Shell Group of companies during World War I". Cited in Swanson on page 158. The first edition is very scarce.

**C128. (SINCLAIR) Spence, Hartzell (editor), A GREAT NAME IN OIL: SINCLAIR THROUGH FIFTY YEARS, NY, McGraw-Hill Book Company, 1966, 104pp.**
This is a commemorative volume describing fifty years of Sinclair Oil as a major factor in the oil industry. In 1966, Sinclair was tenth in size of U.S. oil companies. As a young man, Harry Sinclair lost the family owned drugstore he had inherited through some misdealings and began work in the oil fields. He had an uncanny ability to find oil and attract serious investors. By 1907 he was a millionaire by virtue of several significant oil finds, culminating at Glenn Pool, south of Tulsa. By 1916, Sinclair was the largest independent oilman in the Mid-continent area. He started his company in 1916 at Independence, Kansas with the idea of becoming a fully integrated oil company. Subsequent expansion was made over the U.S. and into foreign operations. One of the best assemblages of old photos from Sinclair's first major discovery for the company at Garber Field to the well-designed promotion of the company logo, brontosaurus. Excellent book. The hardbound first edition is common.

**C129. (OHIO-MARATHON) Spence, Hartzell, PORTRAIT IN OIL: HOW THE OHIO OIL COMPANY GREW TO BECOME MARATHON, NY, McGraw-Hill book Company, Inc., 1962, 357pp.**
Hartzell Spence started out to write a simple company history for Marathon, primarily celebrating their first seventy-five years along with the name change from The Ohio Oil Company to Marathon. What developed was a major oil company propaganda tool funded by the company directors. The company history is comprehensive and interesting, the rest is propaganda. The Ohio Oil Company was formed by Henry Ernst and four partners who pooled their prospects in 1887 at Findlay, Ohio. Numerous other participants bought $1,000,000 worth of stock. The company design was to fight Standard Oil which was depressing prices all over the country by monopolizing the refining and

marketing of crude and its products. Ohio grew by leaps and bounds, confounding Standard. They succeeded largely through shrewd management, aided by timing and luck. Standard bought them out in 1889. James Donnell who had been in the oil business since age sixteen was a wealthy Standard retainer, drilling for them in Ohio at the time of the takeover. He became a company director and field manager. Ohio expanded and shortly had 10,000 producing wells with much undeveloped acreage to drill. Ohio was cut free from Standard by the government order in 1911. Donnell was made president of the biggest crude producer out of the entire Standard group, but they remained short of transportation, pipelines, refineries and marketing outlets. Donnell management along with more acquisitions and mergers took Ohio to integrated world-wide major company status. At this writing a third generation Donnell is in charge. Great stories on the Yates Field of West Texas discovery along with others. The hardbound first edition is common.

**C130. (SINCLAIR) Spencer, P.C., OIL - AND INDEPENDENCE! THE STORY OF THE SINCLAIR OIL CORPORATION , Noplace, American Newcomen Society, 1957, 28pp.**

This pamphlet was not available for review by Neff, but it is regarded as important. Quoting from Swanson: "Activities and accomplishments of Harry F. Sinclair, growth of Sinclair Oil Corporation, leasing of Teapot Dome, resulting developments. Delivered as Newcomen Society address". Cited in Swanson on page 172. The softbound first edition is scarce.

**C131. (STANDARD OIL) Standard Oil Company (New Jersey) , THE LAMP 75TH ANNIVERSARY OF JERSEY STANDARD , NY, The Company, (1957), 80pp.**

Over half of this seventy-fifth anniversary of Jersey Standard magazine is dedicated to history of the company and the oil industry. The first five pages are a well organized chronology of major company and historically pertinent events. The history section opens with The Age of Kerosene 1882-1911. It is a beautifully illustrated look at Jersey and the industry. Kerosene was king as its lighting allowed families to extend their days with ease. Airplanes and cars were becoming popular and ushered in the second section, The Age of Gasoline 1911-1939, which coincided with Jersey's rebuilding period after the Supreme Court dissolution decree. Oil was a weapon helping to win WW I and Standard found oil around the world as gasoline demand skyrocketed. Part three, The Age of Expanding Energy 1939-1957 has much on the role of oil in WW II with the complex mobilization and logistical problems overcome there. Several other well illustrated sections describe company operations with emphasis on their global influence. In the back, large fold-out maps show location of Jersey's facilities and operations around the world. Intriguing sidelight, this is published the year before Castro took control of Cuba and the maps show an operating refinery and active oil exploration work taking place there. Not mentioned in Swanson. The hardbound first edition is scarce.

**C132. (STANDARD OIL) Standard Oil Company (New Jersey);, THIS IS STANDARD OIL COMPANY (NEW JERSEY), NY, The Company, (1967), 32pp.**

Very compact and thorough company booklet. Fold-out color maps at front and back show location of all Standard facilities and operations of consequence around the world. Oil and gas production, pipelines, chemical plants, refineries and marketing areas are shown. The booklet answers questions most frequently asked about Jersey's history and its present organization. A brief, concise section on history of Standard shows Rockefeller organizing the new industry which was in chaos. He and his partners did a little too much organizing, according to the Supreme Court, and in 1911, Standard was broken into thirty-three separate entities, many of which still exist today. Jersey had to rebuild after the dissolution order. They had no oil production of their own so they began acquiring. Two major purchases were Humble Oil of the U.S. in 1919 and Creole of Venezuela in 1928. They kept control of Imperial of Canada which became the largest producer in that country. Esso is Jersey's worldwide marketing subsidiary and conducts operations where their other subsidiaries do not. Favorable propaganda. The hardbound first edition is common.

**C133. (SIERRA PACIFIC) Stedham, Austin W., SIERRA PACIFIC POWER COMPANY, Philadelphia, The Newcomen Society, 1989, 27pp.**

This Newcomen pamphlet deals with the history of Sierra Pacific in an address delivered to the society in Reno by the company CEO. One of the earliest companies (1860's) later to evolve into Sierra Pacific was Virginia City Gas Company which provided lighting to the miners and boomtown there. A great number of service companies were providing Nevada with gas and water, later electricity. The company family tree is very impressive and is presented as a chart consisting of almost 100 companies in Nevada and California. Two of these are oil and gas exploration and development companies: Great Basin Energy and Sierra Energy Company. The softbound first edition is common.

**C134. (UNOCAL) Stegemeier, Richard J., A CENTURY OF SPIRIT: THE HISTORY OF UNOCAL, Philadelphia, The Newcomen Society, 1990, 23pp.**

This Newcomen pamphlet covers an address dealing with the history of Unocal. It was delivered at a society meeting in Los Angeles by company CEO, Stegemeier. He is an engineer who started with the company in 1951 and became CEO in 1988. The original Union Oil was founded in 1890 by Lyman Stewart, Thomas Bard and Wallace Hardison at San Paula, just north of L.A. They had many ups and downs but were successful overall. Bard was tossed out in 1899 in a proxy fight. Union oil gained fame for bringing in the world's greatest well, the Lakeview Gusher, southwest of Bakersfield in 1910. The well flowed an estimated 125,000 barrels of oil per day. The company diversified into the giant integrated major of today and operates worldwide. They have done outstanding pioneering work in geothermal energy and in oil shale technology. The softbound first edition is common.

**C135. (SUN) Sun Oil Company, CENTENNIAL CELEBRATION: THE STORY OF SUN COMPANY, Noplace, The Company, (1986), 96pp.**

A brief history of the company from its beginning as Joseph Newton Pew and partner Edward O. Emerson formed Sun Company in Ohio in 1886 with the intention of exploiting the Ohio oil fields. They were already involved in producing and marketing natural gas in Pennsylvania. Pew directed his oil experienced nephew, Robert, to acquire leases in the Findlay Township, near Lima. This small start led to drilling, producing, pipelining, refining, storage and tank cars within a few years. Consolidating, Sun Oil Company was formed in 1890 to bring the diverse interests together. Sun was just another small independent until the Spindletop boom. Edgar Pew hit Beaumont shortly after the gusher came in and proceeded to involve Sun in drilling, producing and storage within a short time, forming the basis for the major, fully integrated oil company that exists today. Along with historic photos, numerous profiles of top executives are included. Snapshots of the times are interwoven with profiles of Joseph Pew Sr., J. Howard Pew, J. Howard Pew, Jr., Robert G. Dunlap, H. Robert Sharbaugh, Theodore A. Burtis and Robert McClements, Jr. Many Sun advertising logos and gimmicks are shown as they changed throughout the years. The merger with Sunray DX in 1968 and acquisition of Texas Pacific's oil and gas properties in 1980 are highlighted. The hardbound first edition is becoming scarce.

**C136. (STANDARD OIL) Supreme Court of the United States, THE STANDARD OIL COMPANY OF NEW JERSEY ET AL., APPELLANTS vs. THE UNITED STATES, Washington, Government Printing Office, 1911, 43pp.**
This is the final judicial opinion which broke up Standard Oil. Justice White rendered the famous, May 15, 1911 opinion for the court as he upheld a lower court ruling which found Standard in violation of the Antitrust Act statutes and being in restraint of trade. Action recommended for solving the problem: 1) To forbid the doing in the future of acts like those which we have found to have been done in the past which would be violative of the statute; 2) the exertion of such measure of relief as will effectually dissolve the combination found to exist in violation of the statute, and thus neutralize the extension and continually operating force which the possession of the power unlawfully obtained has brought and will continue to bring about". Historic document. Not mentioned in Swanson. The softbound first edition is scarce.

**C137. (GULF) Swensrud, Sidney A., GULF OIL: THE FIRST FIFTY YEARS, NY, Newcomen Society of N.A., 1951, 31pp.**
Swensrud was president of Gulf Oil at this writing. He had a science degree from Minnesota and an MBA from Harvard and worked his way up in Standard Oil of Ohio to Executive VP in 1946. He left to become president of Gulf in 1947. He first tells a general history of the oil industry then begins a story of the Mellon family involvement in it. The Mellons had some oil success in their own right and owned the major petroleum bank. They financed Guffey & Galey in their Spindletop, Texas venture and when it hit big, the development and transportation costs were so enormous Andrew Mellon bought out the partners and formed Gulf Oil. In 1901, the great flow of oil from Spindletop and its independent producers put a real dent in the Standard oil monopoly. Gulf

built refineries and pipelines and sought oil from Oklahoma and Texas. They merged and expanded and by 1950 were a solid top five major oil company. Cited in Swanson on page 173. The softbound first edition is becoming scarce.

**C138. (STANDARD OIL) Tarbell, Ida M., THE HISTORY OF THE STANDARD OIL COMPANY, NY, McClure Phillips, 1904, 2 Volumes. xxx, 815pp.**

Tarbell's series of articles about Standard Oil Company began appearing in McClure's Magazine, of which she was editor, in 1902 and continued for the next seventeen months. They are reedited and produced in this classic two volume work. Of this original two volume 815 page book, only 554 pages are text. The remainder is back up material, being appendices with testimony, contracts, charters and tables for documentation. Tarbell opens with the history of the oil regions of northwest Pennsylvania and the many problems encountered as the operators tried to transport crude oil and refined product (kerosene) from the oil fields to the refiners or from the refiners to the market. First mule teams, then boats and barges and then the railroad. Rail transportation dominated oil movement through the 1860's and 70's and was not threatened by pipe lines until the 1880's. Tarbell gives an early biography of Rockefeller and tells of his start in the oil industry as he invested in a refinery in Cleveland and plunged his entire fortune and effort into that new industry by 1865. She does not complain much about Rockefeller's methods of acquiring his competitors, but she does complain of his methods of combining the refineries into South Improvement Company and negotiating the infamous railroad rebates and drawbacks. Tarbell is not against monopoly as much as she is against the illegal and unfair methods used to create them. She follows Rockefeller and Standard through dissolution of the Standard Oil Trust which changed very little. She gives Rockefeller credit for brilliant organization and timing while calling him miserly and cruel at the same time. This book is a landmark work in two areas, first Tarbell became recognized as the first effective muckraker, setting the stage for years of more and better investigative journalists covering complex and taboo subjects. Secondly, she made the public and government more aware of the magnitude of Standard and the influence one old man had on their lives and the country. She went a long way towards causing the eventual breakup of Standard. Reprinted by Macmillan in 1925. Reprinted by Macmillan (2 volumes in 1) in 1933 and

1937. Reprinted by Smith in 1950. Reprinted in 1969 (Abridged) in softbound by Heinemann. Cited in Swanson on page 166; Hardwicke page 40. The hardbound first edition is very scarce. Other hardbound editions are scarce.

**C139. (UNION OIL) Taylor, Frank J. and Earl M. Welty, BLACK BONANZA: HOW AN OIL HUNT GREW INTO THE UNION OIL COMPANY OF CALIFORNIA, NY, McGraw-Hill Book Company, Inc., (1950), 279pp.**

Taylor and Welty provide a historically accurate and entertaining story of the beginning of the oil industry, the early participation of Lyman Stewart in it, along with his formation and subsequent growth of Union Oil of California. Only a few oil company histories (besides Standard) had been published prior to this book, however the 1950's saw a plethora of them on the market. This is a large, comprehensive book with numerous black and white photos of the men, events and facilities which make up the industry and the company. It was most impressive for the day. Stewart was born near Titusville in 1840. Immediately after the Drake discovery there, young Stewart caught oil fever and made his first of many plunges into the business. He lost his life savings, $125.00 on an oil lease which later produced for others. Two years later he had another near miss, actually finding oil but losing the lease because the price collapsed and he could not produce the well. Most of the other operators were devastated also. After fighting in the Civil War, he found the Pennsylvania oil business to be booming with expanding markets. Lyman and his older brother became prudent oil operators by playing the expo sure game and taking small participations in several prospects thus reducing risk. He and Harbison began patenting oil field inventions. They got in the Bradford Field area early, but the following glut of oil production plus the Rockefeller monopoly collapsed the market one more time. Oil sold for eight cents a barrel in 1881. Stewart struck out for California with a good deal of money and a lot of oil knowledge in 1882. California's oil industry was just beginning at Pico Canyon. Stewart and partner Harbison acquired leases from friend Scofield at California Star Oil and drilled six straight expensive dry holes. Both out of money, they borrowed funds to drill #7 at Pico, called Star #1. At 1,620 feet they hit oil. For several years they seemed to be always on the edge of disaster.

They merged their several companies with the conservative Tom Bard in 1890 to form Union Oil. At that time, California oil industry went wild and Union was right in the middle of it. Union success in California is given in a chronology at the end of the book. Subsequent to publication of this first Union Oil Company history a second edition was issued in 1958, with Welty as the lead editor. It was retitled THE BLACK BONANZA and presented in a smaller, more compact format, viii, 255pp. It included fewer black and white photos, and added more text particularly about new discoveries in Louisiana and Oklahoma. The third edition has not been seen. A fourth edition in 1966 still shows Welty and Taylor as editors. It concentrates on the merger between Pure Oil and Union and is presented in 8.5x11" format with much color, becoming a handsome coffee table book. Title is THE 76 BONANZA with 351pp. A fifth edition titled THE SIGN OF THE 76 was published in 1976 with 424pp, much as the fourth, although no editor is noted. Cited in Swanson on page 173. The hardbound first edition is becoming scarce.

**C140. (OIL WELL SUPPLY) The Oil & Gas Journal, THE "OIL WELL STORY 90th ANNIVERSARY 1862-1952, Tulsa, The Journal, 1952, 271pp.**

Pages 88 through 177 of this special issue are dedicated to Oilwell Supply Company on their 90th anniversary. Excellent short history of the company along with historic photos and numerous ads. Quoting from Swanson: "History of firm, beginning 1862 at Oil City, Pa., when John Eaton started oil well supply business and began five years later to produce equipment specifically for oil industry, made according to specifications and standardized. Company growth in conjunction with oil industry; now a division of U.S. Steel Corporation". By the time Oil Well published their first supply catalog for the industry in 1884, they had already established themselves as the premier supplier, not just to the Oil Regions of Pennsylvania where their two manufacturing plants were, but around the world. From the Oil City and Bradford plants, Oil Well could supply any normal drilling or producing equipment, but was also well known for being able to fabricate anything for special use. The company was owned by the Eaton, Cole & Burnham Company of New York. Cited in Swanson on page 172. The softbound first edition is scarce.

**C141 (GULF) Thompson, Craig, SINCE SPINDLETOP: A HUMAN STORY OF GULF'S FIRST HALF-CENTURY, Pittsburgh, Gulf Oil, (1951), 110pp.**

This is a fine picture and history book prepared by the company. The Spindletop, Texas discovery and Mellon family involvement in formation of Guffey Oil which later became Gulf is covered in good detail. Aggressive expansion into refining and transportation as well as exploration in Texas and Oklahoma lead to major integrated company status. Good stories on early lease bidding at the Osage Indian lease sales between Phillips Petroleum and the Gulf subsidiary, Gypsy in 1922. Astronomical prices were paid on speculation which really paid off later. Early drilling in water in Louisiana was also an important part of history for Gulf. Most of the book presents the Gulf Oil Company of 1950 and brags on the innovations that got them where they were. A bright future! A useful chronology and maps are included. The hardbound first edition is common.

**C142. (SIGNAL) Tompkins, Walter A., LITTLE GIANT OF SIGNAL HILL: AN ADVENTURE IN AMERICAN ENTERPRISE, Englewood Cliffs, NJ, Prentice-Hall, Inc., (1964), xiv, 258pp.**

Tompkins had fortunate timing in undertaking this company history as the founder, Sam Mosher and many of the original employees were still alive and available for interviews. Mosher grew up in Pasadena, California, graduated with a degree in agriculture at U.C. Berkeley in 1916. He became a farmer and barely scraped by with his seventeen heavily indebted acres of lemon and avocado trees. Shell Oil's #1 Alamitos discovery well at Signal Hill (1200 barrels per day at 3,114 feet) in 1921 was located only a few miles from his farm. He visited the hill six months later to find pandemonium with derricks everywhere over the small dome. Wells were producing, rigs were drilling, gas was being vented and flared. Shell had leased part, but not all of the dome. Speculators and promoters ran rampant, buying, selling and trading leases and royalty. Mosher was astounded by the total waste of the gasoline laden natural gas going to the atmosphere. He borrowed $4,000 from his Dad and along with Bob and Louis Berring, founded Signal Gasoline Company to retrieve gasoline from the vented natural gas. With the money and a government pamphlet on how to construct a gasoline plant, they rigged a make-shift plant. They negotiated a favorable deal with Shell for their gas and were in business. It worked so good they incorporated for $100,000 and used these funds to expand further. Mosher's father bought into the company and helped guide it for the next twenty years. Signal in 1964 was a worldwide producer-refiner with an excellent exploratory drilling record. This is a well told story. The hardbound first edition is common.

**C143. (BROKEN HILL) Trengove, A., WHAT'S GOOD FOR AUSTRALIA....! THE STORY OF BHP, Australia, Cassell Australian Limited, (1975), x, 263pp.**

BHP is Australia's largest company and author Trengove claims to have created an objective story of the company from its old mining days beginning in 1885 to the present day (1975). In a remote part of Australia called the Barrier Ranges, Broken Hill with its incredible supply of silver and lead, lay waiting to be discovered. A German named Rasp formed a syndicate of seven men to look for economic mineral deposits. They staked several claims to the hill and proceeded to sink shafts. Large quantities of lead were found but nothing else at first. Persistence by deeper shafting brought high quality silver to the surface and they formed BHP Ltd in 1885 to exploit what now appeared to be the greatest mineral deposit in the country. Rapid development was difficult but proceeded with one major problem after another. A town was built in the remote area to house and take care of the workers. Living conditions were very poor and the workers caused constant trouble over this. Many strikes and much violence took place over the next thirty-five years. Before 1900 BHP expanded into nearby iron deposits and successfully exploited them and began making steel, all the while fighting severe labor problems and a poor public image. By 1915 BHP moved into steel making as their major business. By the early 1960's they again changed emphasis and began exploring for oil. Commercial oil was not actually produced in Australia until the early 1954 and Australia in the early 1960's was a net 90% importer. To get a good start, they hired recently retired Lewis Weeks, who had been chief geologist for Standard Oil (New Jersey) as a consultant. Weeks went directly to the Bass Straits, a difficult drilling area, and had aeromagnetic and seismic surveys done. Several drillable prospects quickly developed. Needing some expertise, BHP negotiated a 50/50 partnership with Standard and proceeded to drill the prospects. The first well found a significant gas field. Subsequent deep drilling discovered four giant oil fields giving the combine 2.5 billion barrels and 7 trillion cubic feet of gas reserves. This success set a pattern for BHP and they began a program of expansion around the world by merger and acquisition and exploration, eventually becoming a significant world oil producer. Difficult to accomplish with such a late start. Much backup material in appendices and source notes. The hardbound first edition is becoming scarce.

**C144. (UNCLE SAM OIL COMPANY) Tucker, H.H., Jr., STANDARD AGAINST UNCLE SAM, MACHINERY OF INJUSTICE LUBRICATED BY STANDARD OIL, Kansas City, KS, The author, 1907, 600pp.**
This book was not available for review by Neff, but it is regarded as important. Quoting from Swanson: "Attacks Standard Oil Company; contains numerous photographs of early oil operations in Kansas, biographical sketches, etc. Author was general manager of Uncle Sam Oil Company". Cited in Swanson on page 166. The hardbound first edition is very scarce.

**C145. (THE McMAN) Tyson, Carl N. et al, THE McMAN: THE LIVES OF ROBERT M. McFARLIN AND JAMES A. CHAPMAN, Oklahoma, Oklahoma Heritage Association, (1977), xiii, 224pp.**
Born in Texas in 1866, McFarlin had a college education but left school to farm. In 1895, he moved to Holdenville, Oklahoma where he was very successful as a farmer and rancher. James A Chapman was born in Texas in 1881 and moved to Holdenville in 1901 to farm. The two got together and after the big 1905 oil discovery at Glenn Pool, south of Tulsa, they bought forty acres and decided to drill. Incredible success. McMan Oil Company was formed by them in 1912 to develop acreage in the Cushing field. They later exploited Healdton field and many others in Kansas and Texas as well. McMan Oil sold its Oklahoma and Texas production to Magnolia (Mobil) for thirty-nine million dollars in 1916. The remainder was sold to Standard of Indiana (Amoco) in 1930 for an additional twenty million dollars. Both men went back to farming, but added banking to their previous success. They continued in and out of the oil business very successfully until the early 1960's. These men are good examples of shrewd and successful exploitationists, as opposed to wildcatters. They evidently never considered taking unusual risk, but were on the scene as soon as someone else found the oil. Very good reading. The hardbound first edition is common.

**C146. (STANDARD OIL) Truth's Investigator, THE GREAT OIL OCTOPUS, London, T. Fisher Unwin, 1911, 255pp.**

The British were not pleased with John D. Rockefeller and his pervasive worldwide influence. Most of the meat of this book concerns U.S. happenings as they had been well advertised in the press and in other print media, although many Standard wrong- doings in Europe and Asia are also documented. Quoting from Swanson: "Based on articles covering many aspects of Standard Oil investigations which appeared in TRUTH, the British magazine". The color pictorial book cover depicts an octopus emerging from an oil barrel with its tentacles encircling the earth. Probably a widely circulated book in England as it was about the same time as the Supreme Court broke up Standard Oil. Cited in Swanson on page 166; Hardwicke page 41. The hardbound first edition is very scarce.

**C147. (WEST TEXAS GEOLOGICAL SOCIETY) Vertrees, Charles, HISTORY OF THE WEST TEXAS GEOLOGICAL SOCIETY: 1926-1969, Midland, TX: WTGS, (1973), 125pp.**

This comprehensive history of WTGS opens with early 1920's oil history in the Permian Basin which leads to the 1926 formation of the society. The presidents annual report along with committee reports for each year are presented through the 1967-68 term. Lots of old photos with several maps and references. Vertrees was a charter member and early president of the society, later awarded a Life Membership. Includes sources and references. The hardbound first edition is common.

**C148. (STANDARD OIL OF NEW JERSEY) Wall, Bennett H., GROWTH IN A CHANGING ENVIRONMENT: A HISTORY OF STANDARD OIL (NEW JERSEY) 1950-1972 AND EXXON CORPORATION 1972-1975, NY, McGraw-Hill Book Company, (1988), xlv, 1020pp.**

This is the fourth of four volumes of the history of Jersey Standard and covers the period from 1950 and the Korean War to 1975. The company reorganized several times during this period and even changed the corporate name to Exxon Corporation in 1972. They increased overseas operations dramatically as the world was becoming dependent on Middle East oil. Foreign oil growth was phenomenal and the Organization of Petroleum Exporting Companies (OPEC) was formed in 1960 to protect the rights of their countries, some of which were being abused by the major oil companies of the world. OPEC presence was not really felt until the early 1970's and at that time it changed the entire complexion of the oil industry. The Teheran and Tripoli agreements of 1971, where those countries acquired ownership of parts of company operations put the majors in a bind and was the beginning of the end of major company oil price control. Chapter 24 is entitled "The Wheels Came Off" and follows the OPEC success from one country to another, setting the stage for the great oil boom which peaked in 1981. There is a good history of the Humble Oil and Richfield joint venture leasing and prospecting in Alaska which lead to the 1968 discovery of Prudhoe Bay oil field, one of the worlds largest. Ninety pages of appendices and notes. The hardbound first edition is common.

**C149. (TEXAS GAS TRANSMISSION) Waters, L.L., ENERGY TO MOVE: TEXAS GAS TRANSMISSION CORPORATION, KY, The Company, (1985), vii, 240pp.**
In the late 1920's America's cities were going to natural gas for lighting, heating and cooking in a big way. Local companies were formed to handle the service and supply for individual cities. Many went bankrupt during the depression and many failed because of lack of supply. Kentucky Natural Gas and Memphis Natural Gas were acquired in 1945 by J.H. Hillman who believed that he could solve their supply problems by building a pipe line from Texas to Ohio and deliver along the way....it worked, with the formation of Texas Gas Transmission Company. The line was 840 miles long and twenty-six inches in diameter. It was completed in 1900 after countless logistical and political problems. In 1954 they formed Texas Gas Exploration which had tremendous success developing reserves in Louisiana and the Texas gulf coast. Generally joint venturing with established operators, they explored offshore and even participated in some foreign exploration. Texas Gas diversified and expanded. By the early 1980's was a two billion dollar corporation and ripe for takeover during the era of rampant takeovers in the oil industry. CSX Corporation won Texas Gas after a protracted bidding war with Coastal States. Good History. The hardbound first edition is common.

**C150. (SHELL) Wells, Barbara, SHELL AT DEERPARK: THE STORY OF THE FIRST FIFTY YEARS, Houston, 1979, 139pp.**
During 1928, Roxana Petroleum (Wholly owned Royal Dutch-Shell owned company) began construction of a large refinery on Buffalo Bayou at the deserted town of Deer Park. The facility in 1979 covers 1,500 acres

and lies on the south bank of the Houston ship channel. Roxana had tremendous oil exploration success in West Texas and planned to pipeline the crude to the Houston area for refining and later distribution by water to the East Coast. The Shell name was shortly adopted for all domestic company ventures. The plant barely weathered the East Texas field oil glut and price collapse and effects of the Depression. As this book was written, during the late 1970's oil boom, the facility employed 3,700 people and was the twelfth largest refinery complex in the U.S. Good story with lots of historic photos. The hardbound first edition is becoming scarce.

**C151. (WESTERN GEOPHYSICAL) Western Geophysical Company, WESTERN GEOPHYSICAL COMPANY OF AMERICA, Houston, TX, The Company, 1979, 130pp.**
The company was formed in 1933 and has operated all over the world. This book commemorates the company's accomplishments through excellent color photographs of their operations, featuring their work in the most difficult areas. It is dedicated to all of their seismic crews, past and present. The company is well known for tackling the most difficult seismograph areas and successfully obtaining usable data where others failed. Alaska, the toughest, then offshore, then various desert and jungle assignments are highlighted. Although little past company history is given, its uniqueness makes this one of the best photo-histories. The hardbound first edition is becoming scarce.

**C152. (STANDARD OIL OF CALIFORNIA) White, Gerald T., FORMATIVE YEARS IN THE FAR WEST: STANDARD OIL COMPANY OF CALIFORNIA, NY, Appleton-Century-Crofts, (1962), xv, 694pp.**
This volume is the definitive history of the company through 1919. White presents a company pedigree on front and back endpapers which shows a complex of eighteen entities coming together under the Standard of California banner by 1906. Taylor, Scofield, Tevis and Felton were entrepreneurs who controlled most of the eventually merged companies. Early California oil history was largely frustration through the 1850's, 1860's and early 1870's. Union Oil and its predecessors drilled and dug in the Ojai area where active seeps and asphalt were abundant on the surface, with little commercial success. Pico Canyon suffered the same fate. Frederick Taylor with Demetrius Scofield wanted an integrated

industry and began exploring as well as refining and marketing oil from their Standard Oil Company of California in 1877. Developing new oil production for expanding markets with a fluctuating price commodity was very difficult, but they became involved in the first truly commercial oil production at Pico Canyon largely through persistence. By the late 1870's Charles Fenton and Lloyd Tevis entered the scene with their Pacific Coast Oil Company. By 1879, Standard Oil of Ohio was dominating the California market through refining and distribution. Standard, being a stabilizing influence across the U.S. by monopoly, took advantage of an erratic supply and demand crude market and acquired the bulk of their competition including Pacific Coast Oil in 1900. By 1906 they had a severe lack of crude supply (many would not sell to them) so they began to break away from Standard philosophy and explore for their own oil. Exploratory success was such that by the 1911 dismemberment of Standard, the California company was one of the few which stood on its own as integrated. They found several large oil fields to feed their refineries. By 1919, Standard of California had a massive marketing area covering the west coast. Much on political hassles and employee relationships along with the WW I war effort. Large appendices and notes section. The hardbound first edition is common.

**C153. (DOW) Whitehead, Don, THE DOW STORY: THE HISTORY OF THE DOW CHEMICAL COMPANY, NY, McGraw-Hill Book Company, (1968), 298pp.**

This is a well researched and typical company history. Herbert Henry Dow was born about 1866. He moved to Midland, Michigan from Cleveland after studying chemistry in college and developing an idea. He believed he could extract bromine and chlorine from salt water by using electricity. He recruited men in Cleveland and Midland to help begin his company, Dow Chemical in 1897. The chlorine was used as a bleach and bromine was used in photography. A couple of years of technical and logistical problems did not slow him down. He died in 1930 after achieving great success in research, development and marketing and seeing his small venture became a giant company. He had hired the best people available in their fields and followed their guidance to success. Today, a multi-billion dollar company with major subsidiaries such as Dow Corning and Dowell Schlumberger, they are a top chemical company. Dowell and Schlumberger have produced much oil field technology and dominate cementing, acidizing and well logging on a worldwide scale. Dow also became a successful oil operator and producer shortly after this 1968 history was written. The hardbound first edition is common.

**C154. (ANGLO-PERSIAN) Williamson, J.W., IN A PERSIAN OIL FIELD: A STUDY IN SCIENTIFIC AND INDUSTRIAL DEVELOPMENT, London, Ernest Benn Ltd, 1927, 189pp.**

This fine book is what might be called an early scholarly oil company propaganda work. Much had been written about the birth of Anglo-Persian, from the Knox D'Arcy concession and Lord Cadman's brilliant directing of the company. This effort takes a look at its operations and its record of scientific and social achievement. The record is displayed here as an example of how to run a giant company. Williamson only spent a month traveling in the oil field area of Persia, but with the help of experts was able to grasp the situation. The center of the field area at this time was 150 miles north of Abadan in a valley within a small mountain range. It centers on a pool called Suleiman. Operations are covered starting with the geologists who are doing extensive and detailed surface geological and topographical mapping as they look for likely structures to drill. Well drilling is by cable tool and rotary rigs, utilizing the most current equipment available. Much on the engineering and other field research continually going on. The company had a massive refinery at Abadan which was capable of producing oil, gasoline and kerosene. The last half of this book concentrates on the social aspects of the company, particularly local labor problems and how they were handled, integrating that work force into their own. Essentially the company created a city of 30,000 people (employees) from barren ground to handle the giant oil development. Top management and local laborers, all with their families and everything it took to make it livable. It was a major undertaking. Cited in Swanson on page 60; Hardwicke page 65. The hardbound first edition is becoming scarce.

**C155. (QUAKER STATE) Wood, Quentin E., QUAKER STATE ROOTS GO DEEP INTO THE WORLD'S FIRST OILFIELD, Philadelphia, The Newcomen Society, 1986, 24pp.**

Wood is Quaker State's Chairman of the Board and presented this address to a meeting of the Newcomen Society at Erie, Pennsylvania in 1986. Company roots go back to the James B. Berry Sons Company, a Titusville lumber company which began drilling oil wells shortly after

Drake brought in his 1859 pumper. Berry was one of nineteen companies merged by Charles Pope in 1931 to form the current Quaker State. The group was diversified from drilling to refining and marketing. The key product is their automobile engine oil, of exceptional quality. T.G. Phinney originated the "Quaker State" name in 1912 and the quality became their by-word. The company produces about 20% of its own crude stock from Pennsylvania fields, largely in the Bradford area. It is now diversified into coal, mining, insurance and Minit-Lube service stations, but even though it has grown to worldwide exposure, the headquarters remain in Oil City. The softbound first edition is common.

**C156. (CANADA GEOLOGICAL SURVEY) Zaslow, Morris, READING THE ROCKS: THE STORY OF THE GEOLOGICAL SURVEY OF CANADA 1842-1972, Ottawa, Canada, The MacMillan Company, 1975, 599p.**
Zaslow is a history professor and was selected for this massive undertaking without regard to his knowledge of geology. His job, which was to present the 130-year history of the Survey, was aided by a plethora of geological experts, particularly Rice and Henderson. Zaslow is certainly an expert now as this scholarly work is incredibly thorough. The original survey was formed to help those in the mining industry as great deposits were being discovered in Canada. In 1842, the government funded a two-man independent organization to map the Dominion, which at that time consisted of only Ontario and Quebec. The other provinces were added to the Survey's responsibility after 1870. Mapping the wilderness, resolving the rocks and supplying statistics has been the assignment, and it has been well done. Areas of concentration changed over the years under the directorship of William Logan first, through Digby McLauren, currently in charge. Politics and economics created the usual amount of turmoil. As technology changed, the Survey kept pace, utilizing the latest laboratory and field methods to record Canada's geology and appraise its mineral wealth. The book includes profiles of many of the directors and important contributors to the success of the Survey. It is well illustrated with photos and drawings. The definitive history. The hardbound first edition is becoming scarce.

**C157. (PENN SQUARE BANK) Zweig, Philip L., BELLY UP: THE COLLAPSE OF THE PENN SQUARE BANK, NY, Crown Publishers, Inc., (1985), x, 500pp.**
This is a thorough and comprehensive history of the captioned debacle with fifty pages of detailed sources and notes. Zweig tries to find motives

and methods in the madness that took a small Oklahoma shopping mall bank and changed it into a monster loan machine for the independent oil man. Greed, power and inattention are the culprits. Chairman of the board, Bill Jennings turns his inexperienced loan officer, Bill Patterson loose during the frenzy of oil and gas drilling activity of the late 1970's and early 1980's. Patterson was completely out of control from the day he was hired, never conducting himself as a normal banker, more as a merchandiser, almost forcing loans on the independent oil people who were always looking for capital. Jennings touted the larger U.S. banks into taking participations in these loans as a way of achieving high returns from their interest in a rapidly expanding industry with a rosy future. A key guru of the boom was Robert Hefner III who is a geologist with a vision of almost unlimited, untapped natural gas reserves in the Anadarko Basin of Western Oklahoma. He is a charismatic salesman for himself and the industry who rode the wave of natural gas price rises from less than a dollar per thousand cubic feet to over ten dollars per thousand within a few short years. Hefner was caught in the middle of the loan debacle when the price of gas crashed from oversupply, owing millions. When Penn Square failed, it took along some very large, well-respected banks around the country in what was a late wake up call to their unconscious lenders. Some of Penn Square's bank victims were hit harder than others, but Chase Manhattan, Seattle First, Continental Illinois, Michigan National and Northern Trust were all substantially injured. Many characters are featured in this saga, some victims, some perpetrators, some innocent and some guilty, but they were all caught in a money feeding frenzy which suddenly crashed like a house of cards. Oil history fans must read this one. The hardbound first edition is common.

ABC Publicity Service . . . A1
Abels, Jules . . . . . . . . . . . B1
Abramson, Louis . . . . . . . . B2
Abrams, Kathleen Crotty . A2
Acheson, Edward Goodrich . B3
Akin, Edward N. . . . . . . . . B4
Alaska Geographic . . . . . . A3
Alcock, F.J. . . . . . . . . . . C1
Alexander, Nancy . . . . . . . B5
Allaud, Louis A. . . . . . . . C2
Allen, Frederick Lewis . . . A4
Allen, Lawrence J. . . . . . A5
Allen, William Harvey . . . . B6
Allstorm, Oliver . . . . . . . . B7
Alvarez, A. . . . . . . . . . . A6
American Bar Association . A8
Ames, Mary Lesley . . . . . . B8
Anderson, Allan . . . . . . . A13
Anderson, G.S. . . . . . . . . A14
Anderson, Irvine H. . . . . . A15
Andrasko, Kenneth . . . . . A16
Anglo-Iranian . . . . . . . . . C4
Aramco . . . . . . . . . . . . C5
Armitage, Merle . . . . . . . . B9
Arnold, Ralph . . . . . A17,A18
Asbury, Herbert . . . . . . . A19
Atlantic . . . . . . . . . . . . C6
Atlantic Richfield . . . . . . A20
Auletta, Ken . . . . . . . . . . C7
A.A.P.G. . . . . . . . . . . . . A7
A.A.P.L. . . . . . . . . . . . . C3
A.P.I . . . . A9,A10,A11,A12
Bacon, Raymond Foss . . . A21
Ball, Max W. . . . . . . . . . A22
Baroid . . . . . . . . . . . . . A23
Barron, Clarence W. . . . . A24
Bartlett, Anderson G. . . . . C8
Bartlett, Donald L. . . . . . B10
Bartlett, Kim . . . . . . . . . B11
Bartley, Ernest R. . . . . . . A25
Bass, Rick . . . . . . . . . . . B12
Bates, J. Leonard . . . . . . A26
Bayne, Samuel Gamble . . . B13
Beaton, Kendall . . . . . . . C9
Benton, Minnie King . . . . A27
Bermudez, Antonio J. . . . . A28
Bey Essad . . . . . . . . . . . B14
Bille, Ed . . . . . . . . . . . . A29
Black, F. Michael . . . . . . A30
Blakey, Ellen Sue . . A31,A32
Blandford, Linda . . . . . . . A33
Blumay, Carl . . . . . . . . . B15
Boatright, Mody C. . . . . . B16
Bone, J.H.A. . . . . . . . . . A34
Bosworth, Allen . . . . . . . A35
Botsford, Harry . . . . . . . . A36
Bowden, J.J. . . . . . . . . . A37
Bowles, Charles E. . . . . . A38
Bowman, Bob . . . . . . . . . A39
Brady, Kathleen . . . . . . . B17
Brantley, J.E. . . . . . . . . . A40
Brostow, Arch . . . . . . . . A41
Brown, Dee . . . . . . . . . . B18
Brown, George W. . . . . . A42
Brown, Marcus M. . . . . . B19
Brown, Stanley . . . . . . . . B20
Brunson, Howard . . . . . . B21
Bryant, Keith L. . . . . . . . B22
Bryson, John . . . . . . . . . B23
Buckley, Priscilla L. . . . . B24
Buente, Frances M. . . . . . C10
Burst, Ardis . . . . . . . . . . B25
Burton, William M. . . . . . B26
Byers, D.E.L. . . . . . . . . . B27
Byles, Axtell J. . . . . . . . . A43
Byrd, David Harold . . . . . B28

Cadman, Wilson K. . . . . . . C11
Cameron Iron . . . . . . . . . . C12
Campbell, Murray . . . . . . B29
Carl, Ella Lane . . . . . . . . . B30
Carr, Albert Z. . . . . . . . . . B31
Carson, Ira . . . . . . . . . . . . A46
Carter, Hodding . . . . . . . A44
Carter, Worrall Reed . . . . A45
Cartwright, Gary . . . . . . . B32
Castaneda, Christopher J. . C13
Champion, Don F. B33,B34,B35
Chandler, David Leon . . . B36
Chasan, Daniel Jack . . . . . A47
Chase, Mary Ellen . . . . . . B37
Cheney, Michael Sheldon . B38
Chesnutt, N.P. . . . . . . . . C14
Chevron . . . . . . . . . . . . C15
Chiles, H.E. . . . . . . . . . . C16
Cities Service . . . . . C17,C18
Clark, James A . . . . . . . . . . . . . . . . . . . . A49,A50,A51,A52
Clark, James A. . . . . . . . . . . . . . . . . B39,B40,B41,B42,B43
Clark, J. Stanley . . . . . . . A48
Clark, Mabel K. . . . . . . . A53
Clark, Opal Bennefield . . . B44
Clinton, Fred S. . . . . . . . A54
Cochran, Mike . . . . . . . . B45
Cocke, W.A. . . . . . . . . . B46
Cohen, Stan . . . . . . . . . . A55
Coleman, Charles M. . . . . C19
Collier, Peter . . . . . . . . . B47
Collings, Ellsworth . . . . . A56
Collins, George B. . . . . . . A56
Coll, Steve . . . . . . . . . . . C20
Cone, Andrew . . . . . . . . A58
Conger, Mrs. Joe . . . . . . A59
Connelly, William L. . . . . B48
Conniff, James C.G. . . . . C21
Connor, Seymour V. . . . . B49
Conn, Frances G. . . . . . . A60
Considine, Bob . . . . . . . . B50
Continental Oil . . . . . . . . C22
Continental Oil . . . . C23,C24
Cook, Lodwrick M. . . . . . C25
Cornell, Doris . . . . . . . . . B51
Corner, George W. . . . . . B52
Cotner, Robert C. . . . . . . B53
Cotten, Kathryn . . . . . . . A61
Cox, Ian . . . . . . . . . . . . C26
Cox, L.W. . . . . . . . . . . . B54
Creole Petroleum . . . . . . C27
Crew, Benjamin J. . . . . . . A62
Cross, Roy . . . . . . . . . . . B55
Crowell, Evelyn Miller . . . B56
Crum, A.R. . . . . . . . . . . A63
Cubage, Granville . . . . . . B57
Culbreath, H.L. . . . . . . . C28
Cummings, Jim . . . . . . . . B58
Curtin, D. Thomas . . . . . A64
Darrah, William Culp . . . . A65
Davis, Ralph E. . . . . . . . B59
Davis, Wallace . . . . . . . . B60
Davis, William . . . . . . . . A66
Dean, George F. . . . . . . . B61
DeChair, Somerset . . . . . B62
Dedmon, Emmett . . . . . . C29
DeGolyer, Everette . . . . . A67
DeGolyer, Everette . A68,A69
Dennison, Charles Stewart C30
Desmond, James . . . . . . . B63
Destler, Chester McArthur B64
Deterding, Sir Henry . . . . B65
Dias, Earl J. . . . . . . . . . . B66
Dodd, S.C.T. . . . . . . . . . B67
Dodd, S.C.T. . . . . . . . . . C31

Dolson, Hildegarde . . . . . A70
Donahue, Jack . . . . . . . . B68
Donahue, Jack . . . . . . . . C32
Donnell, J.C. . . . . . . . . . C33
Drosnin, Michael . . . . . . B69
Dugan, Haynes W. . . . . . B70
Eason, Al . . . . . . . . . . . A71
Easton, Robert . . . . . . . . A72
East, Lrecia . . . . . . . . . . B71
Eaton, Reverend S.J.M. . . A73
Eby, Brian J. . . . . . . . . . B72
Edinger, David H. . . . . . C111
Edmonds, I.G. . . . . . . . . A74
Egloff, Gustav . . . . . . . . A75
El Mallakh, Ragaei . . . . . A76
Elder, Jack . . . . . . . . . . . A77
Ellis, William Donohue . . . . . . . . . . . . . . . . . . . . . B74,C34
Ellis, W. Erby . . . . . . . . B73
Elmer, Isabel Lincoln . . . . B75
Englebrecht, Curt E. . . . . B76
Erni, Henri . . . . . . . . . . A78
Estes, Pam . . . . . . . . . . . B77
Ethyl Corporation . . . . . . A79
European Gas . . . . . . . . . C35
Ewing, Jefferson S. . . . . . B78
Exxon . . . . . . . . . . . . . . C36
Ezell, John Samuel . . . . . C37
Fairbanks, Helen R. . . . . . B79
Fairhill, David . . . . . . . . A80
Fanning, Leonard M. . . . . . . . . . . . . . . . A81,A82,183,C38
Farley, Joseph M. . . . . . . C39
Faulk, Odie B. . . . . . B80,B81
Fay, Stephen . . . . . . . . . B82
Fears, Jerry . . . . . . . . . . A84
Fellows, Kenneth . . . . . . C40
Fenton, Carroll Lane . . . . B83
Fineran, John Kingston . . . B84
Finney, Guy W. . . . . . . . C41
Finnie, Richard . . . . A85,C42
Finstad, Suzanne . . . . . . . B85
Flemmons, Jerry . . . . . . . B86
Fletcher, Grady W. . . . . . B87
Floherty, John J. . . . . . . . A86
Flores, Giovanni . . . . . . . B88
Flynn, John T. . . . . . . . . B89
Fogarty, Charles F. . . . . . C43
Forbes, Gerald . . . . . . . . A87
Forbes, R.J. . . . . . . A88,C44
Fosdick, Raymond B. . . . . B90
Foster, Peter . . . . . . C45,C46
Franks, Kenny A. . . . . . . . . . . . . . A89,A90,A91,A92,A93
Franks, Kenny A. . . . . . . . . . . . . . . . . . . A94,A95,A96,A97
Fraser, Duncan . . . . . . . . A98
Frey, John W. . . . . . . . . A99
Friends of Flagler . . . . . . B91
Fuermann, George . A100,A101
Fulton, John F. . . . . . . . . B92
Galbreath, Frank . . . . . . A102
Galena Oil . . . . . . . . . . . C47
Gale, Thomas A. . . . . . . A103
Gard, Wayne . . . . . . . . A104
Garfias, Valentin R. . . . . . B93
Garrison, Omar V. . . . . . B94
Geiser, S.W. . . . . . . . . . B95
Gentry, Mickey . . . . . . . . C48
Gentry, Winalee . . . . . . . B96
Gerber, Albert B. . . . . . . B97
Gerretson, F.C. . . . . . . . C49
Gervasi, Frank . . . . . . . . B98
Getty, J. Paul . . . . . . . . . . . . . . . . . . . B99,B100,B101,B102
Getty, J. Paul . B103,B104,C50

Gibb, George Sweet . . . . . C51
Gibson, George R. . . . . . . A105
Giddens, Paul H.
. . . . . . . . . . . A106,A107,A108
Giddens, Paul H. . . . . . . A109,
. . . . . . . . . . . . . . . . A110,C52
Gilbert, Miriam . . . . . . . B105
Gillispie, Ora . . . . . . . . . A111
Glasscock, C.B. . . . . . . . A112
Glasscock, Lucille . . . . . B106
Goralski, Robert . . . . . . A113
Gould, Charles N. . B107,B108
Grable, Donovan B. . . . . B109
Grad, Morton . . . . . . . . . C53
Graham, Hal . . . . A114,A115
Graham, Hugh Davis . . . B110
Gray Earle . . . . . . . A116,C54
Greenburg, Jonathan . . . B111
Gregory, Robert . . . . . . A117
Griffin, John Howard . . . A118
Gulbenkian, Nubar . . . . B112
Gwin, Lucy . . . . . . . . . B113
G.S.I. . . . . . . . . . . . . . . . C55
Haas, Paul R. . . . . . . . . . C56
Hadlock, Adah . . . . . . . B114
Haigh, Berte R. . . . . . . . . C57
Halasz, Nicholas . . . . . . B116
Halbouty, Michel T. . . . B117
Haley, James L. . . . . . . A119
Haley, J. Evetts . . . . . . . . C58
Hamill, Curtis G. . . . . . A120
Hamilton, Charles W. . . A121
Hammer, Armand . . . . . B118
Hammons, Terry . . . . . . B119
Hanna, Betty Elliott . . . . A122
Hansl, Proctor W. . . . . . A123
Hardwicke, Robert E. . . . . . .
. . . . . . . . . . . . . . A124,A125
Hari, William Ivy . . . . . B115
Harris, Charles H. . . . . . . A126
Harris, Kenneth . . . . . . B121
Harris, Leon . . . . . . . . . . B120
Harter, Harry . . . . . . . . . A127
Hassman, Heinrich . . . . A128
Hatch, Vernelle A. . . . . A129
Hatfield, Reginald W. . . A130
Hatley, Allen G. . . . . . . A131
Hawke, David Freeman . B122
Heiman, Monica . . . . . . B123
Helgesen, Sally . . . . . . . A132
Henriques, Robert . . . . . B124
Henry, James Dodds . . . A133
Henry, J.T. . . . . . . . . . . A134
Heritage Magazine . . . . . A135
Herrick, John P. . . A136,A137
Hersh, Burton . . . . . . . . B125
Hewins, Ralph. . . . . . . . . . . .
. . . . . . . . . . A138,B126,B127
Heydrick, L.C. . . . . . . . . A139
Hidy, Ralph W. . . . A140,C59
Higgins, Patillo . . . . . . . A141
Hill, George P. . . . . . . . B128
Hill, Margaret Hunt . . . . B129
Hines, Gordon . . . . . . . B130
Hoffman, William . . . . . . . . .
. . . . . . . . . . A142,B131,B132
Hogg, James Stephen . . . B133
Holbrook, Stewart H. . . . A143
Hollyman, Thomas . . . . A144
House of Representatives . C60
House, Boyce . . . . . . . . . . . .
. . . . . A145,A146,A147,A148
House, Boyce..A149,A150,B134
House, J.D. . . . . . . . . . A151
Howland, Harold J. . . . . . C61
Hubbard, Elbert . . . . . . . C62

Hubbard, Silas . . . . . . . B135
Hughes, Dudley J. . . . . . A152
Hughes, Howard . . . . . . B136
Humble Oil . . . . . . . . . A153
Humphrey, William D. . . A154
Hunt, H.L. . . B137,B138,B139
Hunt, H.L. . . . . . . B140,B141
Hurley, Marvin . . . . . . . A155
Hurt, Harry . . . . . . . . . B142
Hutchinson, W.H. . . . . . B143
Hutson, William Gardiner..B144
Ickes, Harold L. . . A156,B145
Ironside, Roberta Louise . . C63
Irvin, Samuel P. . . . . . . A158
I.P.A.A. . . . . . . . . . . . A157
Jackson, Elaine . . . . . . . . C64
Jacobson, Norman H. . . . B146
James, Marquis . . . . . . . . C65
Jarman, Rufus . . . . . . . . . C66
Johnes, Richard Lloyd . . A163
Johnson, Arthur Menzies...A159
Johnson, Arthur M. . . . . . C67
Johnson, J.K. . . . . . . . . A160
Johnson, Robert L. . . . . . C68
Johnson, T.E. . . . . . . . . B147
Johnston, Marguerite . . . A161
Jones, Billy M. . . . . . . . B148
Jones, Charles S. . . . . . . . C69
Jones, Howard Mumford A162
Jordan, Constantin N. . . . A164
Josephson, Emanuel M. . B149
Josephson, Matthew . . . . A165
Karasick, Norman M. . . B150
Kay County Gas . . . . . . A166
Keats, John . . . . . . . . . B151
Keeble, John . . . . . . . . A167
Keller, David Neal . . . . . C70
Kelly, John S. . . . . . . . . B152
Kelly, Louise . . . . . . . . A168
Kelsay, R.E. . . . . . . . . B153
Kelso, J. Richard . . . . . . . C71
Kennedy, donald S. . . . . . C72
Kennedy, Tom . . . . . . . A169
Keplinger, H.F. . . . . . . B154
Kerr, Aubrey .A170,A171,A172
Kerr, Robert S. . . . . . . . B155
Kilman, Ed . . . . . B156,B157
King, John O. . . . . . . . . B158
King, Kerren . . . . . . . . . C73
Kistler, Ron . . . . . . . . . B159
Kittle, Kit . . . . . . . . . . A173
Klappenbach, E.W. . . . . B160
Klein, Henry H. . . . . . . . C74
Knowles, Ruth Sheldon . . . . . . . . . . . . . . . . . . . . . A174,A175
Koskoff, David E. . . . . . B161
Kramer, Michael . . . . . . B162
Kutz, Meyer . . . . . . . . . B163
Lambert, Paul F. . . . . . . A176
Larson, Elwin S. . . . . . . . C75
Larson, Henrietta M. C76,C77
Latham, Earl . . . . . . . . B164
Latta, F.F. . . . . . . . . . . A177
Lawrence, Albert A. . . . A178
Lawson, Thomas W. . . . A179
Lay, Kenneth L. . . . . . . . C78
Leeston, Alfed M. . . . . . A180
Lenzner, Robert . . . . . . B165
Leonard, Charles C. . . . A181
LeVane, Ethel . . . . . . . . B166
Lief, Alfred . . . . . . . . . . C79
Lilly, Doris . . . . . . . . . B167
Lloyd, Henry Demarest . . C80
Lloyd, John . . . . . . . . . B168
Lobsenz, Norman M. . . . B169
Lockhart, Charles . . . . . B170

Lodwick, John . . . . . . . . B171
Logan, Chauncey . . . . . A182
Longhurst, Henry . . . . . . C81
Longrigg, Stephen Hemsley . . .
. . . . . . . . . . . . . . . . . . . A183
Look Magazine . . . . . . . A184
Loos, John L. . . . . . . . . . C82
Love, Philip H. . . . . . . . . B172
Lummus Company . . . . . . C83
Lundberg, Ferdinand . . . . . . .
. . . . . . . . . . A185,B174,B175
Lund, Robina . . . . . . . . B173
Lynch, Gerald . . . . . . . A186
Lyon, Jim . . . . . . . . . . . C84
Mabry, Mrs. Grant . . . . A187
Machlin, Milt . . . . . . . . . B176
Mallison, Sam T. . . . . . . B177
Malsch, Brownson . . . . . B178
Mangan, Frank . . . . . . . . C85
Marcosson, Isaac F. . . . . A188
Markmann, Charles Lam B179
Marshall, J. Howard . . . B180
Martin Sidney Walter . . . B181
Martinez, Anibal R. . . . . A189
Massie, Joseph L. . . . . . B182
Masters, John A. . . . . . . . C86
Mather, John A. . . . . . . A190
Mathews, John Joseph . . B183
Mathison, Richard . . . . . B184
Matthews, Elizabeth Turner . . .
. . . . . . . . . . . . . . . . . . B185
May, Lioy . . . . . . . . . . A191
McBeth, Reid Sayers . . . . . . .
. . . . . . . . . . . . . . ..A192,B186
McClintick, David . . . . . . C87
McComb, David G. . . . . A193
McConnell, Burt M. . . . A194
McCoy, Dorothy Abbott . A195
McDaniel, Robert W. . . . B187
McDaniel, Ruel . . . . . . . . A196
McGee, Dean A. . . . . . . . . C88
McGowen, Norris Cochran . . .
. . . . . . . . . . . . . . . . . . . A197
McGuire, Paul F. . . . . . A198
McKay, Seth S. . . . . . . . A199
McLaughlin, Roy P. . . . B188
McLaurin, John J. . . . . . A200
McLean, John G. . . . . . . C89
McPhee, John . . . . . . . . B189
McWorkman, Lee G. . . . B190
Mead, Robert Douglas . . A201
Mellon, William Larimer B191
Meyer, Lewis . . . . . . . . B192
Meyer, Lorenzo . . . . . . A202
Mikesell, Raymond F. . . A203
Miles, Leo . . . . . . . . . . B193
Miller, Ernest C. . . . . . . . . . .
. . . . . A204,A205,A206,A207
Miller, Ernest C. . . . . . . . . . .
. . . . . . A208,A209,A210,B194
Miller, Max . . . . . . . . . A211
Miller, Russell . . . . . . . B195
Mills, Charles E. . . . . . . . C90
Mills, N. Elizabeth . . . . B196
Milsten, David Randolph .B197
Mineau, Wayne . . . . . . . A212
Miner, Craig . . . . . . . . A213
Miner, Craig . . . . . . . . A214
Mintrop Ludger . . . . . . A215
Mitchell, Johnny . . . . . . B198
Molk, Isador . . . . . . . . B199
Montague, Gilbert Holland .C91
Moore, Austin Leigh . . . B200
Moore, Richard R. . . . . A216
Moreno, Federico . . . . . A217
Morgan, Anne Hodges . . B201

Morgan, Jonnie R. . . . . . . A218
Morley, Harold T. . . . . . . . C92
Morris, Edmund . . . . . . A219
Morris, Joe Alex . . . . . . B202
Morris, John W. . . A220,A221
Morton, Frederick . . . . . B203
Moscow, Alvin . . . . . . . B204
Murrah, David J. . . . . . B205
Murray, William H. . . . . B206
Myers, Elisabeth P. . . . . B207
Myres, Samuel D. . A222,A223
Naifeh, Steven . . . . . . . B208
Nash, Robert . . . . . . . . B209
Neal, Harry Edward . . . A224
Neely, Harold G. . . . . . B210
Neff, E. Richard . . . . . . B211
Nelson, Ed . . . . . . . . . . B212
Nevins, Allan . . . . A225,B213
Newsom, D. Earl . A227,A228
Newton, J.H. . . . . . . . . A226
Nicholson, Patrick . . . . . B214
Nichols, Allen G. . . . . . A229
Noggle, Burl . . . . . . . . A230
O'Connor, Harvey . . . . . . . . . . . . . . . . . . . . A231,B215,C93
O'Connor, Richard . . . . A232
O'Green, Fred W. . . . . . . C94
Oil and Gas Journal . . . . . . . . A310,A311,A312,A313,A314
Oil City Derrick . . . . . . . . . . . A233,A234,A235,C95,C140
Oklahoma Natural . . . . . . C96
Olien, Roger M. . . . . . . . . . . . . . . A236,A237,A238,A239
Orcutt, Mary Logan . . . . B216
Owen, Edgar Wesley . . . A240
Owen, Roderic . . . . . . . B217
Pahlavi, Shah . . . . B218,B219
Pan American . . . . . . . . . C97
Parker Drilling . . . . . . . . C98
Parker, George . . . . . . . B220
Patrick, Kenneth G. . . . . . C99
Payne, Darwin . . . . . . . C100
Peckham, S.F. . . . . . . . A241
Peebles, William K. . . . . C101
Peeples, Malcolm W.H. . A242
Perryman, W.L. . . . . . . C102
Persico, Joseph E. . . . . . B221
Person, Harlow S. . . . . . A243
Peto, Sir S. Morton . . . . A244
Petroleum County . . . . . A245
Petroleum Producers . . . C109
Petty, O. Scott . . . . . . . C103
Petzinger, Thomas . . . . . C104
Phelan, James . . . . . . . . B222
Philips, Shine . . . . . . . . A246
Phillips Petroleum . . . . . C105
Phillips, David Atlee . . . B223
Pickens, T. Boone . . . . . B224
Poling, James . . . . . . . . B225
Ponca City D.A.R. . . . . A247
Pope, Clarence . . . . . . . B226
Popple, Charles Sterling . C106
Porterfield, Bill . . . . . . . . . . . . . . . . . . . . . B227,B228,B229
Porter, J.W. . . . . . . . . . C107
Powell, J. Richard . . . . . A248
Pratt, Wallace E. . . . . . . A249
Pratt, Wallace E. . . . . . . B230
Presley, James . . . A250,C108
Price, W.C. . . . . . . . . . B231
Purcell, Ann . . . . . . . . . B232
Pyle, Tom . . . . . . . . . . B233
Quaker State . . . . . . . . . A251
Rabe, Stephen G. . . . . . A252
Ramsey, Robert H. . . . . C110

Ravage, M.E. . . . . . . . . . A253
Redpath, Lionel V. . . . . A254
Redwood, Sir Boverton . . A255
Reed, Ed . . . . . . . . . . . B234
Reinholt, Oscar H. . . . . . A256
Rhoden, Harold . . . . . . B235
Rintoul, William . . . . . . . . . . . . . . . . . . . . A257,A258,A259
Rister, Carl Coke . . . . . A260
Roberson, Courtney Ann A261
Roberts, Glyn . . . . . . . . B236
Roberts, Harold D. . . . . A262
Robert, Joseph C. . . . . . C112
Robinson, Jeffrey . . . . . B237
Robinson, Spencer W. . . A263
Rockefeller, John D. . . . B238
Rockefeller, Nelson A. . . B239
Rodgers, William . . . . . B240
Rodgers, W.S.S. . . . . . . C113
Roff, Charles L. . . . . . . B241
Rogers, John William . . . A264
Roosevelt, Kermit . . . . . A265
Roscow, James P. . . . . . A266
Ross, Victor . . . . . . . . . A267
Rountree, Russs . . . . . . A268
Rowe, Melodia B. . . . . . B242
Rowland, John . . . . . . . B243
Royal Dutch . . . . . C114,C115
Rundell, Walter, Jr. . . . . . . . . . . . . . . . . . . . . . . A269,A270
Sadler, Jerry . . . . . . . . . B244
Sailor, Merlin F. . . . . . . B245
Sampson, Anthony . . . . . C116
Sanders, Christine Moor . C117
Sanders, Sue . . . . . . . . . B246
Sargent, M.P. . . . . . . . . A271
Sarnoff, Paul . . . . . . . . B247
Schlumberger, Anne . . . C118
Schooley, John S. . . . . . A272
Schruben, Francis W. . . . A273
Schweitzer, P. . . . . . . . A274
Schweizer, Peter . . . . . . A275
Schwettmann, Martin W. A276
Scofield, Donald F. . . . . B248
Scott, Charles C. . . . . . . A277
Scott, Harold W. . . . . . . B249
Scott, John David . B250,B251
Scott, Otto J. . . . . B252,C119
Sennholz, Mary . . . . . . B253
Seward, William B. . . . . C120
Shannon, JAmes . . . . . . C121
Shaplen, Robert . . . . . . B254
Shawcross, William . . . . B255
Sheehy, Sandy . . . . . . . B256
Shell Union . . . . . . . . . C122
Shelton, Hooper . . . . . . A278
Shield, Charles F. . . . . . B257
Shulman, Morton . . . . . C123
Sidwell, Eugene Chenoweth . . . . . . . . . . . . . . . . . . . . B258
Siegel, Stanley . . . . . . . A279
Silliman, Benjamin A280,A281
Simpson, Colin . . . . . . . C124
Singerman, Philip . . . . . B259
Singer, Mark . . . . . . . . . C125
Sinise, Jerry . . . . . . . . . A282
Smalheer, Calfin V. . . . . C126
Smiley, Alfred Wilson . . B260
Smith, Corita Jean . . . . . B261
Smith, P.G.A. . . . . . . . C127
Solberg, Carl . . . . . . . . A283
Spencer, P.C. . . . . . . . . C130
Spence, Hartzell . . C128,C129
Spender, John A. . . . . . . B262
Spratt, John S. . . . . . . . A285
Stadler, Walter . . . . . . . A286

Standard Oil .A287,C131,C132
Stanley, F. . . . . . . . . . . . . . . .
. . . . . A288,A289,A290,A291
Stanley, F. . . A292,A293,B263
Stedham, Austin W. . . . . C133
Steele, John Washington . B264
Stegemeier, Richard J. . . C134
Stegner, Wallace . . A294,B265
Steinhart, Carol E. . . . . . A295
Stephens, Robert W. . . . B266
Sternberg, C.W. . . . . . . B267
Stewart, Roy P. . . . . . . . A296
Stillwagon, C.K. . . . . . . A297
Stotz, Louis . . . . . . . . . A298
Stowe, Estha Briscoe . . . B268
Straton, Florence . . . . . . A299
Stratton, David H. . . . . . B269
Sullivan, Mark . . . . . . . A300
Summers, Festus P. . . . . B270
Sun Oil . . . . . . . . . . . . C135
Supreme Court . . . . . . . C136
Swanson, E.B. . . . . . . . A301
Sweetman, John . . . . . . A302
Sweet, George Elliott . . . B271
Swensrud, Sidney A. . . . C137
Szalewicz, Steve S. . . . . A303
S.W. Texas Oil Scouts . . A284
Tait, Samuel W. . . . . . . A304
Talbot, Frederic A. . . . . A305
Tarbell, Ida M. . . . B272,C138
Tavoulareas, William P. . B273
Taylor, Frank J. . . . . . . C139
Tennent, James C. . . . . . A306
Terry, Pamela . . . . . . . . A307
Texas Parade Magazine . A308
Texas Permian Historical...A309
Thompson, Arthur Beeby. .B274
Thompson, Craig . . . . . C141
Tinkle, Lon . . . . . A315,B275
Tinnin, David B. . . . . . . B276
Toernudd, GUstaf . . . . . B277
Tolbert, Frank X. . . . . . B278
Tolf, Robert W. . . . . . . B279
Tompkins, Walter A. . . . C142
Tower, Walter S. . . . . . A316
Trafzer, Clifford Earl . . . B280
Trainer, George Francis . B281
Trengove, A. . . . . . . . . C143
True, H.A. . . . . . . . . . . B282
Truth's Investigator . . . . C146
Tucille, Jerome . . . . . . . B283
Tucker, H.H. . . . . . . . . C144
Tugendhat, Christopher . A317
Twitchell, K.S. . . . . . . . A318
Tygiel, Jules . . . . . . . . . B284
Tyson, Carl N. . . . . . . . C145
U.S. Congress . . . . . . . B285
U.S. Senate . . . . . A319,A320
Verrees, Charles . . . . . C147
Vincent, Mordelo L. . . . B286
Wade, Henry F. . . . . . . A321
Walker, James P. . . . . . A322
Wallis, Michael . . . . . . . B288
Wall, Bennett H. . . B287,C148
Ward Edward . . . . . . . . A323
Ward, John L. . . . . . . . B289
Warner, C.A. . . . . . . . . A324
Warner, Thor . . . . . . . . B290
Watchorn, Robert . . . . . B291
Waters, L.L. . . . . . . . . C149
Watkins, Lew . . . . . . . . A325
Weaver, Bobby D. . . . . . A326
Weeks, Lewis G. . . . . . . B292
Weinberg, Steve . . . . . . B293
Wells, Barbara . . . . . . . C150
Wells, Laura Lou . . . . . A327

Wells, W.B. . . . . . . . . . . B294
Welsh, Louise . . . . . . . . . A328
Wendt, Lloyd . . . . . . . . . B295
Werner, M.R. . . . . . . . . . A329
Western Geophysical . . . C151
Whitehead, Don . . . . . . C153
Whiteshot, Charles A. . . A333
White, Ben P. . . . . . . . . B297
White, Gerald T. . . . . . . . . . .
. . . . . . . . . . A330,B296,C152
White, John H. . . . . . . . A331
White, Wayne E. . . . . . . A332
Whitney, Caspar . . . . . . B298
Wickware, Potter . . . . . B299
Wilkie, Franc Bangs . . . A334
Williamson, Chilton . . . . B301
Williamson, Harold . A335,A336
Williamson, J.W. . . . . . C154
Williams, Harry T. . . . . B300
Wilson, Robert E. . . . . . B302
Wilson, Terry P. . . . . . . A337
Winkleman, B.F. . . . . . . B303
Winkler, John K. . . . . . . B304
Winslow, Ron . . . . . . . . A338
Witteman, A. . . . . . . . . A339
Wolfe, Jane . . . . . . . . . B305
Woodward, Guy H. . . . . A340
Wood, John M. . . . . . . . B306
Wood, Quentin E. . . . . . C155
Wright, William . . . . . . A342
W.P.A. Writers . . . . . . A341
Yergin, Daniel . . . . . . . A343
Youle, W.E. . . . . . . . . . B307
Young, Bob . . . . . . . . . A344
Zaslow, Morris . . . . . . . C156
Zinman, David . . . . . . . B308
Zonis, Marvin . . . . . . . . B309
Zweig, Philip L. . . . . . . C157

Abercrombie . . . . . . . . B214
Abramson . . . . . . . . . . . . B2
Acheson . . . . . . . . . . . . . B3
Adair . . . . . . . . . . . . . B259
Alabama Power . . . . . . . . C39
Alaska . . . . . . . . . . . . . . .
. . . . A3,A5,A16,A47,A55,A84
Alaska. A167,A201,A266,A300
American Meter . . . . . . . C79
AMPOL . . . . . . . . . . . C124
Anderson . . . . . . . . . . . B121
Anglo-Iranian . . . . . C4,C154
ARAMCO . . . . . . . . . . . C5
Archbold . . . . . . . . . . . B200
ARCO . . . . . . . . . . . . . . C25
Ashland . . . . . . . . . . . . C119
Atlantic . . . . . . . . . . . . . C6
A.A.P.G. . . . . . . . . . . . . C92
A.A.P.L. . . . . . . . . . . . . C3
A.P.I. . . . . . . . . . . . . . . C38
Bailey . . . . . . . . . . . . . B46
Bard . . . . . . . . . . . . . . B143
Barnsdall, A. . . . . . . . . B150
Barret . . . . . . . . . . . . . B278
Bartlett . . . . . . . . . . . . . B11
Bass . . . . . . . . . . . . . . . B12
Bayne . . . . . . . . . . . . . . B13
Bechtel . . . . . . . . . . . . . C42
Benedum . . . . . . . . . . . B177
Bey . . . . . . . . . . . . . . . . B14
Bibliography . . . . . . A67,A124
Blazer . . . . . . . . . . . . . B182
Booth . . . . . . . . . . . . . B194
Bow Valley . . . . . . . . . . C45
British Petroleum . . . . . . . C81
Broken Hill . . . . . . . . . C143
Brooklyn Union . . . . . . . C75
Brown, Claude . . . . . . . B209
Brown, Dee . . . . . . . . . . B18
Brunson . . . . . . . . . . . . . B21
Buckley . . . . . . . . B24,B179
Bunn . . . . . . . . . . . . . . . B51
Burton . . . . . . . . . . B26,B302
Byers . . . . . . . . . . . . . . B27
Byrd . . . . . . . . . . . . . . . B28
Cabot . . . . . . . . . . . . . B120
Cadman . . . . . . . . . . . . B243
California . . . . . . . . . . . .
. . . A72,A89,A130,A157,A177
California . . . . . . . . . . . .
. A229,A254,A257,A258,A259
California . . . . . . . . . . . .
. A277,A280,A286,A295,A330
Camden . . . . . . . . . . . . B270
Cameron . . . . . . . . . . . C12
Canada. . A13,A85,A116,A151
Canada . . . . . . . . . . . . . .
. . . . . A169,A170,A171,A172
Canada Superior . . . . . . C107
Canada Survey . . . . C1,C156
Canadian Hunter . . . . . . . C86
Canfield . . . . . . . . . . . B298
Carl . . . . . . . . . . . . . . . B30
Carter . . . . . . . . . . . . . . B86
Champion . . . . B33,B34,B35
Cheney . . . . . . . . . . . . . B38
Cities . . . . . . . C17,C18,C34
Connelly . . . . . . . . . . . . B48
Conoco . . . . . . C22,C23,C24
Cooper . . . . . . . . . . . . . C70
Cox, Jack . . . . . . . . . . . B54
Creole . . . . . . . . . . . . . . C27
Cross . . . . . . . . . . . . . . B55
Cullen . . . . . . . . . B156,B157
Cullinan . . . . . . . . . . . B158
Cummings . . . . . . . . . . . B58

Davis, Cullen . . . . . . . . . . .
. . . . . . . . B32,B45,B208,B223
Davis, Ralph . . . . . . . . . B59
Dean . . . . . . . . . . . . . . . . B61
Degolyer . . . . . . . B230,B275
Delta Drilling . . . . . . . . C108
Deterding . . . . . . . . B65,B236
Devitt . . . . . . . . . . . . . . B205
Dodd . . . . . . . . . . . . . . . B67
Dome . . . . . . . . . . C46,C84
Dow . . . . . . . B29,C120,C153
Dresser . . . . . . . . . . . . . C100
Drummond . . . . . . . . . . B119
Dugan . . . . . . . . . . . . . . . B70
Dysart . . . . . . . . . . . . . . . B9
East . . . . . . . . . . . . . . . . B71
East Ohio . . . . . . . . . . . . C71
Eby . . . . . . . . . . . . . . . . B72
El Paso . . . . . . . . . . . . . C85
Ellis . . . . . . . . . . . . . . . . B73
Elmer . . . . . . . . . . . . . . . B75
Enron . . . . . . . . . . . . . . . C78
Estes . . . . . . . . . . . . . . . . B77
Ethyl . . . . . . . . . . . . . . . C112
Ewing . . . . . . . . . . . . . . . B78
Exxon . . . . . . . . . . . . . . . C36
Fall . . . . . . . . . . . . . . . B269
Farfias . . . . . . . . . . . . . . B93
Flagler . . . B4,B36,B91,B181
Fletcher . . . . . . . . . . . . . B87
Flores . . . . . . . . . . . . . . . B88
Foster . . . . . . . . . . . . . . . B74
Franklin . . . . . . . . . . . B271
Galena Oil . . . . . . . . . . . C47
Gates . . . . . . . . . . . . . B295
General . . . . . . . . . . . . . . .
. . . . . A2,A8,A9,A10,A11,A12
General . . . . . . . . . . . . . . .
. . A14,A17,A21,A22,A23,A31
General . . . . . . . . . . . . . . .
. . A32,A34,A38,A40,A43,A45
General . . . . . . . . . . . . . . .
. . A48,A50,A62,A64,A66,A68
General . . . . . . . . . . . . . . .
. . A69,A75,A78,A79,A81,A82
General . . . . . . . . . . . . . . .
. A83,A86,A87,A88,A99,A105
General . . . . . . . . . . . . . . .
. A113,A123,A125,A131,A140
General . . . . . . . . . . . . . . .
. A143,A144,A156,A158,A159
General . . . . . . . . . . . . . . .
. A162,A165,A174,A175,A180
General . . . . . . . . . . . . . . .
. A184,A185,A188,A192,A200
General . . . . . . . . . . . . . . .
. A211,A215,A224,A225,A231
General . . . . . . . . . . . . . . .
. A232,A236,A240,A241,A242
General . . . . . . . . . . . . . . .
. . As44,A249,A255,A256,A267
General . . . . . . . . . . . . . . .
. A274,A275,A283,A287,A297
General . . . . . . . . . . . . . . .
. A298,A301,A302,A306,A308
General . . . . . . . . . . . . . . .
. A312,A314,A316,A317,A323
General . . . . . . . . . . . . . . .
. A335,A336,A338,A343,A344
General American . . . . . C102
Gentry . . . . . . . . . . . . . . B96
Geosource . . . . . . . . . . C103
Getty . . . . . . . . . . . . . . . .
. . . . B62,B99,B100,B101,B102
Getty . . . . . . . . . . . . . . . .
. . B103,B104,B126,B165,B166
Getty . . B169,B173,B195,C50
Gilcrease . . . . . . . . . . . B197

Gladys City . . . . . . . . . C117
Glasscock . . . . . . . . . . B106
Gonzaullas . . . . . . B178,B266
Gould . . . . . . . . . B107,B108
Grable . . . . . . . . . . . . . B109
Great Britain . . . . . . . . . .
. . . . . A6,A56,A80,A98,A340
Gulbenkian, Calouste B127,B171
Gulbenkian, Nubar . . . . B112
Gulf . . . . . . . . . . C137,C141
Gwin . . . . . . . . . . . . . B113
G.S.I . . . . . . . . . . . . . . C55
Hadlock . . . . . . . . . . . . B114
Halbouty . . . . . . . . B68,B117
Hammer . . . . . . . . . . . . . .
. . . . . B15,B23,B50,B118,B293
Hass Oil . . . . . . . . . . . . C56
Hefner . . . . . . . . . . . . B280
Hefner, Jr. . . . . . . . . . . . B80
Higgins . . . . . . . . . . . . B187
Hill, George . . . . . . . . . B128
Hill, Robert . . . . . . . . . . . B5
HNG . . . . . . . . . . . . . . C40
Hogan . . . . . . . . . . . . . B281
Hogg . . . . . . . . . . B53,B133
Home-stake . . . . . . . . . . C87
House . . . . . . . . . . . . . B134
Houston Club . . . . . . . . . C32
Hughes . . . . . . . . . . . . . .
. . . B10,B69,B82,B85,B94,B97
Hughes B136,B151,B159,B184
Hughes . . . . B222,B235,B276
Humble . . . . . . . . . . . . . C76
Hungary . . . . . . . . . . . A303
Hunt . . . . . . . . . . . . . . . .
. . . . B20,B25,B129,B137,B138
Hunt . . B139,B140,B141,B142
Hunt . . . . . . . . . . B247,B283
Ickes . . . . . . . . . . . . . . B145
Interstate . . . . . . . . . . . C82
Jacobson . . . . . . . . . . . B146
Jones . . . . . . . . . . . . . B242
Julian . . . . . . . . . . . . . B144
Julian Petroleum . . . . . . . C41
Kansas . . . . A213,A214,A273
Kansas Gas . . . . . . . . . . C11
Kavanaugh . . . . . . . . . . . B95
Kelly . . . . . . . . . . . . . . B152
Kelsay . . . . . . . . . . . . . B153
Keplinger . . . . . . . . . . . B154
Kerr . . . . . . B155,B201,B285
Kerr-McGee . . . . . . C37,C88
Lafortune . . . . . . . . . . . . B81
Lechner . . . . . . . . . . . . . B40
Lee . . . . . . . . . . . . . . . B248
Lesley . . . . . . . . . . . . . . . B8
Litton . . . . . . . . . . . . . . C94
Lloyd . . . . . . . . . . . . . . B57
Lockhart . . . . . . . . . . . B170
Long . . . B84,B110,B115,B234
Long . . . . . . . . . . B300,B308
Louisiana . . . . . . . . . A44,A91
Love . . . . . . . . . . . . . . B189
Lubrizol . . . . . . . . . . . C126
Lucas . . . . . . . . . . . . . B186
Lufkin . . . . . . . . . . . . . . C64
Lummus . . . . . . . . . . . . C83
Mapco . . . . . . . . . . . . . . C66
Marathon . . . . . . . . C33,C129
Marland . . . . . . . . . . . . B183
Marshall . . . . . . . . . . . B180
Matthews . . . . . . . . . . . B185
McCarthy . . . . . . . . . . . . B60
McKee & Co. . . . . . . . . . C90
McLaughlin . . . . . . . . . B188
McLean . . . . . . . . . . . . . B42

McWorkman . . . . . . . . B190
Mellon Family . . . . . . . . . . . . . . . . . . . . B125,B161,B191
Mellon, Andrew . . B172,B215
Mellon, Paul . . . . . . . . B132
Mexico . A24,A28,A121,A194
Mexico A202,A243,A248,A325
Meyer . . . . . . . . . . . . . B192
Middle East . . . . . . . . . . . . A15,A33,A74,A76,A138,A183
Middle East . . . . . . . . . . . . A203,A212,A265,A294,A318
Miles . . . . . . . . . . . . . B193
Mills . . . . . . . . . . . . . . B196
Mississippi . . . . . . A152,A197
Mitchell . . . . . . . . . . . . B198
Molk . . . . . . . . . . . . . . B199
Moncrief . . . . . . . . . . . B250
Montana . . . . . . . . A1,A245
Morgan . . . . . . . . . . . . . B16
Murchison . . . . . . . . . . B305
Murray . . . . . B22,B130,B206
Neely . . . . . . . . . . . . . B210
Neff . . . . . . . . . . . . . . B211
Nelson . . . . . . . . . . . . B212
New York . . . . . . A136,A137
Newmont . . . . . . . . . . . C110
Nobel Family . . . . . . . . B279
Nobel, Alfred . . . . . . . . B116
Ohio . . . . . . . . . . . . . . A154
Oil Well Supply . . . . . . C140
Oil Workers Union . . . . . C93
Oklahoma . . . . . . . . . . . . . . A54,A56,A90,A92,A93,A94
Oklahoma . . . . . . . . . . . . . . . . A95,A96,A97,A102,A112
Oklahoma . . . . . . . . . . . . . A117,A139,A163,A166,A176
Oklahoma . . . . . . . . . . . . . . A198,A220,A221,A227,A228
Oklahoma . . . . . . . . . . . . . . A247,A261,A296,A315,A320
Oklahoma . . . . . . . . . . . . . . A321,A322,A327,A328,A332
Oklahoma . . . . . . . . . . A341
Oklahoma Natural . . . . . . C96
Onassis . . . . . . . . . . . . B167
Orcutt . . . . . . . . . . . . . B216
Owen, Jimmie . . . . . . . B261
Owen, Roderic . . . . . . . B217
O.G.& E. . . . . . . . . . . . C72
Pacific Petroleums . . . . . . C54
Page . . . . . . . . . . . . . . . B44
Pan American . . . . . . . . . C97
Parker Drilling . . . . . . . . C98
Pearson . . . . . . . . . . . . B262
Peckham . . . . . . . . . . . B296
Penn Square . . . . . C125,C157
Pennsylvania . . . . . . . . . . . . . A7,A19,A36,A41,A42,A53
Pennsylvania . . . . . . . . . . . . A58,A60,A63,A65,A70,A73
Pennsylvania . . . . . . . . . . . A103,A106,A107,A108,A109
Pennsylvania . . . . . . . . . . . A110,A126,A129,A134,A178
Pennsylvania . . . . . . . . . . . A181,A190,A191,A204,A205
Pennsylvania . . . . . . . . . . . A206,A207,A208,A209,A210
Pennsylvania . . . . . . . . . . . A219,A226,A233,A234,A235
Pennsylvania . . . . . . . . . . . A251,A271,A272,A280,A304
Pennsylvania . . . . . . . . . . . A307,A333,A334,A339,A342

Pennzoil . . . . . . . . . . . . C48
Penrose . . . . . . . . . . . . . B79
Pertamina . . . . . . . . . . . C8
Peru . . . . . . . . . . . . . . A217
Pew . . . . . . . . . . . . . . B253
Phillips . . . . . . . . . . . . C105
Phillips, Frank . . . . . . . B288
Phillips, L.E. . . . . . . . . B148
Pickens . . . . . . . . . . . . B224
Pope . . . . . . . . . . . . . . B226
Porterfield . . . . . . B228,B229
Powell . . . . . . . . . . . . . B265
Price . . . . . . . . . . . . . . B231
Purcell . . . . . . . . . . . . B232
Pure . . . . . . . . . . . . . . . C95
P.G.& E. . . . . . . . . C19,C111
P.S.E.& G. . . . . . . . . . . C21
Quaker State . . . . . . . . . C155
Richfield . . . . . . . . . . . . C69
Rockefeller . . . . . . . . . . .
. . . . . . B1,B6,B7,B19,B31,B76
Rockefeller . . . . . . . . . . .
. . . B89,B105,B122,B135,B149
Rockefeller . . . . . . . . . . .
. . B164,B207,B213,B238,B303
Rockefeller . . . . . . . . . B304
Rockefeller Family . . . . . . .
. . . . B47,B52,B163,B175,B202
Rockefeller Family . . . . . . .
. . . . . . . . . . B204,B233,B254
Rockefeller, A. . . . . . . . . B37
Rockefeller, David . . . . . B131
Rockefeller, Jr. . . . . . . . . B90
Rockefeller, M. . . . . . . B176
Rockefeller, Nelson . . . . . . .
. . . . . . . . B63,B98,B162,B221
Rockefeller, Nelson . . . . . . .
. . . . . . . . . . B225,B239,B240
Rockefeller, Winthrop . . B289
Roff . . . . . . . . . . . . . . B241
Rogers . . . . . . . . . . . . . B66
Romania . . . . . . . . . . . A164
Rothschild Family . . . . . B203
Royal Dutch . . . . . . . . . . .
. . . . . . . C44,C49,C114,C115
Russia . . . . . . . . . . . . . A133
Sadler . . . . . . . . . . . . . B244
Sailor . . . . . . . . . . . . . B245
Samuel . . . . . . . . . . . . B124
Sanders . . . . . . . . . . . . B246
Saunders . . . . . . . . . . . B252
Schlumberger . . . C2,C7,C118
Scott . . . . . . . . . . . . . . B249
Shah . . . . . . . . . . . . . . B218
Shah . . . . . . . . . . . . . . B219
Shah . . . . . . . . . . . . . . B255
Shah . . . . . . . . . . . . . . B309
Shamrock . . . . . . . . . . . C58
Shell C9,C26,C122,C127,C150
Sherman . . . . . . . . . . . . B64
Shield . . . . . . . . . . . . . B257
Sidwelladair . . . . . . . . . B258
Siegfried . . . . . . . . . . . B160
Sierra Pacific . . . . . . . . C133
Signal . . . . . . . . . . . . . C142
Silliman . . . . . . . . . . . . . B92
Simmons . . . . . . . . . . . B111
Sinclair . . . . . . . . C128,C130
Skelly . . . . . . . . . . . . . . C63
Smiley . . . . . . . . . . . . . B260
South Improvement . . . . C109
Southern Union . . . . . . . . C14
Standard . . . . . . . . . . . . .
. . A179,C31,C35,C51,C53,C59
Standard . . . . . . . . . . . . .
. . C60,C61,C62,C74,C77,C80

Standard . . . . . . . . . . . . . .
. . C91,C106,C131,C132,C136
Standard . . . C138,C146,C148
Standard Cal . . . . . . C15,C152
Standard Indiana . . . C29,C52
Steele . . . . . . . . . . . . . B264
Sternberg . . . . . . . . . . . B267
Stoker . . . . . . . . . . . . . B220
Stowe . . . . . . . . . . . . . B268
Sun . . . . . . . . . . . . C67,C135
Tarbell . . . . . . . . . B17,B272
Tavoulareas . . . . . . . . . B273
Teagle . . . . . . . . . . . . . B287
Teco . . . . . . . . . . . . . . . C28
Texaco .C20,C30,C65,C73C104
Texaco . . . . C104,C113,C121
Texas .A20,A25,A27,A30,A35
Texas . . . . . . . . . . . . . . . .
. . . . . A37,A39,A436,A49,A49
Texas A51,A52,A59,A61,A71
Texas . . . . . . . . . . . . . . . .
. . A77,A100,A101,A104,A111
Texas . . . . . . . . . . . . . . . .
. A114,A115,A118,A119,A120
Texas . . . . . . . . . . . . . . . .
. A122,A127,A132,A135,A141
Texas . . . . . . . . . . . . . . . .
. A142,A145,A146,A147,A148
Texas . . . . . . . . . . . . . . . .
. A149,A150,A153,A155,A160
Texas . . . . . . . . . . . . . . . .
. A161,A168,A173,A182,A186
Texas . . . . . . . . . . . . . . . .
. A187,A193,A195,A196,A199
Texas . . . . . . . . . . . . . . . .
. A216,A218,A222,A223,A237
Texas . . . . . . . . . . . . . . . .
. A238,A239,A246,A250,A260
Texas . . . . . . . . . . . . . . . .
. A263,A264,A269,A270,A263
Texas . . . . . . . . . . . . . . . .
. A264,A269,A270,A276,A278
Texas . . . . . . . . . . . . . . . .
. A279,A282,A284,A285,A288
Texas . . . . . . . . . . . . . . . .
. A289,A290,A291,A292,A293
Texas . . . . . . . . . . . . . . . .
. A299,A309,A310,A311,A324
Texas . . . . . . . . . A326,A331
Texas Eastern . . . . . . . . . C13
Texas Gas . . . . . . . . . . C149
Texasgulf . . . . C43,C99,C123
The McMan . . . . . . . . . C145
Thompson, Arthur . . . . . B274
Thompson, Ernest . . B43,B147
Thornton . . . . . . . . . . . B263
Tide Water . . . . . . . . . . . C10
Toernudd . . . . . . . . . . . B277
True . . . . . . . . . . B251,B282
Tygiel . . . . . . . . . . . . . B284
T.P.& L. . . . . . . . . . . . . . C68
Udden . . . . . . . . . . . . . B123
Uncle Sam Oil . . . . . . . C144
Union Oil . . . . . . . . . . C139
University Lands . . . . . . . C57
UNOCAL . . . . . . . . . . C134
USSR . . . . . . . . . . . . . A128
Various Bios . . . . . . . . . . .
. . . . . . . . . B41,B49,B56,B83
Various Bios . . . . . . . . . . .
. . . . . . . . . . B174,B227,B256
Various Companies . . . . . . .
. . . . . . . . . . . C89,C101,C116
Venezuela . . A18,A189,A252
Vincent . . . . . . . . . . . . B286
Warner . . . . . . . . . . . . B290

Watchorn . . . . . . . . . . . B291
Weeks . . . . . . . . . B168,B292
Welch . . . . . . . . . . . . . . B39
Wells . . . . . . . . . . . . . B294
Western . . . . . . . . . . . . . C16
Western Geophysical . . . C151
White . . . . . . . . . . . . . B297
Wickware . . . . . . . . . . B299
Williamson . . . . . . . . . . B301
Wood . . . . . . . . . . . . . B306
Wyoming . . . . . . . . . . . . . .
. . . . . A4,A26,A29,A230,A253
Wyoming . . . . . . . . . . . . . .
. . . . . . A262,A268,A319,A329
W.T.G.S. . . . . . . . . . . C147
Yamani . . . . . . . . . . . . B237
Youle . . . . . . . . . . . . . B307